High-Performance Cement-Based Concrete Composites

Technical Resources

Journal of the American Ceramic Society

www.ceramicjournal.org

With the highest impact factor of any ceramics-specific journal, the *Journal of the American Ceramic Society* is the world's leading source of published research in ceramics and related materials sciences.

Contents include ceramic processing science; electric and dielectic properties; mechanical, thermal and chemical properties; microstructure and phase equilibria; and much more.

Journal of the American Ceramic Society is abstracted/indexed in Chemical Abstracts, Ceramic Abstracts, Cambridge Scientific, ISI's Web of Science, Science Citation Index, Chemistry Citation Index, Materials Science Citation Index, Reaction Citation Index, Current Contents/ Physical, Chemical and Earth Sciences, Current Contents/Engineering, Computing and Technology, plus more.

View abstracts of all content from 1997 through the current issue at no charge at www.ceramicjournal.org. Subscribers receive full-text access to online content.

Published monthly in print and online. Annual subscription runs from January through December. ISSN 0002-7820

International Journal of Applied Ceramic Technology

www.ceramics.org/act

Launched in January 2004, *International Journal of Applied Ceramic Technology* is a must read for engineers, scientists,and companies using or exploring the use of engineered ceramics in product and commercial applications.

Led by an editorial board of experts from industry, government and universities, *International Journal of Applied Ceramic Technology* is a peer-reviewed publication that provides the latest information on fuel cells, nanotechnology, ceramic armor, thermal and environmental barrier coatings, functional materials, ceramic matrix composites, biomaterials, and other cutting-edge topics.

Go to www.ceramics.org/act to see the current issue's table of contents listing state-of-the-art coverage of important topics by internationally recognized leaders.

Published quarterly. Annual subscription runs from January through December.
ISSN 1546-542X

American Ceramic Society Bulletin

www.ceramicbulletin.org

The *American Ceramic Society Bulletin*, is a must-read publication devoted to current and emerging developments in materials, manufacturing processes, instrumentation, equipment, and systems impacting the global ceramics and glass industries.

The *Bulletin* is written primarily for key specifiers of products and services: researchers, engineers, other technical personnel and corporate managers involved in the research, development and manufacture of ceramic and glass products. Membership in The American Ceramic Society includes a subscription to the *Bulletin*, including online access.

Published monthly in print and online, the December issue includes the annual *ceramicSOURCE* company directory and buyer's guide. ISSN 0002-7812

Ceramic Engineering and Science Proceedings (CESP)

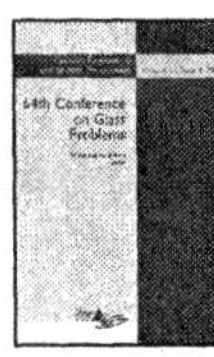

www.ceramics.org/cesp

Practical and effective solutions for manufacturing and processing issues are offered by industry experts. CESP includes five issues per year: Glass Problems, Whitewares & Materials, Advanced Ceramics and Composites, Porcelain Enamel. Annual subscription runs from January to December.
ISSN 0196-6219

ACerS-NIST Phase Equilibria Diagrams CD-ROM Database Version 3.0

www.ceramics.org/phasecd

The ACerS-NIST Phase Equilibria Diagrams CD-ROM Database Version 3.0 contains more than 19,000 diagrams previously published in 20 phase volumes produced as part of the ACerS-NIST Phase Equilibria Diagrams Program: Volumes I through XIII; Annuals 91, 92 and 93; High Tc Superconductors I & II; Zirconium & Zirconia Systems; and Electronic Ceramics I. The CD-ROM includes full commentaries and interactive capabilities.

High-Performance Cement-Based Concrete Composites

Materials Science of Concrete, Special Volume

Proceedings of the Indo-U.S. Workshop on High-Performance Cement-Based Concrete Composites, Chennai, India (2005)

Editors
Joseph J. Biernacki
Surendra P. Shah
N. Lakshmanan
S. Gopalakrishnan

Published by
The American Ceramic Society
PO Box 6136
Westerville, Ohio 43086-6136
www.ceramics.org

For information on ordering titles published by The American Ceramic Society, or to request a publications catalog, please call 614-794-5890, or visit our website at www.ceramics.org

ISBN 1-57498-199-4

Contents

Preface ix
List of Participants xii

Defining High Performance and Today's State-of-the-Art

Hybrid Fiber Reinforced Cement Composites 3
S.P. Shah and T. Voigt

High Performance Concrete—Present Scenario and Future Prospects in Indian Context 21
S. Gopalakrishnan, J.A. Peter, and K. Balasubramanian

Application of High Performance Concrete in India—Some Case Studies . . . 33
S.A. Reddi

Fracture Characteristics of High Strength High Performance Concrete. 43
B.K. Raghuprasad and B.H. Bharatkumar

Fiber-Based Systems for High Performance

High Performance Hybrid Composites for Thin Cementitious Products: the Next Generation. 55
A.E. Naaman, T. Wongtanakitcharoen, and V. Likhitruangsilp

Fibre Based Systems for High Performance 73
S.K. Kaushik

Guidelines for Design of Reinforced Concrete Structural Elements with High Strength Steel Fibres in Concrete Matrix. 83
N. Lakshmanan and T.S. Krishnamoorthy

Steel Fiber Reinforced Concrete—Applications in India 93
V.S. Parameswaran, K. Balasubramanian, and S. Gopalakrishnan

Flexural Cracks in Beams with Glass FRP Rebar and Fiber Reinforced Concrete. 101
D.C. Jansen and W.K. Lee

Investigation of Wood Pulp Fiber Reinforcement for Ready Mixed Concrete Applications 113
H.J. Brown and J.H. Morton

Confinement Analogy—Towards Toughness-Based Design for Reinforced Fiber Reinforced Concrete Beams 123
V.S. Gopalaratnam, Z. El-Shakra and H. Mihashi

Enhanced Performance of Fiber Reinforced Concrete with Low Fiber Volume Fractions 137
M. Lopez de Murphy, T. Hockenberry, and A. Achenbach

Next-Generation Cement Blends for High Performance

Hydration Kinetics of Portland Cement Containing Supplementary Cementitious Materials 149
Y. Peng, W. Hansen, C. Borgnakke, and J J. Biernacki

Next-Generation Cement Blends for High Performance Concrete 165
A.K. Jain

Fly Ash Based High Performance Cementitious Composites. 171
K. Ganesh Babu and P. Dinakar

Multi-Component Cementitious Systems and Their Influence on Durability of Concrete 183
N. Bhanumathidas, N. Kalidas, and G.A.B. Suresh

New Generation Admixtures for Enhanced Performance of Concrete 193
R. Shridhar

Influence of Fine Aggregate Lithology on Delayed Ettringite Formation in High Early Strength Concrete 199
A.M. Amde, Richard A. Livingston, and Kenneth Williams

Potential Use of Beneficiated Fly Ash in High Performance Concrete 211
K.A. Riding and M.C. Garci Juenger

Tools for Modern and Next Generation High Performance Research

Comparison of Different Methods for Characterization of Cement-Based Materials Subjected to Sulfate Attack. 223
K.E. Kurtis, A.C. Jupe, N.N. Naik, S.R. Stock, and P. Stutzman

Sustainable Development: Approach for Research on Next Generation High Performance Concrete with Fly Ash in India 235
P.C. Basu

Rheological Measurements and Very Early Age Viscoelastic Property Measurements of Concrete 247
M. Neelamegam, N.P. Rajmane and J.K. Dattatreya

Non-Destructive and Partially-Destructive Test Methods for Condition Assessment of Corrosion Affected Structures 255
H.G. Sreenath

Fiber Optic Instruments for Monitoring Service Life Performance of In-place Concrete 269
K. Ravisankar

Considering Moisture Gradients and Time-Dependent Crack Growth in Restrained Concrete Elements Subjected to Drying 279
N. Neithalath, B. Pease, J.H. Moon, F. Rajabipour, J. Weiss, and E. Attiogbe

Meso-Scale Strain Measurements Using Synchrotron X-Rays 291
J.J. Biernacki, C. Parnham, J. Bai, T. Watkins and C. Hubbard

Investigation of Early Age Material Property Development in Cementitious Materials Using One-Sided Ultrasonic Technique 301
K.V. Subramaniam and Jaejun Lee

Author Index 311
Keyword Index 312

Preface

The common need for infrastructure binds all nations together, yet the resource availability and specific needs differ. India is in great need of housing and basic infrastructure for utility systems, roadways and urban development while the United States of America is, effectively, in the process of rebuilding its deteriorated systems. While basic practices may be similar in both countries, specific practices regarding issues, such as design for earthquake and disaster management, use of by-products and waste materials, hot weather curing and formulation with advanced admixtures, vary.

The world's infrastructure is largely built of concrete. With growing populations in the developing countries of the world, such as India, and with the decay of existing infrastructure in developed countries, such as the U.S., the need for new materials with improved properties has become as imperative now as ever. High-performance concrete composites that include fibers and particulate matter along with advanced chemical admixtures and complex ternary and even quaternary cement blends represent a growing proportion of the concrete being used and would possibly be the future norm.

Numerous obstacles, however, continue to block the widespread implementation of high-performance cement-based materials in both the U.S. and India. These include a lack of fundamental understanding of the interactions between complex ternary and quaternary blends of cement and other cementitious materials and their even more complex interaction with chemical admixtures and long-term performance and durability of composite materials containing metal and organic fibers. As a result, understanding the fundamental interactions between cementitious material constituents and reinforcing elements in these new world concretes is necessary and a plan to realize their potential is needed.

Thanks to a grant from the U.S. National Science Foundation (NSF Grant Award No. INT-0352838) and financial support from the Council of Scientific and Industrial Research (CSIR), India, a joint Indo-U.S. workshop on ***high-performance cement-based concrete composite*** has been made possible. This workshop endeavors to isolate key enabling research topics that may accelerate the advancement of next generation high performance concrete composite materials for broad-scale global applications in both developed and developing countries. It is expected that the outcome of the workshop will promote an exchange of knowledge that will lead to greater interaction between research centers and scientists of the two nations.

The Indo-U.S. workshop was held at the Structural Engineering Research Center (SERC), CSIR Campus, in Chennai, India. Thirteen U.S. and 14 Indian resource persons participated in the two-day activity as well as special invitees from Indian academia, research institutions and industry. The workshop agenda was organized into five thematic sessions: (1) Defining high performance and today's state-of-the-art; (2) Fiber-based systems for high performance; (3) Next generation cement-blends for high performance; (4) Tools for modern and next generation high performance research; and (5) Planning for collaborative and follow-on activities. This edition of The Materials Science of Concrete contains 27 papers that include review articles as well as research manuscripts summarizing new research results and represents contributions from each of the 27 participants and their collaborating co-authors.

Joseph J. Biernacki
Surendra P. Shah
N. Lakshmanan
S. Gopalakrishnan

List of Participants: U.S. Contingent

A.M. Made	University of Maryland
J.J. Biernacki*[&]	Tennessee Technological University
H.J. Brown	Middle Tennessee State University
M.C. Garci Juenger	University of Texas at Austin
S. Gopalaratnam	University of Missouri, Columbia
W. Hansen*	University of Michigan
D.C. Jansen*	California Polytechnic State University
K.E. Kurtis	Georgia Institute of Technology
M. Lopez de Murphy	Pennsylvania State University
A.E. Naaman*	University of Michigan
S.P. Shah*[#]	Northwestern University
K.V. Subramaniam	City College of the City University of New York
W.J. Weiss	Purdue University

List of Participants: Indian Contingent

K.G. Babu	Indian Institute of Technology Madras
P.C. Basu*	Atomic Energy Regulatory Board
N. Bhanumathidas	Institute for Solid Waste Research & Ecological Balance
S. Gopalakrishnan*[#]	Structural Engineering Research Centre
A.K. Jain	Grasim Industries Ltd.
S.K. Kaushik	Indian Institute of Technology Roorkee
N. Lakshmanan*[&]	Structural Engineering Research Centre
M. Neelamegam	Structural Engineering Research Centre
V.S. Parameswaran	Intl Centre for FRC and Design Technology Consultants
B.K. Raghuprasad*	Indian Institute of Science
K. Ravisankar	Structural Engineering Research Centre
S.A. Reddi*	Gammon India Limited
H.G. Sreenath	Structural Engineering Research Centre
R. Shridhar	Fosroc Chemicals (India) Pvt Limited

Invited Guests

K. Balasubramanian	Structural Engineering Research Centre
B.H. Bharatkumar	Structural Engineering Research Centre
J.K. Dattatreya	Structural Engineering Research Centre
P. Dinakar	IIT - Department of Ocean Engineering
N. Kalidas	Institute for Solid Waste Research & Ecological Balance
T.S. Krishnamoorthy	Structural Engineering Research Centre
J.A. Peter	Structural Engineering Research Centre
N.P. Rajmane	Structural Engineering Research Centre
G.A.B. Suresh	Ramco Research & Development Centre

* Organizing Committee member, [&] Organizing Committee chair, [#] Organizing Committee co-chair.

Defining High Performance and Today's State-of-the-Art

HYBRID FIBER REINFORCED CEMENT COMPOSITES*

S.P. Shah and T. Voigt
Northwestern University
Center for Advanced Cement-Based Materials
2145 Sheridan Road,
Evanston, IL 60201

ABSTRACT

This paper provides an overview about recent research on the field of hybrid and high performance fiber reinforced cement composites conducted at the Center for Advanced Cement-Based Materials, Northwestern University, USA. The presented investigations focus on the fiber dispersion, tensile and shrinkage cracking behavior and water permeability of hybrid fiber reinforced cement composites. It is shown how advanced experimental techniques and constitutive modeling can be used to effectively relate the properties of the fiber reinforcement and the internal composite structure to the performance of the tested fiber-cement composites.

INTRODUCTION

The most common motive for the inclusion of fibers in concrete is to increase ductility under tensile or flexural loading. This is possible because fibers influence the fracture mechanism within the concrete structure[1]. To understand which shapes, sizes, and types of fibers are most effective, one must consider the mechanisms through which the fibers interact with the concrete matrix. As soon as load is applied to concrete, short and discontinuous microcracks begin to develop in a distributed manner throughout the material[2]. Very fine fibers, also known as microfibers (typically less than 50 μm in diameter), may delay the process by which the microcracks coalesce to form large, macroscopic cracks known as macrocracks and thereby increasing the tensile strength of the matrix. With conventional fiber contents, the coalescence of the first macrocrack coincides with the peak load and initiates localization, where subsequent deformation is concentrated in the opening of the crack (Fig. 1). Upon further material deformation, the widening of the macrocrack is resisted by coarser fibers, known as macrofibers, which bridge the localized cracks. The postpeak degradation of the concrete is prolonged and the toughness of the material is increased as these macrofibers, which carry nearly all the load on the cracked composite, either pull out or break. There has been much enthusiasm recently in the field of fiber-reinforced concrete for the development of hybrid fiber systems where two or more types of fibers are combined. This is done with the intent of conferring the best performance characteristics of each of the constituent fiber types to the composite material.
The mechanism of the mechanical performance of plain, conventional and high performance fiber reinforced cement composites is compared in Fig. 2. The plain and conventional fiber reinforced cement composites exhibit a clear strain softening behavior after the maximum tensile load is reached. In contrast to this, high performance fiber reinforced cement composites are characterized by an increased value of the proportional limit and a strain-hardening behavior in the post-peak region. This effect can be achieved by combining the right type of fibers with proper processing techniques and strong matrices[4].

* This paper was originally presented at the Symposium "Erfahrung und Zukunft des Bauens" (Experiences and Future of Construction) on September 28, 2004 in Leipzig, Germany

This paper is meant to provide an overview about recent research on the field of hybrid and high performance fiber reinforced cement composites conducted at the Center for Advanced Cement-Based Materials, Northwestern University, USA. The presented investigations focus on key material properties and the fracture process of fiber reinforced cement composites. To effectively determine these parameters and establish their relation to the mechanical performance of the composites advanced experimental techniques and constitutive models were employed.

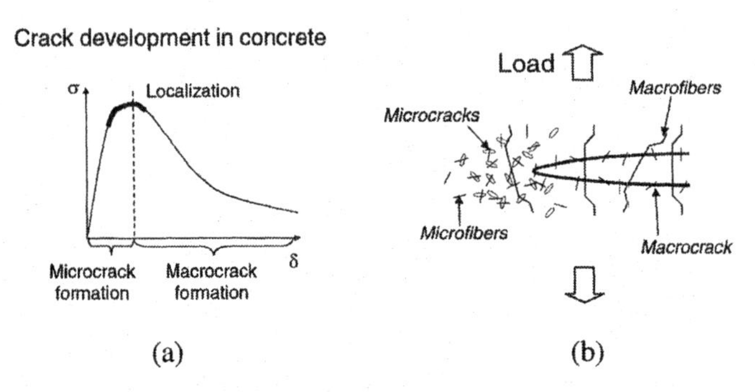

Fig. 1: Relationship between: (a) mechanical response and crack development; and (b) relative scale of crack at which fiber interaction occurs[3]

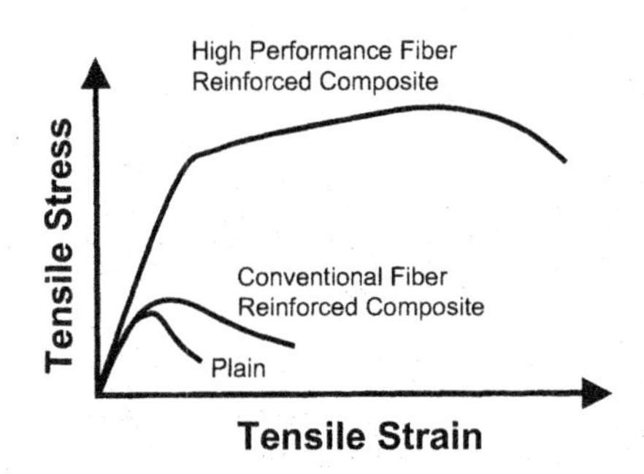

Fig. 2: General behavior of plain, conventional and high performance fiber reinforced cement composites

ANALYSIS OF FIBER DISPERSION

Motivation

The fiber dispersion and the size and number of fiber-free areas are believed to play an important role in the initiation and sequence of the composite cracking. As the fibers clump and the size and number of matrix areas that are not supported by fibers increase, the initiation of a crack requires less energy, and once the crack forms, it can advance easily through the fiber-free areas in the matrix.

Experimental Program and Test Procedure

The effect of fiber dispersion on the behavior of fiber reinforced cement composites was studied by means of flexural beam tests. The beam specimens were produced by extrusion of fresh cement-fiber composites with a ram type extruder. Details of the extruder can be found elsewhere[5]. After steam curing for two days, the extruded beams (size 25.4 x 4 mm) were tested in oven dried condition at the age of six days. The basic mix design of the composite was (by volume) 45% cement, 12% silica fume, 3% fibers, 1% superplasticizer, and 39% water, with a 0.29 water/cement ratio (by weight). Polyvinyl alcohol (PVA) fibers with a diameter of 14 micron and a length of 2 mm were used for all beams.

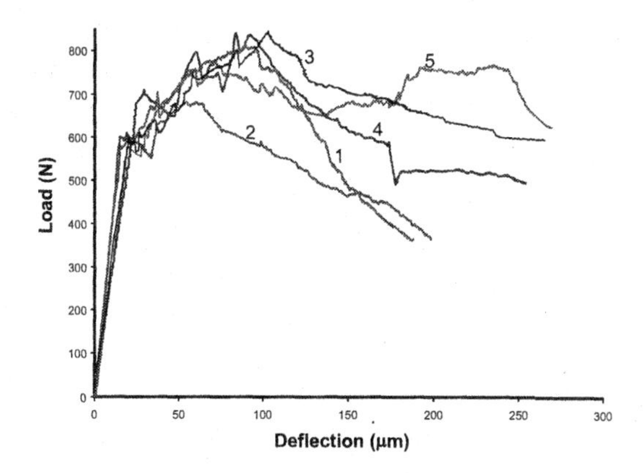

Fig. 3: Four-point flexure test results of identical composites[6]

The beams were tested with a four-point flexural test setup in displacement-controlled (closed loop) mode. Five different specimens with identical mix design were tested and the resulting load versus displacement curves are given in Fig. 3. The electronic speckle pattern interferometry (ESPI) technique is used to record the location of crack initiation and the sequence of the multiple cracking of the cement composites during the flexure tests. The ESPI technique is a highly accurate displacement measurement method, which allows mapping of the crack propagation at the microscale. The experimental setup includes a helium-neon laser as a light source, an interferometer, a loading unit, image acquisition and processing units, an analog/digital interface, and an image storage and display (Fig. 4). The immediate results of the ESPI test is a fringe pattern that allows determining the location of microcracks. Details about the technique and fringe formation can be elsewhere[7].

Experimental Results of Bending Tests and Fiber Distribution

The ESPI pictures of cracks observed at one of the tested samples are given in Fig. 5. From this figure the crack initiation locations and corresponding stresses and displacements at the beam surface can be derived. For this beam, a total of six cracks have been observed and examined. The crack locations can be identified by discontinuities in the fringes of the ESPI pictures. To evaluate the relationship between fiber dispersion and cracking sequence the cross sections of the beams at the locations of the cracks were investigated. SEM pictures of the polished crack surfaces were digitized and each fiber marked and represented as a point for further treatment by an image analysis program. The distributions of the fibers over the cross sections at the locations of six consecutive cracks is given in Fig. 7. Each single cross section represents the area of 10 mm^2 at the point where the individual crack were initiated.

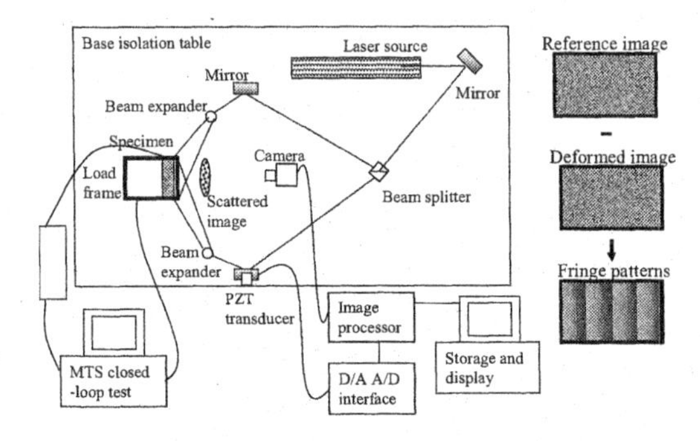

Fig. 4: Principle and setup of the ESPI test

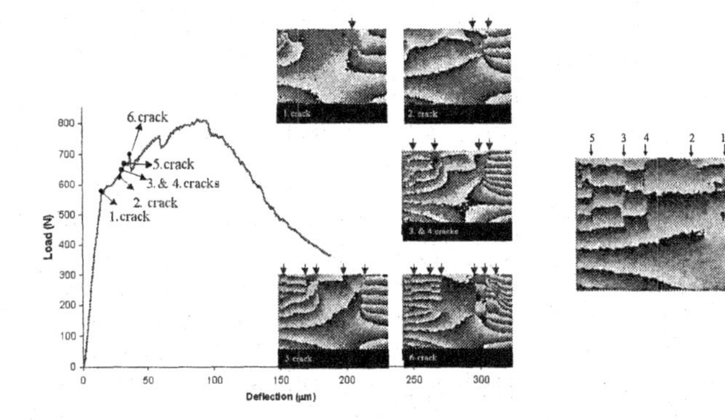

Fig. 5: ESPI pictures of specimen, taken during test, and locations and sequence of cracks[6]

Fig. 6: Sequential multiple cracking as observed by ESPI fringe pattern

Statistical Analysis

The availability of the fiber distribution in digitized form offers the possibility to analyze the fiber dispersion with statistical means. Two statistical functions are calculated to quantitatively describe the fiber dispersion, the K-function and the F-function. The K-function is a standard measure of the expected number of fibers within a certain distance of a given fiber location. The K-function is calculated to describe the tendency of fibers to clump. The F-function is the distribution of distances between arbitrary points to the nearest fiber, which can be used to describe the size of the fiber-free areas. The F-function gives the probability of any point in the cross section being at least a distance *r* away from the nearest fiber. More details of the two statistical functions and their calculation are given elsewhere[6].

In the following it will be shown how the calculated point process statistics can be used to quantify the dispersion of the fibers and how the obtained parameters can be related to the cracking behavior of the tested composites. The radius of the largest fiber free area (derived from the F-function) at the crack location for six consecutive cracks are given in Fig. 8. It is clear from the figure that this parameters dramatically decreases with the number of the crack.

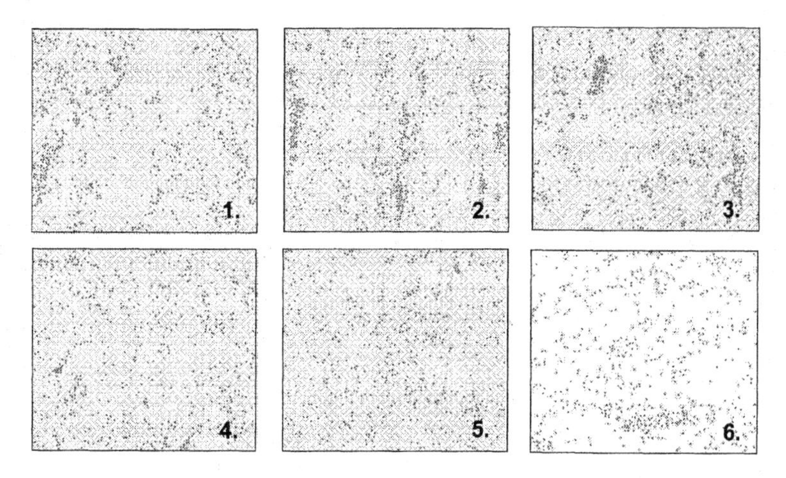

Fig. 7: Fiber distribution over the cross section at the different cracking locations[6]

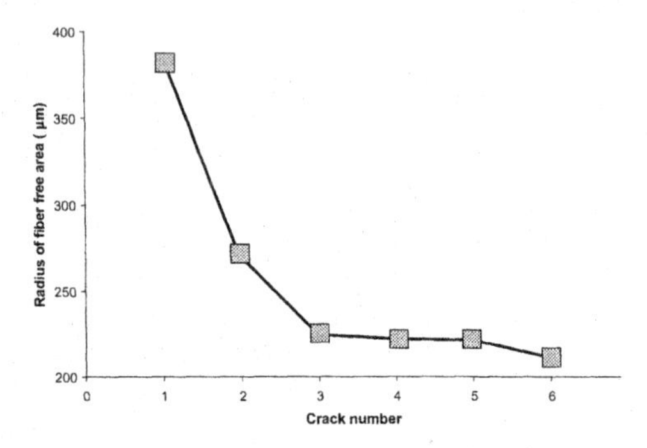

Fig. 8: Comparison of largest fiber-free area at crack locations[6]

The radius of the fiber free area can also be taken as the defect (flaw) size that caused the given system to fail. The data presented in Fig. 9 show that a general relationship exist between this flaw size and the cracking load of the considered composite. The figure also shows that this relationship can be explained with the critical stress intensity factor K_{IC}. More details are given elsewhere[6].

Another parameter describing the mechanical performance of a composite is the toughness defined as the ratio of displacement at the maximum stress to the displacement at the first cracking stress. The relationship between the toughness of the five tested beam specimens and the percentage of fiber clumping at the first crack location is given in Fig. 10. The percentage of fiber clumping is calculated from the K-function. It can be seen that both parameters are closely related. An efficient fiber bridging and transfer of the load to other parts of the composite can be achieved if there is a better fiber dispersion and less fiber clumps at the first crack location. Better fiber dispersion, in turn, increases the toughness of the composite.

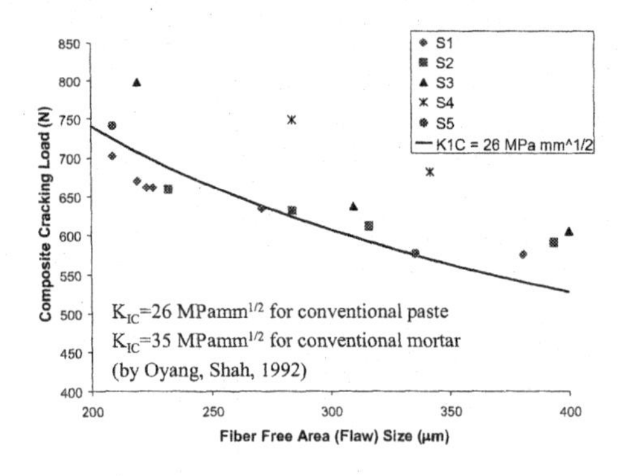

Fig. 9: Comparison of theoretical and experimental cracking stress versus defect size[6]

Conclusions of the Fiber Dispersion Investigations

Microstructural parameters such as the size of the fiber-free areas and the fiber dispersion in the composite are associated with the mechanical performance and sequential multiple cracking of the fiber-reinforced cement-based composites. It is seen that fiber-free areas reduce the cracking stress of the composite by acting as defects in the material. Sequential cracks form depending on the size of the fiber-free areas at the composite cross section.

The fiber dispersion affects the toughness of the composites by its role in transferring the load to the other parts of the specimen. An effective crack bridging and increase in the toughness of the composite can be achieved if the fiber dispersion is better at the first crack location.

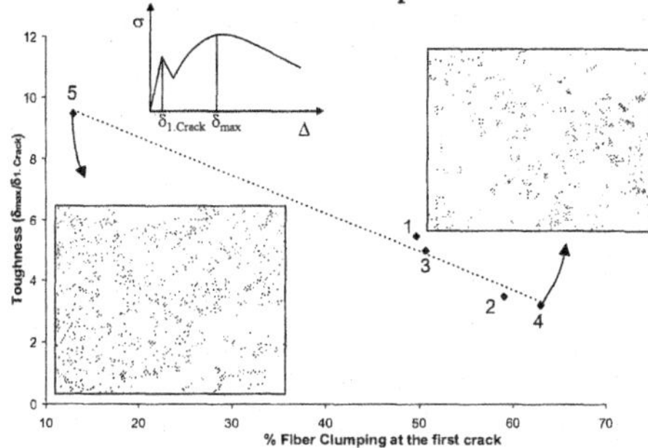

Fig. 10: Effect of fiber clumping at first crack location on specimen toughness[6]

ANALYSIS OF FRACTURE PROCESS

Motivation

Hybrid fiber systems, where two or more types of fibers are combined with the objective of conferring the best performance characteristics of each to the composite material, are a promising approach for making efficient use of fibers. However, attempts to achieve a complementary and additive blend have been only sporadically successful in practically applicable concrete mixes. The research presented in this section seeks to further the understanding of fiber reinforcement in general and the development of hybrid blends in particular. The most elementary method for understanding the mechanism of fracture in fiber-reinforced concrete is to observe the process as it occurs. Since the direct observation is difficult with microfibers because of the small scale at which they participate, an advanced technique, Subregion Scanning Computer Vison (SSCV), capable to perform this type of observations will be employed.

One important property affected by the cracking behavior of fiber reinforced concrete is crack induced water permeability. Concrete mixtures containing hybrid fiber blends have the ability to develop multiple cracking patterns with reduced widths of the single cracks. This makes these mixes especially beneficial to reduce the water permeability since permeability strongly depends on the crack width. Under the assumption of laminar flow, the rate of flow through a parallel sided slot, such as a crack, is directly proportional to the crack width cubed, multiplied by the length of the crack (see Fig. 11). Therefore, for similar overall displacement, a multiple-cracked material will demonstrate a lower permeability than one exhibiting a single crack.

Table I: Types and properties of used fibers

Type	Designation	Shape	Diameter (μm)	Length (mm)
steel	macrofiber	hooked	500	30
	microfiber A	straight	22	6
PVA	microfiber B	monofilament	14	12

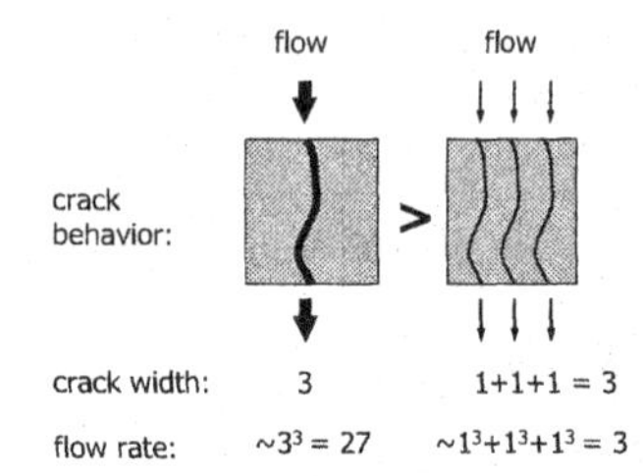

Fig. 11: Dependency of flow rate on the cracking behavior (single versus multiple)

Observation of the Fracture Process

The investigations described in the following were conducted with mortar specimens with the proportions 1:0.45:1 (cement : water : fine aggregate) by weight. Three different fiber types were used in different combinations. The properties of the fibers are given in Table 1.
The development of cracking in the cement composites was determined by means of uniaxial tensile tests performed on specimens with a rectangular cross section. The tensile load was imparted to the specimen via steel plates that transferred shear stress through epoxy-glued surfaces. The experimental setup for the uniaxial tensile tests is given in Fig. 12.

To monitor the cracking process during the uniaxial tensile tests the SSCV method is used. This technique utilizes a charged-coupled device (CCD) camera that is mounted on a precisely controlled two-dimensional motion stage. By moving vertically and horizontally along the surface of the specimen the CCD camera captures a set of 56 images, called subregions, which together represent the full surface of the specimen (see Fig. 13). Two white light sources were placed on the left and right of the specimen to produce sufficient lighting of the specimen.

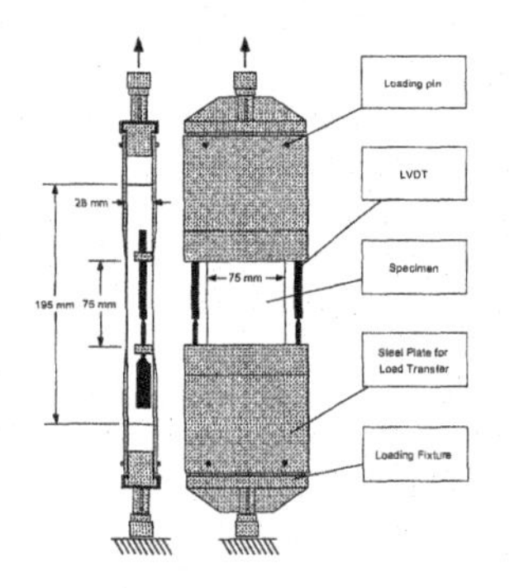

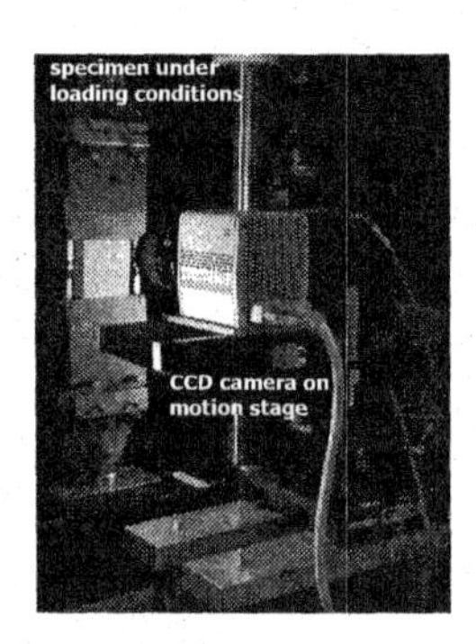

Fig. 12: Uniaxial tension test and subregion scanning computer vision (SSCV) setup[8]

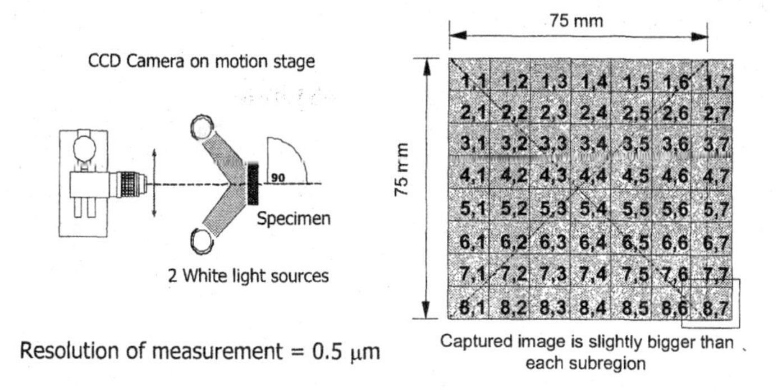

Fig. 13: Principle of subregion scanning computer vision (SSCV) [8]

To record the images the loading process of the specimen is stopped and the tensile load is maintained on a constant level for about two minutes. After acquiring the images the uniaxial tensile test is continued until the next images are to be collected. Given this procedure, the SSCV technique can monitor the crack growth at numerous selected stages in the loading history without interfering with the fracture process itself.

The acquired sub-images are processed with a computer vision technique, called Digital Image Correlation (DIC). This technique compares successive digital images to measure two-dimensional deformations on the surface of the specimen (see Fig. 14). This process of measuring deformations is based on tracking a portion of an image, known as the subimage, as the pattern of pixels described by this subimage moves in a sequence of images of the specimen surface. The displacement field is calculated for a regularly spaced grid of nodes arbitrarily positioned on the reference image. Further details about the SSCV and DIC techniques are given elsewhere[8-10].

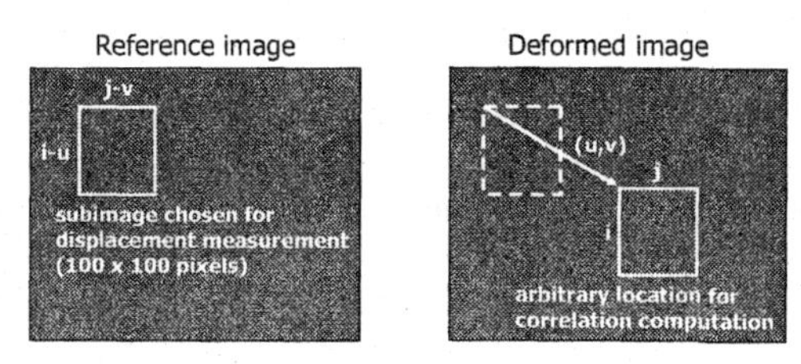

Fig. 14: Principle of digital image correlation (DIC)

The effect of the addition of micro- and macrofibers on the cracking behavior is shown in Fig. 15. The figure shows the images of the specimen surface as obtained from DIC at three different deformation stages. Two mortar mixtures containing macrofibers and microfibers only are compared. It can be seen that for each of the three deformations stages the specimen containing the micro fibers shows a more distributed crack formation with smaller crack widths of the single crack. This example verifies the benefit the addition of microfibers for the inducing multiple cracking behavior of cement composites. More detail on the results of this study are given elsewhere[8].

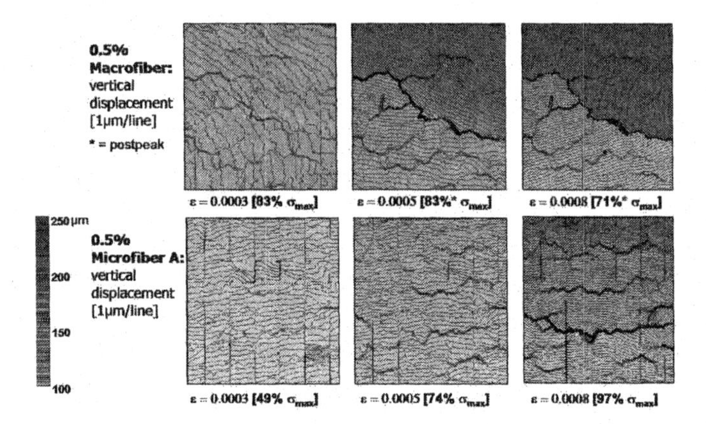

Fig. 15: Influence of macro- and microfibers on the cracking behavior at different deformation stages

Relationship Between Cracking Behavior and Flow Rate

To investigate the influence of hybrid fiber reinforcement on the crack induced water permeability uniaxial tensile tests in combination with water permeability tests were conducted. The principle of this test is given in Fig. 16. A pressure box, whose edge is sealed with a flexible rubber gasket and which contains water, is fixed to the specimen and clamped in place with threaded rods. A pressure of 13.8 kPa, which is approximately equivalent to the pressure head of 1.4 m of water, is generated inside the box to induce flow. During testing, this pressure is controlled by a regulator and measured directly at the specimen. The water traveling through the specimen is captured in an identical box sealed to the other side of the specimen, which has an outlet spout at a height equal to the top of the box. This is filled with water before testing so that any flow through the concrete that causes an overflow is caught and measured. More details about this testing procedure are given elsewhere[3]. The water permeability tensile tests were conducted with the same materials that were described in the previous section.

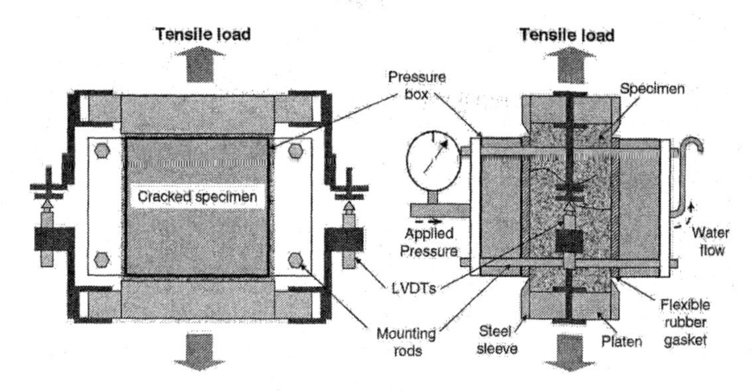

Fig. 16: Water permeability tensile testing setup[3]

The uniaxial tensile stress is plotted versus the overall axial displacement in Fig. 17(a) for plain mortar, mortars containing macro- and microfibers only and mortar containing both fiber types. It can be seen that the improvement in mechanical performance resulting from the inclusion of PVA microfibers takes the form of increased strength. The larger steel macrofibers, on the other hand, provide postpeak resistance and a lower increase in strength. The hybrid fiber-reinforced mortars containing PVA microfibers and steel macrofibers demonstrated increased strength and improved ductility. The clear contribution from both micro- and macrofibers is observable, demonstrating the effect of blending fibers of two sizes.

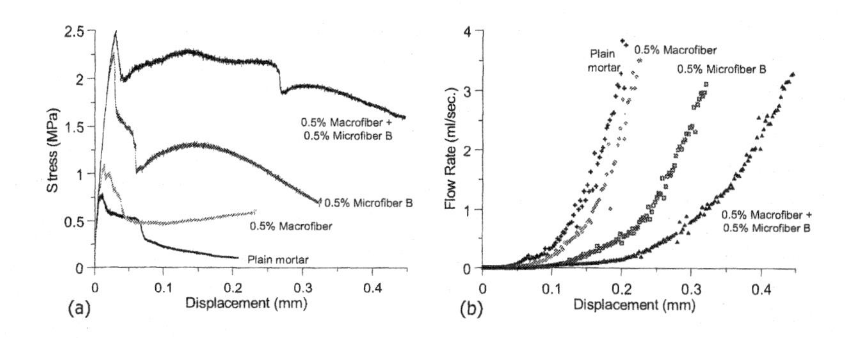

Fig. 17: Uniaxial tensile performance of mortars; and (b) corresponding flow rates for same displacements (from 3).

The flow rate of water through the mortars versus overall displacement for the same specimens is plotted in Fig. 17(b). A lower flow rate at a given displacement is an indication of reduced water permeability. The steel macrofibers decrease the permeability slightly compared with the unreinforced mortar. The inclusion of PVA microfibers individually lessens the permeability further, with the hybrid blend of steel macrofibers and PVA microfiber showing the greatest beneficial effect on the permeability of the cracked material. For the specimens containing the PVA hybrid blend, the overall specimen displacement reaches over 200 μm before flow becomes appreciable, while a similar flow rate was visible through the unreinforced mortar at a displacement of 100 μm. In addition, subsequent increases in the flow rate occur more slowly in the hybrid-reinforced material than in the unreinforced mortar. Further details on the evaluation of these results are given elsewhere[3].

Conclusions of the Fracture Process Investigations

Through the use of the SSCV technique, clear differences in cracking behavior were observed in mortars containing no fibers, microfibers, macrofibers and hybrid blends. Compared to the cracking behavior of macrofiber-reinforced mortars the use of microfiber-reinforcement modified the mechanism by which failure occurred, reducing the widening of coalesced cracks and inducing multiple cracking before the peak load is reached. This is likely the result of the smaller inter-fiber spacing of the microfibers, which increased the probability that the crack path would intersect with the fibers.

The permeability of cementitious materials was a determining factor in its durability. For the mixtures examined herein, when fibers were included, an improvement in resistance to permeation was seen for cracked mortars. While a marginal improvement was seen with conventional macrofibers, microfiber reinforcement produced a significant reduction in cracked permeability, and the blending of these fibers with macrofibers further decreased the water flow through the cracked material.

ANALYSIS OF FIBER TYPE AND INTERNAL COMPOSITE STRUCTURE

Motivation

Restrained shrinkage cracking is a severe problem for concrete structures. This section describes investigations on concrete mixtures reinforced with various steel and polypropylene fibers as well as welded wire fabric that have yielded an important comparison of the performance of the different reinforcement types in preventing and controlling the drying shrinkage cracking. This kind of information is essential for the effective and durable design of concrete structures.

Experimental Program and Test Procedure

To evaluate the performance of the different fiber types in preventing and controlling drying shrinkage cracking, restrained shrinkage ring tests were performed. The principle of the test consists of casting an annulus of concrete that is to be tested around a rigid steel ring. While the wooden base automatically seals the bottom surface of the concrete ring, the top surface is sealed by a layer of silicone rubber. This procedure creates symmetrical drying conditions for the specimen. Due to the drying of the concrete from the outer, circumferential surface the material shrinks. Since this shrinking is prevented by the steel ring, a condition of internal uniform pressure in the concrete is created: tensile stress parallel and compressive stress radial to the steel-concrete interface. After a certain time the tensile stresses inside the concrete will exceed its tensile strength. A crack will develop at the weakest point of the concrete ring. Due to further shrinkage, eventually the tensile stresses will again increase up to levels high enough for a second or third crack to occur. Further details on the ring test are given in 11.

The experiments described in this paper were conducted with six different types of steel fibers, three different polypropylene fibers, a welded wire fabric (WWF) and plain concrete. The fibers have various sizes and geometries. Dimensions and material of the fibers and the WWF can be found in Table 2. The fibers were tested in single-fiber mixtures in different fiber volumes and additionally in two fiber blends (see Table 3). The mix proportion of the concrete by weight was 1:0.5:2:2 (cement : water : coarse aggregate : sand). No additives or admixtures were used and the concrete composition was the same for all tested mixtures.

Table II: Dimensions of the tested fibers and WWF[11] (figures not to scale)

Shape/ Material	Length	Equivalent Diameter	Aspect Ratio	View
Flat End (Steel)	30 mm	0.60 mm	50	
	50 mm	1.00 mm		
Hooked End (Steel)	50 mm	1.00 mm	50	
Crimped (Steel)	38 mm	1.14 mm	33	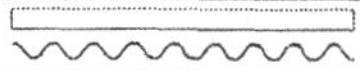
	50 mm		44	
Profiled (Steel)	20 mm	0.78 mm	26	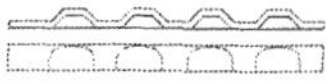
Crimped (Polypropylene)	50 mm	0.90 mm	56	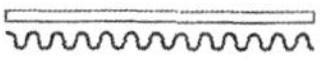
Fibrillated (Polypropylene)	20 mm	< 0.05 mm	≈ 400	
Multifilament (Polypropylene)	5-15 mm	< 0.05 mm	≈ 100-300	
WWF* (Steel)	Spacing: 152 mm	3.4 mm	–	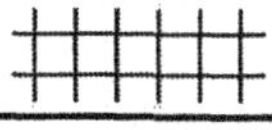

Table III: Volumes* and combinations of tested fiber mixtures[11] (in volume percent)

Steel Fibers

Flat End 30 mm	0.125 %	0.25 %	0.5%
Flat End 50 mm	0.125 %	0.25 %	0.5%
Profiled 20 mm	0.125 %	0.25 %	0.5%
Hooked End 50 mm		0.25 %	0.5%
Crimped 38 mm		0.25 %	0.5%
Crimped 50 mm		0.25 %	0.5%

Polypropylene (PP) Fibers

Crimped 50 mm	0.3%	0.6%	1.0%

Fiber Blends

Crimped Steel, 38 mm + Multifilament (PP)	0.17% + 0.07%
Crimped (PP) + Fibrillated (PP)	0.26% + 0.17%

The ring specimens were placed on turntables in an environmental chamber to ensure controlled and repeatable curing conditions at a temperature of 22°C and a relative humidity of

50%. The rings were checked by eye every 24 hours for visible cracks and the age of occurrence of each crack was recorded. The width of each crack was measured using a microscope at 42 days after casting.

Influence of Fiber Type

The experimental results of the maximum crack width for the different reinforcements are presented in Fig. 18. The maximum crack width of the plain concrete is 1.07 mm. The maximum crack width discussed here represents the maximum width of all cracks measured on a group of three ring specimens for each fiber type and content.

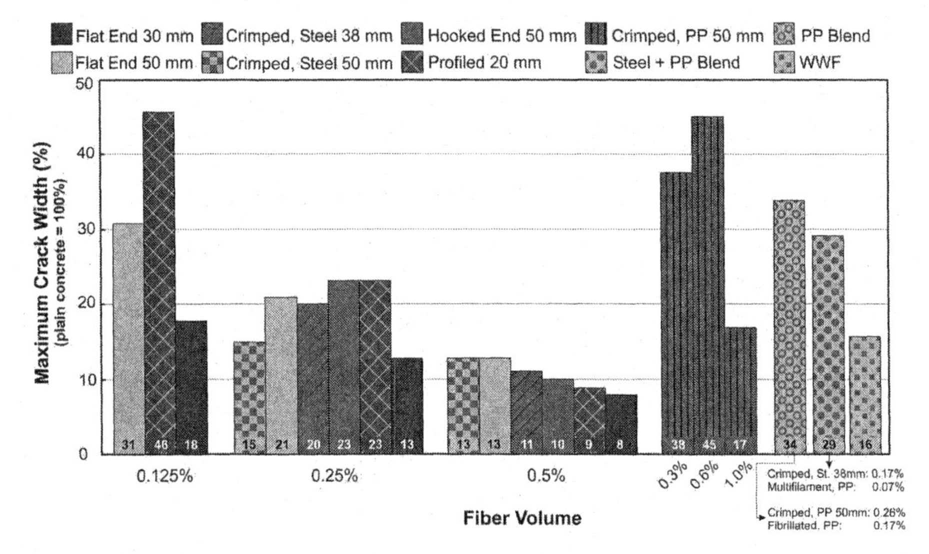

Fig. 18: Maximum crack width for all tested materials relative to plain concrete (crack width for plain concrete = 100%)[11]

The maximum crack width decreases with increasing fiber volume for an individual fiber type. This trend is general except for the fiber Crimped, PP 50 mm. This deviation can be explained by the experimental error. The best performing reinforcement is the fiber Flat End 30 mm. The mixtures containing this fiber have the smallest crack width for all three fiber volumes. The fiber Crimped, PP 50 mm is most effective in the high fiber volume (Vf=1%), but the lower fiber volumes still effectively reduce the crack width to less than 50% compared to the plain concrete mix. In case of the PP Blend the partial replacement of the macro fibers by the micro fibers did not alter the maximum crack width when compared to the corresponding single fiber mixture (Crimped, PP 50 mm for Vf=0.3%). Improvements can be expected in the ability of this fiber blend to reduce the plastic shrinkage cracking. The WWF is effective in reducing the crack width. The maximum crack width could be reduced to 16% of that of the plain mixture. The performance of the WWF is similar to that of the tested steel fibers in the fiber volume of 0.25%.

Combined Effect of Aspect Ratio and Fiber Volume

Another approach to analyze the shrinkage cracking behaviour is to relate the performance of the composite to the dimensions and geometry of its fiber reinforcement. In the following this approach the combination of the volume and the aspect ratio of the individual fibers is considered. The aspect ratio (A) of a fiber is calculated by the ratio of its length and diameter. The aspect ratios of the used fibers are given in Table 2. Figure 19 shows the relationship between the product of fiber volume and aspect ratio (V_f A) and the maximum crack width for all the tested steel fibers.

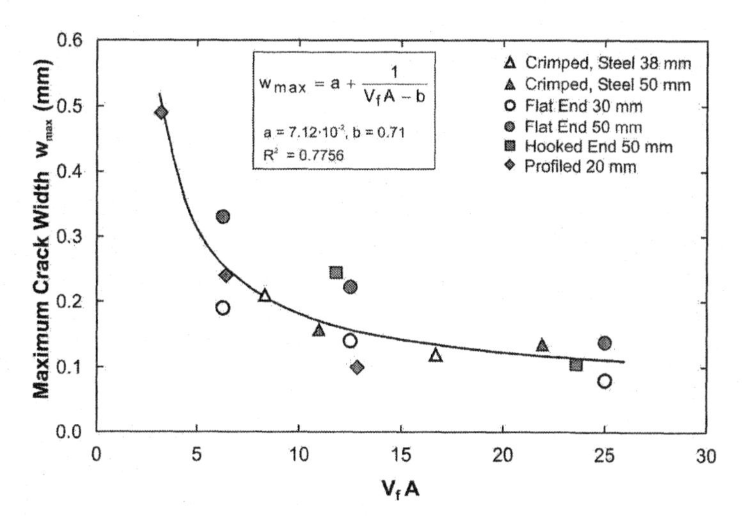

Fig. 19: Relationship between the product of fiber volume and aspect ratio to the maximum crack width for tested steel fibers (from 11)

The parameter V_f A incorporates two properties of the used fiber reinforcement: fiber volume and fiber geometry. It can be seen from Fig. 19 that V_f A and maximum crack width have a relatively close correlation (R^2=0.7756). This result suggests that including more parameters that specify the type and the level of the fiber reinforcement is beneficial for the potential of such parameters to explain the shrinkage cracking behavior of fiber reinforced concrete.

Internal Structure of the Cement Composite

It is now attempted to relate the performance of the reinforcements to the parameters that uniquely characterize the internal structure of the entire composite, not only certain phases. A concept is applied that distinguishes between fibers and coarse aggregates as one phase and matrix as another phase of the composite. By considering matrix volume, void content and total surface area of the constituents, the interaction of fibers, gravel and matrix can comprehensively be analyzed and effectively related to the performance of the individual test mixtures.

Figure 20(a) shows a schematic of the internal packing of fibers and gravel (without the matrix). Under the assumption of maximum compaction, this mix of fibers and gravel contains a certain volume of air voids that solely depends on the volume, the size distribution and the shape of the fibers and the gravel. If now a certain volume of matrix (dependent on the mixture proportion) is added to this conglomerate, the model assumes that the matrix has to fill up exactly that volume of air voids that was derived from Fig. 20(a). The volume of matrix that exceeds this void volume (excess matrix) is equally used to cover the surface of the fibers and the gravel. Figure 20(b) shows the composite consisting of gravel, fibers and matrix with the distinction between matrix filling voids and matrix covering fibers and gravel. Based on these assumptions the average thickness of the matrix layer (t_m) that envelops fibers and gravel can be specified (Fig. 20c). It results from the ratio of the volume of the excess matrix and the total surface area of fibers and gravel. Further details on the determination of the matrix thickness can be found elsewhere[9].

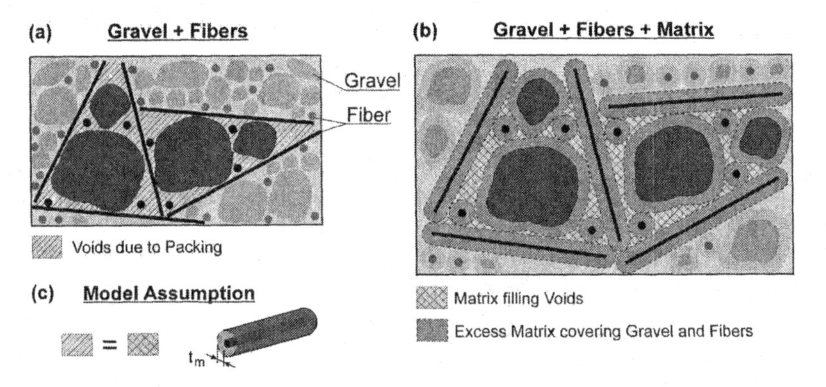

Fig. 20: Schematic explanation of multi-aspect concept[11]

(a) Void content as a results of maximum packing of gravel and fibers (without matrix)
(b) Distinction between matrix enveloping fibers and gravel and matrix filling voids
(c) Model assumption and thickness of matrix layer covering fibers and gravel

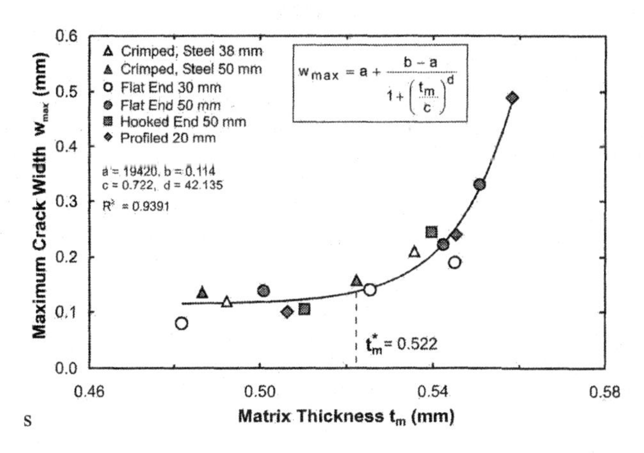

s

Fig. 21: Relationship between matrix thickness t_m and maximum crack width for tested steel fiber mixtures[11]

In Fig. 21 the relationship between the matrix thickness t_m and the maximum crack width for all tested steel fiber mixtures is presented. The plotted data points exhibit a very strong trend and show a clear improvement in the ability of the used concept to explain the shrinkage cracking behaviour of the tested materials. The R^2-value for the shown trend line (R^2=0.9391), describing the quality of the correlation between the single data points and the plotted trend clearly exceeds the R^2-value for the trend given in Fig. 19 (R^2=0.7756 for V_fA).

The trend given in Fig. 21 can be divided into two parts: a part with a very steep slope (right side) and a part with a very small slope (left side). This pattern suggests the existence of an optimal value for the matrix thickness (t^*_m = 0.522 mm) that marks an optimum composition of the composite in regard to the maximum crack width. Values of t_m smaller than t^*_m achieved for example by an increase in fiber volume do not significantly reduce the crack width. The higher fiber volume of such mixes (and thereby higher material costs) would not be justified by the reduction in crack width.

Conclusions of the Investigation of Fiber Type and Internal Composite Structure

In conclusion of this section it is emphasized that the shrinkage cracking behavior is governed by the internal structure of the entire composite, not just by the properties of the fiber reinforcement. It is shown that the characteristics and the interaction of all the composite constituents, comprehensively described by the parameter matrix thickness t_m , are the governing factors of the shrinkage cracking behavior of a fiber reinforced cement composite.

References

[1]P.N. Balaguru and S.P. Shah, "Fiber-Reinforced Cement Composites," McGraw-Hill, Inc., Singapore, 1992, 530 pp.

[2]Z. Jia and S.P. Shah, "Two-Dimensional Electronic-Speckle-Pattern Interferometry and concrete fracture processes," *Experimental Mechanics*, **34**, [3] 262-270 (1994)

[3]J.S. Lawler, D. Zampini, and S.P. Shah, "Permeability of cracked hybrid fiber-reinforced mortar under load," *ACI Materials Journal*, **99** [4] 379-385 (2002)

[4]S.P. Shah, "Betontechnologie der Zukunft (Material technology of concrete in the future). Betonwerk + Fertigteil-Technik (Concrete Precasting Plant and Technology)," **59** [2] 39-45 (1993)

[5]R. Shrinavasan, D. DeFord, and S.P. Shah, "The use of extrusion rheometry in the development of extruded fiber reinforced cement composites," *Concrete Science and Engineering*, **1** [1] 26–36 (1999)

[6]Y. Akkaya, S.P. Shah, and B. Ankenman, "Effect of fiber dispersion on multiple cracking of cement composites," *Journal of Engineering Mechanics*, **127** [4] 11-316 (2001)

[7]M. Ghandehari, S. Krishnaswamy, and S.P. Shah, "Technique for evaluating kinematics between rebar and concrete," *ASCE Journal of Engineering Mechanics*, **125** [2] 234–241 (1999)

[8]J.S. Lawler, T. Wilhelm, D. Zampini, and S.P. Shah, "Fracture processes of hybrid fiber-reinforced mortar," *Materials and Structures*, **36** [257] 197-208 (2003)

[9]S. Choi and S.P. Shah, "Measurement of deformations on concrete subjected to compression using image correlation," *Experimental Mechanics*, **37** [3] 307-313 (1997)

[10]S. Choi and S.P. Shah, "Propagation of microcracks in concrete studied with Subregion Scanning Computer Vision (SSCV)," *ACI Materials Journal*, **96** [2] 255-260 (1999)

[11]T. Voigt, V.K. Bui, and S.P. Shah, "Drying shrinkage of concrete reinforced with fibers and welded wire fabric," *ACI Materials Journal*, **101** [3] 233-241 (2004)

HIGH PERFORMANCE CONCRETE-PRESENT SCENARIO AND FUTURE PROSPECTS IN INDIAN CONTEXT

S. Gopalakrishnan
Director Grade Scientist and Advisor (Management)

J.A. Peter
Scientist

K. Balasubramanian
Scientist

Structural Engineering Research Centre
CSIR Campus, Taramani, Chennai 600 113, INDIA

ABSTRACT

Construction activities account for a major component of the budget in developing countries including India. Portland cement concrete is the most extensively used material for the construction of large infrastructural facilities worldwide. Significant distress or deterioration is being observed in reinforced concrete structures, such as bridges, multi-storeyed buildings, hyperboloid cooling towers and chimneys, particularly in coastal regions even well within their expected life span. Ensuring durability of concrete is one of the important issues to be addressed in evolving strategies to bring about sustainable development. This paper presents the typical results of investigations carried out at Structural Engineering Research Centre, Chennai, India on the development of High Performance Concrete (HPC), including Self Compacting Concrete (SCC) and the present scenario and future prospects on application of HPCs in the construction industry in India.

INTRODUCTION

High performance concrete (HPC) is a concrete in which enhancement of some or all of the properties of concrete, such as placement, compaction, no segregation, long-term mechanical properties, early age strength, toughness, volume stability and durability related properties resulting in extended service life in severe environment, is ensured/achieved[1,2]. Research carried out worldwide has well established that suitable addition of pozzolanic materials in concrete mixtures can lead to improvements in the durability of concrete [3,4]. Such concrete mixtures come under the category of HPC. Pozzolanic materials, such as fly ash (FA), ground granulated blast furnace slag (GGBS), and silica fume (SF), when used in concrete mixtures, due to their reaction with calcium hydroxide (a product of cement hydration) results in calcium syndicate hydrate compounds which, in turn, render the transition zone between the matrix and aggregates in the concrete mixture, denser and stronger and result in refinement of pore structure, thereby high degree of impermeability to ingress of water, air, chlorides, sulphates and CO_2. Pozzolanic materials are also known as mineral admixtures, supplementary cementitious materials (SCMs) and cement replacement materials (CRMs).

Experimental investigations were carried out at the Structural Engineering Research Centre (SERC), Chennai, on the use of SCMs, viz., FA, GGBS and SF, in developing HPC mixtures[5,6,7]. These studies were further extended to develop self-compacting concrete (SCC) mixtures using fly ash as SCM. Typical result of these studies are presented in this paper. The presentation is preceded by a review of Indian scenario on use of SCMs in construction industry during 1990s and during the present decade. It is seen that the performance of concrete mixtures, both in terms of strength and durability is enhanced with the use of SCMs. Details of typical construction in India using HPC and future prospects of HPC in the construction industry are also presented.

INDIAN SCENARIO TILL 1990s

Use of mineral admixture to make blended cements, i.e., Ordinary Portland Cement (OPC, termed as 'cement' in this paper) blended with pozzolanic materials to produce concrete mixtures is not new in India and can be traced back to the 1960s. Use of blended cements, viz., Portland slag cement[8] and Portland pozzolana cement[9] has been permitted by Bureau of Indian Standards (BIS). However, use of the blended cements in concrete construction in the country has been much limited till recent years.

The lack of large scale effort in India in exploiting the durability related improvements that fly ash imparts to concrete mixtures can be attributed to the rather poor quality of fly ash which was being generated in the country till 1980s. However, during the past decade (i.e., 1990s), there has been considerable improvement in the quality of fly ash collected through the mechanism of electrostatic precipitators in the various thermal power plants as reflected in the characteristics of fly ash with reduced loss on ignition (<5%) and increase in its fineness.

The importance of ensuring long term durability of concrete structures using SCMs has been incorporated in the revised Indian Standard IS 456-2000[10], wherein the earlier clause on durability aspects has been revised, permitting use of combinations of OPC with SCMs in the manufacture of concrete.

INDIAN SCENARIO DURING THE PRESENT DECADE

With the provisions made in the IS 456-2000 on the use of SCMs to manufacture structural grade concretes up to M80 grade, use of SCMs, especially fly ash and GGBS, in concrete mixtures, either as an additive/admixture while mixing the concrete or in the form of blended cements is on ever increasing trend in the construction industry. It is pertinent to point out that codes issued by the Indian Road Congress (IRC), which govern the construction of bridges/ flyovers in the country, are yet to be revised to permit use of SCMs in the concrete mixtures. However, efforts have been made to issue guidelines for use of HPC in the construction of bridges by the Ministry of Surface Transport / IRC in India and is in the draft stage. Central Public Works Department (CPWD) in India has recently issued circular permitting the use of fly ash based concrete mixture for constructions in the country only in a restricted manner. Ready Mixed Concrete Plants in India have also been contributing much on the use of SCMs in concrete mixtures. These efforts would lead towards attaining the much-needed sustainability on the use of cements in the construction industry in India.

R&D AT SERC, CHENNAI ON CONCRETE MIXTURES CONTAINING SCMs

Demonstration on use of fly ash in concrete during 1970s

During 1970s, when only 33 grade cement was available in India, SERC, Chennai, was one of the few institutions in the country to demonstrate use of fly ash in concrete mixtures. To demonstrate the use of fly ash in Reinforced Concrete (RC) and Prestressed Concrete (PSC), during 1973-1977, a two-storeyed experimental building in SERC Campus was constructed [11,12,13]. The floor and the roof of this building were built by using RC / PSC precast channel units. The RC channel units were used for spans up to 5m and PSC channel units were used for longer spans. Fly ash from the Basin Bridge Thermal Power Station (North Chennai) was used in cement mortar and concrete. Three different grades of concrete mixtures, M15, M20, and M40 by replacing 20% cement with fly ash were developed and used in producing the channel units. About 20% savings in total cement consumption and about 4% savings in total cost of the building were achieved at that time with the use of fly ash. This building, which is known as the fly ash building, now houses the Experimental

Mechanics Laboratory and is in good condition even after more than two decades of its construction (Figs. 1 and 2). This building serves as a successful model construction using fly ash to achieve cost-effective and durable construction.

(a)

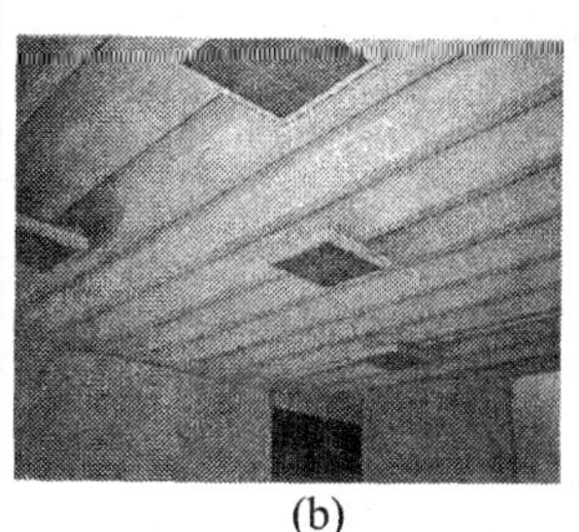

(b)

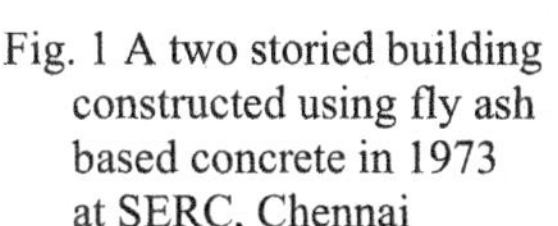

Fig. 1 A two storied building constructed using fly ash based concrete in 1973 at SERC, Chennai

Fig. 2
a) PSC Channel units being cast in the prestressing bed
b) A view of the under deck of the floor constructed using PSC Channel units

Investigations on the Use of SCMs in Concrete Mixtures

Three grades of concrete mixtures having target mean strength (compressive strength on 100 mm cubes) at 28 days of 40MPa, 60MPa and 70MPa, were investigated as follows:

(i) Fly ash (at 30% CRM level) based concrete mixtures and designated as AF30, BF30 and CF30 corresponding to the three target mean strengths respectively,

(ii) GGBS based concrete mixtures, designated as AS40, BS40 and CS40 (for 40% CRM level) and AS70, BS70 and CS70 (for 70% CRM level) corresponding to the three target mean strengths respectively and

(iii) OPC based concrete mixtures (without SCMs and meant as control mixtures for purposes of comparison of results with those of SCM based concrete mixtures) and designated as AO, BO, CO corresponding to the three target mean strengths respectively.

Mechanical properties: The mixture proportions and mechanical properties, viz., compressive and flexural strengths at 28 days for the above concrete mixtures are given in Table I. The standard deviations arrived at from the results of testing 30 cubes for compressive strength, for all the concrete mixtures are in the range of 3.0 to 4.0 MPa.

Table I Mixture proportions and properties of OPC based, fly ash based and GGBS based concrete mixtures

Concrete Series	OPC 53 gr.	SCM	Sand	Aggre-gate	Water l/cu.m	SP % of binder (OPC+SCM)	Compres-sive Strength	Flexural Strength
	Kg/cu.m						at 28 days (MPa)	
OPC Based (Control) concrete mixtures								
AO	300	--	804	1129	165	0.4	41	4.8
BO	375	--	715	1195	150	1.2	62	5.3
CO	450	--	635	1195	150	1.4	71	6.9
Fly ash* based (30% CRM Level) concrete mixtures								
AF30	232	101	782	1085	160	0.6	39	4.1
BF30	286	122	672	1135	155	1.2	58	5.2
CF30	375	161	609	1195	150	1.6	69	5.8
GGBS based (40% CRM level) concrete mixtures								
AS40	198	132	779	1129	165	0.5	43	4.53
BS40	247	166	683	1195	150	1.4	62	5.59
CS40	310	206	595	1195	150	1.45	69	6.20
GGBS based (70% CRM level) concrete mixtures								
AS70	103	241	787	1135	155	0.6	40	4.40
BS70	131	300	714	1135	155	1.2	59	5.20
CS70	161	375	638	1135	150	2.5	67	6.30

* Properties of FA: % retained on 45 micron sieve: 33; LOI: 1.08 %; CaO:1.02%; SiO_2: 59.1%, Specific gravity: 2.15

AO: A Series with OPC as binder; AF30: A Series with FA at 30% CRM level;
AS40: A Series with GGBS at 40% CRM level; AS70: A series with GGBS at 70% CRM level

From Table I, it is observed that by a judicious choice of binder (OPC + SCM) and water-binder ratio, it is possible to produce SCM based concrete mixtures of desired strength levels at 28 days. The flexural strength of SCM based concrete mixtures is also in par with that of OPC based concrete mixture at 28 days.

Stress Strain Characteristics: Typical plot of stress-strain curves at the age of 28 days by testing cylindrical specimens for control concrete mixtures, AO, BO and CO and for control concrete mixture, AO, together with that for SCM based concrete mixtures viz., AF30 and AS70, are given in Figs.3 and 4 respectively. From Fig.3, it is observed that for higher strength concrete mixtures (BO and CO), the failure strain of concrete is less than the value of 0.0035, which is the ultimate strain considered/assumed in the designs. From Fig. 4, it is seen that for 40MPa concrete, stress-strain behaviour of SCM based concrete mixtures is similar to that of control concrete mixture.

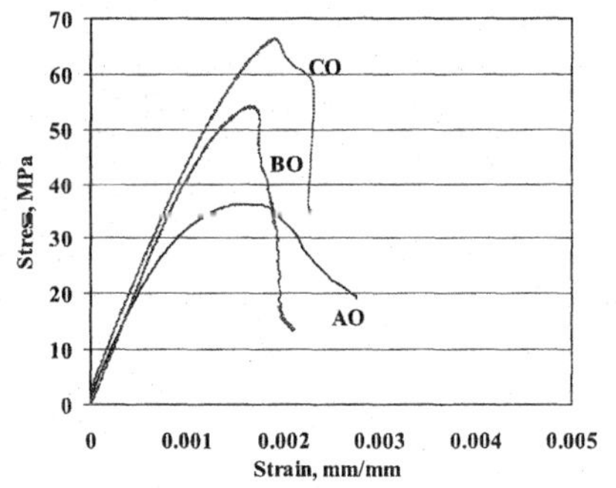

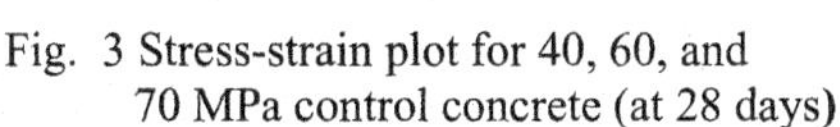

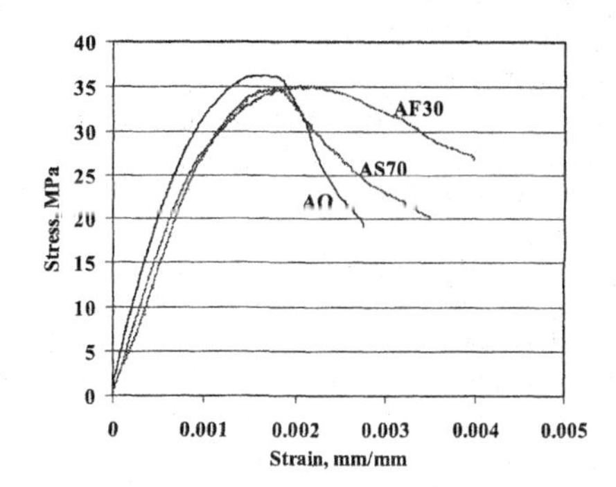

Fig. 3 Stress-strain plot for 40, 60, and 70 MPa control concrete (at 28 days)

Fig. 4 Stress-strain plot for 40 MPa concrete (OPC and SCM based at 28 days)

Study of pore size distribution using Mercury Intrusion porosimetry (MIP): Pore structure of the concrete matrix greatly influences the strength and durability properties of concrete. Pore structure of cement paste and concrete has been studied by various researchers[14,15,16]. Hence, tests were carried out at SERC using MIP to evaluate the pore size distribution of OPC based and SCM based concrete mixtures. Typical results of test carried out on samples of concrete for control concrete mixtures and for slag based (40% CRM level) concrete mixtures, using MIP are presented in Figs.5 and 6. From Fig. 5, it is seen that as w/b decreased (i.e., with increase in strength from A series to C series), incremental intrusion plot shifts towards lower range of pore size and the peak intrusion (critical pore diameter) decreases. From Fig. 6, it is also seen that as w/b decreased, cumulative intrusion descends towards the x-axis indicating reduced porosity level. From Figs. 5 and 6, it is seen that in the case of slag based concrete mixtures, the above pore structure characteristics are further improved than those of OPC based concrete mixtures, which reflect in their improved durability properties. These trends are more pronounced as the age of concrete increases.

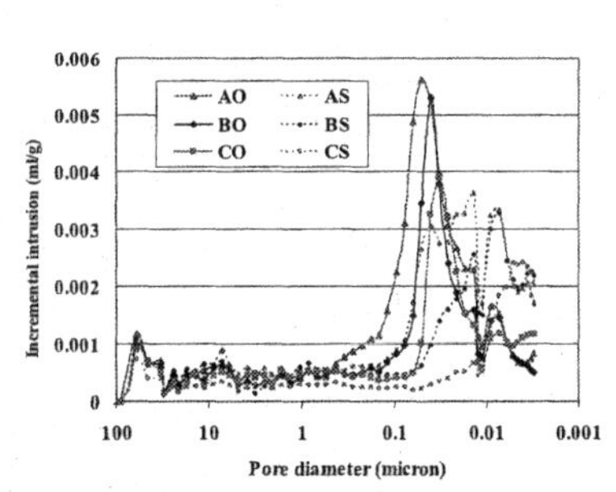

Fig. 5 Incremental intrusion vs pore diameter (at 28 days)

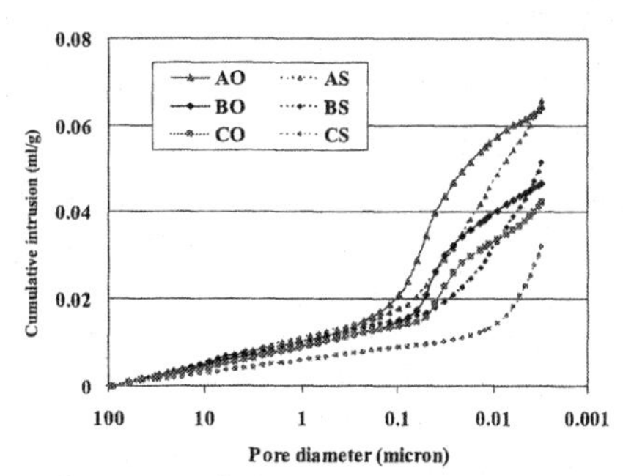

Fig. 6 Cumulative intrusion vs pore diameter (at 28 days)

Flexural Behaviour of Reinforced SCM based Concrete Beams: Flexural behaviour was investigated on a series of rectangular reinforced concrete (RC) beams cast using both OPC based concrete mixtures, fly ash based concrete mixtures and GGBS concrete mixtures as per mixture proportions given in Table 1. All the RC beams were designed, as balanced section as per IS 456:2000. The cross section of the beams for all the concrete test series were chosen suitably with varying percentage of tension reinforcement such that the moment capacity of the beams remained the same (≈ 90 kN-m). Adequate shear reinforcement was provided to avoid premature shear failure. The span of the beams (3200 mm) and the nature of the loading were kept the same for all the beams belonging to each series. Each series comprised of two beam specimens.

The RC beam specimens were tested under two-point loading with the middle 1000 mm length kept under constant bending moment. The load was applied through a hydraulic jack and distributed at two points such that each load point was 1000 mm from support point. Hinge condition was provided at one end of the RC beam, and roller support was provided at the other end. Table II gives the results of the flexural tests.

Table II Results of the flexural test on the RC beams (balanced capacity)

Beam Id.	Cross section (mm)	P_{cr} (kN)	P_u (kN)	P_{sl1} (kN)	P_{sl2} (kN)	Crack Spacing (mm)	Crack Width at P_{sl1} (mm)	Ductility Ratio
AO	150x300	30	192	140	128	76.9	0.14	2.02
AF30		30	187	130	125	83.3	0.16	1.74
AS40		35	187	130	125	71.4	0.18	1.68
AS70		30	187	140	125	83.3	0.16	-
BO	150x250	20	198	116	133	69.7	0.20	1.62
BF30		20	175	108	116	66.6	0.24	1.79
BS40		20	193	118	129	77.4	0.20	1.60
BS70		25	187	110	125	83.3	0.24	-
CO	150x225	20	168	87	109	91	0.19	1.60
CF30		20	170	85	114	100	0.16	1.63
CS40		20	170	82	113	87.1	0.20	-
CS70		20	170	90	113	83.3	0.16	1.55

P_{sl1} - Service load corresponding to deflection = span/250 (12mm); Span : 3200 mm
P_{sl2} - Service load computed as (P_u/1.5); P_{cr} : Load at first cracking; P_u : Load at failure

The flexural behaviour of reinforced control concrete and SCM (i.e., for both fly ash and GGBS) based concrete beam specimens were similar, both at the cracking and service load stages. The stiffness of the RC beams was not affected with the addition of GGBS in the concrete. The stiffness of the RC beams was found to be marginally lower for the fly ash based RC beam. The crack width and crack spacing were found to be similar for the control and SCM based RC beams. It was noted that the deformation ductility, as seen from the ductility ratio, relatively decreased with increase in concrete compressive strength from 40 MPa to 70 MPa. The bending moment values obtained in the present investigation were found to be much closer to the values computed as per IS:456-2000 code than those computed based on the ACI and CEB-FIB model codes. From the values of P_{sl1} and P_{sl2}, it is noted that defining service load **from** deflection criteria is more conservative, especially for higher strength concrete (i.e., for B and C series).

Concrete Mixtures having Ternary Blends: To offset the effect of retardation on strength development at early ages of concrete due to the incorporation of SCMs, such as fly ash, studies were carried out at SERC on triple blended concrete mixtures, i.e., concrete mixtures containing FA and SF as SCMs. Table III gives the mixture proportions and properties of control concrete, silica fume based concrete at 8% CRM level, fly ash based concrete at 25%

CRM level and ternary blends based concrete mixtures (i.e., SF at 8% together with FA at 25% CRM levels).

Table III Mixture proportions and properties of Concrete without SCMs and with SCMs (SF and FA)

Mix Id.	C (kg/m^3)	SF (kg/m^3)	FA (kg/m^3)	Compressive Strength MPa			Water Absorption (%)	Rapid Chloride Permeability (Coulombs)
				3days	7 days	28days		
C2	480.0	-	-	40.1	56.1	70.7	4.85	1672
C2D1	441.6	38.4	-	45.0	57.0	72.4	2.76	576
C2F1	360.0	-	120	33.0	44.6	61.0	4.46	515
C2T1	321.6	38.4	120	40.0	62.0	73.6	3.12	348

The parameters which were kept constant for all the concrete mixtures given in Table IV are w/b = 0.4 ; sand = 672 kg/m^3 ; coarse aggregate = 1105 kg/m^3; water = 192 l/m^3.

From the properties given in Table III, it is seen that drastic reduction in percentage water absorption and chloride permeability values are obtained in concrete mixtures having ternary blends. For the concrete mixtures containing only SF as SCM and both SF with FA as SCMs, the compressive strengths are in par or marginally improved when compared with that of control concrete.

Self Compacting Concrete (SCC): As an extension to the studies on developing HPC using SCMs, a study was performed at SERC, Chennai to develop SCC mixtures using locally available materials, especially fly ash as SCM. SCC requires a very high slump, which could be achieved using superplasticiser. However, to ensure that such concretes remain cohesive during handling operations, special attention needs to be given while proportioning the mix to avoid instability. Viscosity Modifying Agent (VMA) prevents segregation on addition of superplasticiser. Incorporating large quantities of fines, especially SCMs, such as fly ash, as partial cement replacement material, improves cohesiveness, reduces temperature rise in concrete during hydration of cement and enhances the durability besides reduction in cost of the SCC.

Tests for Self Compactibility: The workability of concrete mixtures to qualify as SCC is checked by the tests for self compactibility, viz., filling ability, passing ability and segregation resistance. This is verified by testing for slump flow, slump flow time, and V-funnel flow time and testing in the L-box, U–box, and Filling Ability Box. With reference to SCC mixtures developed at SERC and given in Table IV, the test results for self-compactability are as follows:

Slump flow →Mix S1:700mm; Mix S2:700 mm; Mix S5: 720 mm
Slump flow time → Mix S1: 4Sec.; Mix S2: 3 Sec.; Mix S5: 3 Sec.
V-funnel flow time→ Mix S1 =8 Sec.; Mix S2 = 6 Sec., ; Mix S5= 6 Sec.

The above values, obtained from the tests for self-compactibility for the concrete mixtures, are within the ranges specified by EFNARC-2004[17] for SCCs. Visual observations on the different SCC mixtures indicated that they were cohesive and possessed high stability. Proportions and properties of typical SCC mixtures, developed at SERC, are given in Table IV.

Table IV Mixture proportions and properties of SCC mixtures

Mix Id.	Cement kg/m^3	FA kg/m^3	VMA % of binder	SP % of binder	Sand kg/m^3	CA kg/m^3	Water l/m^3	Comp. Strength at 28days MPa	Chloride permeability-Coulombs	Resistivity kilo ohms-cm
S1	650	-	-	0.4	790	700	220	72	2552	8.0
S2	490	160	-	0.4	790	700	220	68	864	23.6
S5	390	260	0.05	0.5	880	730	202	58	324	62.8

From Table IV, it is observed that a significant reduction in the total charges passed through the concrete matrix was observed during the accelerated chloride permeation test (under 60V DC) for the concrete mixtures with SCMs, especially for the mixture with ternary blend, which could be attributed to the combined effect of particle packing, pozzolanic reaction and reduction in the conductivity of the pore fluid.

For both the SCC mixtures, S2 and S5, which have resistivity values greater than 20 kilo ohms-cm, the likelihood of corrosion risk is low as per guidelines given by J.H. Bungey[18].

Reinforced SCC Beams: RC beams were cast using SCC mixture and tested under flexure (Fig.7). From the test results, it was seen that even though the SCC mixture had higher binder content, lower values for maximum size and quantity of coarse aggregate (CA), the flexural behaviour of RC beam cast using SCC was similar to that of conventional concrete.

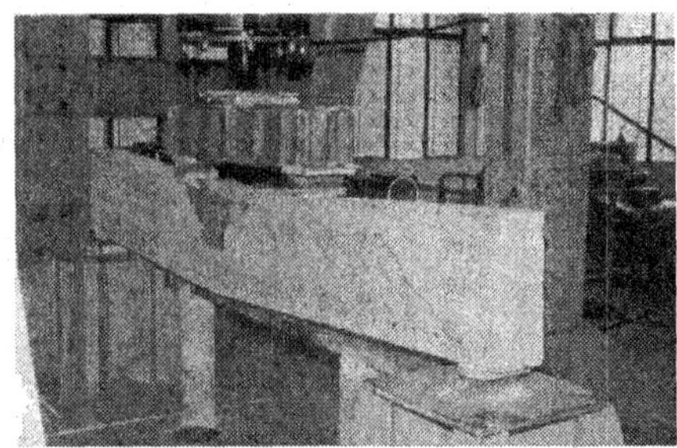

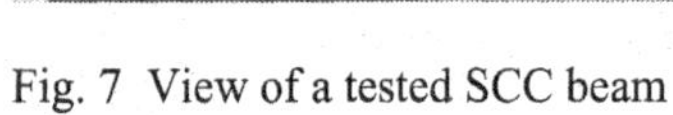

Fig. 7 View of a tested SCC beam

Fig. 8 Under reamed SCC pile ready for dynamic test

Under-reamed piles: The dynamic behaviour of cast-in-situ under-reamed piles (one with SCC and one with conventional concrete having similar strengths), 25 cm diameter and 4m long with two bulbs, was studied by exciting the pile in both horizontal and vertical directions using a Lazon type mechanical exciter (Fig. 8). It was observed that the dynamic characteristics, such as natural frequency, percentage damping, resonant, acceleration response and the normalised resonant displacement response for the SCC pile were almost the same as that of the conventional concrete pile.

Application of HPC Technology in Typical Construction in the Country

Details of typical construction in the country, which have been constructed using HPC technology based on the technical, advise of SERC, are given below.

Construction of Superstructure of the Flyover Bridge Across the Dumper Lines in Vizagapatnam Port Trust: As part of consultancy services, M45 grade GGBS based HPC (with 40% CRM level) for the construction of the prestressed concrete beams for the superstructure of the flyover bridge was designed and demonstrated by SERC. As per the requirements of the construction, the mix was designed for target strength of 57 MPa, which was to be achieved in 14 days. The compressive strength at 28 days was 64 MPa. The mix had a slump of 130 mm with water-binder ratio of 0.35. Naphthalene based superplasticizer at the rate of 0.85% by weight of the binder was added to get the required workability. Fig. 9 shows a view of the prestressed concrete bridge girder cast using M45 grade GGBS based HPC.

Fig. 9 A view of prestressed concrete fly over bridge girder cast using GGBS based HPC.

Fig. 10 Shows the dome unit being tested for strength at SERC

Construction of Bubble Type Dome Units for the Roof structure of the Parliamentary Library Building, New Delhi.: Central Public Works Department, New Delhi, sponsored a project to SERC on development of bubble type dome units for the roof structure of the Parliamentary Library Building, New Delhi. After detailed technical considerations, steel fibre reinforced HPC, M50 grade was selected and designed. The assessment of characteristics of the HPC including the casting and testing the structural capacity of the bubble type dome units were carried out at SERC, as part of the project. The concrete mixture was designed as a ternary blend with total CRM level of 40%, comprising 35% GGBS and 5% Silica Fume and with Dramix steel fibres at 0.75 % by volume of concrete. Naphthalene based superplasticizer was used to obtain the required workability. The water-binder ratio adopted was 0.35. The maximum size of coarse aggregate was limited to 10mm. Steel fibres were added with a view to improve the shrinkage resistance of fresh concrete and the tensile strength of hardened concrete so that the possibility of microcracking of concrete was minimised besides enhancing the toughness i.e., energy of absorption capacity. Fig. 10 shows a view of the dome being tested for strength at SERC.

FUTURE PROSPECTS AND R&D NEEDS

The awareness amongst civil engineering professionals in the country on the benefits of using SCMs in concrete mixtures to ensure long-term durability of RC/PSC structures is growing. This aspect, coupled with the enabling provisions, permitting use of SCMs in concrete mixtures, incorporated in the IS 456-2000 Code has paved the way for use of HPC in the concrete constructions in the country. Already, HPC has been used in some of the important infrastructural constructions in the country. Governmental agencies, such as Ministry of Surface Transport (MOST) and Central Public Works Department (CPWD), who construct most of the infrastructures in the governmental sectors, are taking steps to lift the ban on the use of fly ash based concrete mixtures for structural grade concrete. On the whole, use of HPC in concrete constructions is expected to grow rapidly in India during the coming decade. Ready mix concrete plants in the vicinity of metropolitan cities in India, are well poised towards making HPC. Development of SCC, which is a recent development in the country in the HPC technology, is in a nascent state and holds much potential for use in the construction industry in the near future.

For the application of HPC, especially high-strength HPCs in the RC/PSC constructions, salient R&D requirements are listed below. R&D is in progress at SERC on some of these aspects.

- Verification and extension of the design recommendations of IS 456-2000, which are essentially applicable for concretes up to medium strengths (≤40MPa), to the present day needs of higher strengths (up to 80 MPa).
- Evaluation of ductility aspects of structural elements constructed using high strength concretes and imparting ductility to such concretes to ensure lower damage levels during accidental over loading or during natural disasters.
- Evaluation of performance of reinforced high strength HPC structural elements under cyclic loading.

CONCLUDING REMARKS

In this paper, it is demonstrated through laboratory investigations that SCMs (available in India) could be used to produce HPCs. Through judicious choice of water-binder ratio and use of SCMs conforming to standards, it is possible to produce durable concretes having desired rate of strength development.

Use of HPC, besides ensuring long-term durability of concrete structures would also bring in economy and sustainability in concrete constructions in the country. Further, significant savings on use of cement (OPC), which consumes considerable energy in its production, could be realised by adopting SCMs in concrete mixtures.

ACKNOWLEDGEMENT

This paper is published with the kind permission of the Director, Structural Engineering Research Centre, Chennai.

REFERENCES

[1]Paul Zia, "International Workshop on HPC", held at Bangkok during Nov. 1994, SP-159, ACI, USA, 1996.

[2]Civil Engineering & Research Foundation (CERP,1993),"High Performance Construction Materials and Systems: An Essential Program for America and Its Infrastructure"(Technical Report (# 62939)

[3]Malier Yves (Ed.) "High Performance Concrete from Material to Structure", E&FN SPON, 1993

[4]CEB-FIP, "Application of High Performance Concrete", Report of CEB - FIP Working Group on HS/HPC, 1994.

[5]Gopalakrishnan, S., "Durable Concretes using Mineral Waste Materials - An Overview of SERC's Experience", pp 285-295, *Proceedings of Waste and by-products for Alternate Building Materials*, BMTPC, New Delhi, 14-15, April, 1999.

[6]Gopalakrishnan S, Rajamane N.P, Neelamegam M, J. Annie Peter and Dattatreya J.K., 'Effect of Partial Replacement of Cement with Fly ash on the strength and Durability Characteristics of High Performance Concrete', *Indian Concrete Journal*, 335-341 (2001).

[7]Rajamane, N.P, Annie Peter, J., Neelamegam, M, Dattatreya, J.K and Gopalakrishnan, S., "Improvements in Properties of High Performance concrete with Partial Replacement of cement by GGBS", *Journal of The Institution of Engineers (India)*, **84** (May) 38-42 (2003).

[8]IS:455-1989 "Indian Standard Code of Practice for Portland Slag Cement" (Fourth Revision), Bureau of Indian Standards, New Delhi.

[9]IS:1489 (Part 1)-1991, "Indian Standard Code of Practice for Portland Pozzolana Cement-Flyash based" (Third Revision), Bureau of Indian Standards, New Delhi.

[10] IS 456-2000, "Indian Standard Code of Practice for Plain and Reinforced Concrete (Fourth Revision), Bureau of Indian Standards, New Delhi, July 2000.

[11] Madhava Rao A.G., Parameswaran V.S. and Ramachandramurthy D.S., "Prestressed concrete channel flooring units using high strength deformed beams prestressed by electrothermal method", *Indian Concrete Journal*, December, pp 2-7 (1974)

[12] Venkateswarlu B., Ramaiah M. and Sreenath H.G., "High strength Fly Ash Concrete", *Journal of Structural Engineering*, **1**[3] 1-7 (1973).

[13] Venkateswarlu, B., Ramaiah, M. Sreenath, H.G., "Strength and Serviceability of Precast Channel Units in Reinforced Concrete", *Proceedings of the Symposium* on *Low Cost House, Public Buildings and Industrial Structures,* Institution of Engineers (India), January 1974.

[14] Rossler M and Odler I, "Investigations on the Relationship between Porosity, Structure and Strength of Hydrated Portland Cement pastes, I- Effect of Porosity", *Cement and Concrete Research,* **15**[2] 320-330 (1985)

[15] Lu Cui and Cayhadi, J.H, "Permeability and pore Structure of OPC Paste", *Cement and Concrete Research,* **31** 277-282 (2001)

[16] Rakesh Kumar and Bhattacharjee, B. "Porosity, pore size distribution and in-situ strength of concrete", *Cement and Concrete Composites*, **33** 155-164 (2003)

[17] "Specifications and Guidelines for Self Compacting Concrete", pp 1-32, EFNARC-2004.

[18] Bungey, J.H., "Durability Tests", pp.37-161 in *Chapter 7: The Testing of Concrete in Structures*, Surrey University Press, London, 2nd Edition, 1984.

APPLICATION OF HIGH PERFORMANCE CONCRETE IN INDIA-SOME CASE STUDIES

S.A. Reddi
Dy.Managing Director
Gammon India Ltd.,
Gammon House, Veer Savarkar Marg,
Prabhadevi, Mumbai 400 025

ABSTRACT

High Performance Concrete (HPC) has been used in India for pre-stressed concrete structures since 1940. Zero slump concrete was in extensive use until about 1970 for pre-stressed concrete construction with concrete grades 40 to 50 MPa.

After the advent of mechanization in construction and use of superplasticisers as chemical admixtures for concrete, higher grades of HPC are being adopted. The use of HPC has also been facilitated by introduction of higher grades of concrete up to 80 MPa in the Indian National Code for Concrete (IS : 456 revised in the year 2000).

60 MPa concrete was first adopted for the reconstruction of the collapsed dome of a nuclear power reactor at Kaiga near Goa in the Nineties, followed by selective use in the subsequent construction of a reactor building.

In the field of bridges and flyovers, HPC has been used for the 2.5 km long J.J. Hospital flyover (75 MPa concrete) and the Bandra Worli Sea Link project (60 MPa concrete).

Some building projects have also been using HPC during the last five years. High Performance Spread Concrete for tunnel linings and rock administration has been adopted for a number of hydro-electric projects. HPC includes the use of silica fume, fibre composites, alkali free superplasticizers, fly ash, GGBS to name a few.

Substantial research has been undertaken by number of research institutions in India on the subject of Self Compacting Concrete. For the Kaiga Atomic Power Project Units 3 & 4, some of the walls and columns with heavily congested reinforcement have been constructed using Self Compacting Concrete, along with high volume fly ash (50% of cementitious materials content).

This paper describes the case histories in detail.

DEFINITION OF HPC

High Performance Concrete (HPC) is defined as a concrete specially designed to meet long term strength and durability requirements in addition to the specific performance characteristics required for the particular structure. Though HPC is commonly associated with high strength (60 MPa and above), in reality the HPC, as name implies, is relevant to one or more of the following performance characteristics :

- Concrete in a marine environment
- High 28 day compressive strength
- High early strength
- Very low/high slump
- Underwater concrete
- Abrasion resistance
- High flexural strength
- Slipform concrete
- Water/air tightness
- Shotcrete
- Fibre reinforced concrete
- Ferrocement
- Colcrete

The first pre-stressed, zero slump concrete structure (the Meerut garrage[1]) was constructed in 1944.

The first pre-stressed concrete railway bridge any where in the World was constructed in 1949 for the Assam Rail Link in Eastern India. The first pre-stressed concrete road bridge constructed in 1952 was the Palar Bridge near Chenai which utilized HPC with zero slump and a high characteristic strength of 40 to 50 MPa. Due to the peculiarities of the Indian codes requiring high target mean strengths, concrete actually used in these structures had characteristic strength of 60 to 75 MPa.

UNDERWATER CONCRETE

During the construction of Ganga bridge at Patna[11] (fig.1) in the Seventies, one of the 12 m diameter wells sunk to a depth of about 55 m and developed cracks. As a remedial measure a fresh well of smaller diameter was constructed inside the cracked well, with heavily reinforced grade M-40 underwater concrete. This represents possibly the deepest underwater structural concrete for a bridge anywhere in the world. The high performance requirement relates its ability to be placed underwater and maintain its structural integrity.

Fig.1 Ganga bridge, Patna

Fig.2 Narmada bridge, Gujrat

The main foundations for the Second Hooghly Bridge[2] in Kolkata were made of cellular reinforced concrete caissons with diameters up to 23 m. The design specifies that the caissons should be kept empty during their service life in order to optimize the dead load. This is achieved by providing concrete of liquid retaining grade for well steining. The bottom plug was carried out utilizing special grades of colcrete / concrete. The colcrete mix and the process were tailored to ensure a water tight bottom plug.

WATER RETAINING STRUCTURES

The earliest pre-stressed concrete water retaining structures were constructed in the Fifties at Kilpauk (fig.3) in Chennai and BARC, Mumbai (fig.4). The characteristic strength realized was an average of 60 MPa. The concrete was dense and water tight. These towers are still in service.

Fig.3 Kilpauk Water Tower

Fig.4 Ball Tank, BARC Mumbai

AIRTIGHT NUCLEAR CONTAINMENT STRUCTURES

Already fourteen nuclear power stations are in operation and nine more are under construction. The concrete in the containment structures is required to have high compressive strength, high tensile strength and be airtight. HPC with characteristic strength of 60 MPa had first been adopted in the reconstruction of the collapsed pre-stressed dome of the Kaiga Power Project in the mid Nineties[3] (fig.5), followed by Units 3 & 4 of Tarapur Atomic Power Project (60,000 m^3) (fig.6).

Fig.5 Nuclear reactor dome at Kaiga

Fig.6 Tarapur Atomic Power Plant Units 3 & 4

In a nuclear power plant the primary containment structure (PCS) is the most important structure. Its principal safety function is to contain the radioactive releases, following a nuclear accident, thereby preventing radiation exposure to the plant personnel, public and environment beyond the acceptable limits. A cross-section of Kaiga NPP, unit-2 consists of a 42 m diameter prestressed concrete vertical cylindrical shell capped with segmented spherical shell dome. HPC has been found suitable for the principal design requirements of moderate compressive strength and high tensile strength, leak tightness, durability and high workability for good quality concreting of PCS. HPC is particularly useful in the dome having congestion of reinforcement, prestressing cables and embedded parts.

The performance parameters for Kaiga HPC were:
Compressive strength : 60 N/mm^2
Tensile strength : 3.87 N/mm^2
Permeability : 5 mm (max)
Workability (slump) : 175 mm $\pm$ 25 mm

For concrete grades of M-60 and above it has become common practice to use silica fume as a mineral admixture. Initially silica fume procured in 25 kg bags was directly fed into the batching plant. Some projects are using silica fume in the form of slurry by mixing it with water (Bandra Worli sealink, Fig.7). On some projects, triple blend cement (OPC + GGBS + Micro silica) has been used (LNG Project at Dahej in Gujarat, Fig.8). For the Bandra Worli sea link project M-50 grade was used for piles and M-60 grade was used for piers and the superstructure. Apart from micro silica, fly ash was also used.

Fig.7 Bandra Worli Sea Link

Fig.8 LNG Storage Tanks, Dahej, Gujarat

Fig.9 JJ Flyover, Mumbai

CONCRETE IN MARINE ENVIRONMENT REQUIRING HIGH DURABILITY

JJ flyover in Mumbai[4] (fig.9) is one of the longest elevated roads in India at about 2.5 km. M-75 grade concrete with micro-silica, has been used for the substructure and superstructure. For this project, 2% of superplasticizer (by weight of cement) was used for concrete with a very low water cement ratio of 0.3. With the addition of superplasticizers, it has been possible to increase the slump up to 200mm despite such a low water cement ratio. Though structural elements were designed for 60 MPa, M75 grade has been adopted in order to ensure durability in the coastal environments, without the need for any reinforcement coatings. It has also opened up opportunities for ready-mixed HPC suppliers.

HPC FOR HIGH EARLY STRENGTH

Effective use of pre-tensioning requires very high early strength (25 to 30 MPa) at the time of release of strands. For seven flyovers in Delhi requiring more than 1000 pre-cast pre-tensioned beams, high early strength mix, above 25 to 30 MPa, in about 12 to 16 hours time was realized. Thus a 24 hours time cycle for production of beams was achieved, without steam curing, thanks to the adoption of high early strength concrete.

HIGH PERFORMANCE FIBRE REINFORCED CONCRETE

The Parliament Library Building in Delhi[5] (fig.10) is a four-story RCC framed structure, two storys below ground and two storys above ground. Precast concrete bubbles were used for the dome structure. SERC, Chennai was consulted on the selection of suitable material for these bubbles. The alternatives considered were Ferro cement, Polymer-Impreganated Concrete Composite and Fibre-Reinforced HPC. Fibre-reinforced HPC satisfied characteristics like high strength, durability and impregnability with protection for embedded steel reinforcement against corrosion and was chosen.

The ingredients[5] of high performance fibre reinforced cement concrete used for dome units by weight were in the proportions of

Cement : 60 % of total cementetious materials
GGBS : 35 % of total cementetious materials
Silica fume : 5 % of total cementetious materials
a/c ratio : 3.5
w/c ratio : 0.35
Superplasticiser : 1-2 % of total cementetious materials

Fig.10 Parliament House Library

Steel fibres used were 0.8 % by volume of concrete. The HP-FRC mix had a characteristic compressive strength of 50 N/m^2 at 28 days. A pan type mixer of capacity 200 litres was specially fabricated to suit the mixing requirement of HP-FRC.

HPC FOR BUILDINGS

For some of the commercial buildings constructed by the Raheja Group in Mumbai, concrete mix with 7 day strength of 51 MPa and 28 day strength of 72 MPa has been successfully realized using a cement content of 425 kg and micro silica of 40 kg/m^3 of concrete.

For the two 65 residential complexes under construction in downtown Mumbai, HPC is being extensively used in order to ensure early de-shuttering and maximum re-use of the formwork. The concrete is being produced at the batching plant installed at the site.

For the Taj Hotel in Whitefield, Bangalore, Self Compacting Concrete (SCC) was used for the reinforced concrete walls, beams and slabs, resulting in very good quality concrete finishing. Vibrators are not used nor are they required for SCC. Because of fluidity of SCC, the formwork is required to be specially designed catering to increased hydro static pressure.

The roof of the Ramakrishna Mandir in Pune[6] consists of ferro cement shells with silica fume. The reported strength at 3 days, 7 days and 28 days were 28, 41 and 66 MPa respectively (fig.11).

The dome of the Sathya Sai Hospital at Puttaparthy[6] in Andhra Pradesh was built with M-35 grade concrete. Micro silica was used to increase the workability and high early strength of the concrete.

Fig.11 Ramakrishna Mandir at Pune

Fig.12 Dul Hasti HRTTunnel lining

HYDRO ELECTRIC POWER PROJECTS

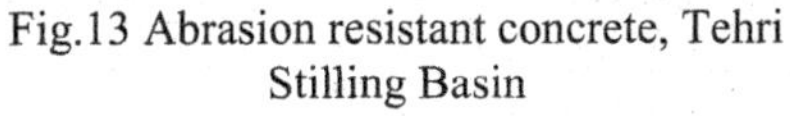

Fig.13 Abrasion resistant concrete, Tehri Stilling Basin

Fig.14 Kinzua dam

HPC with strength up to 60 MPa has been extensively used for tunnel linings, spillways, etc. Outstanding examples include Dul Hasti head race tunnel lining[6] (fig.12), Chamera Stilling Basin[6], Kurichu Dam in Bhutan, head race project at Tala and Tehri Dam Spillway (fig.13). In all these cases, micro silica has been used as a mineral admixture.

Steel fibre reinforced concrete has also been used for a number of hydro electric projects with the following properties:

- Concrete Strength : 36 MPa
- Micro silica : 7 %
- Steel fibres : 7 %
- Accelerator : 8.5 %

Abrasion resistance micro silica concrete has been used for Tehri Dam Stilling Basins.

Shotcreting[6] is adopted for a large number of Hydel projects for initial stabilizing of the rock surface in tunnels prior to concrete lining. Use of micro silica and alkali-free accelerators substantially improves the adhesion and cohesion properties of shotcrete and reduces the rebound resulting in significant savings.

TALA HYDRO ELECTRIC PROJECT IN BHUTAN[7]

Spillways, stilling basins and pressure tunnels are sometimes subjected to very high velocity (40 m/second). Hence HPC with high strength and low abrasion loss was adopted. Abrasion loss typically was reduced from 5.57 % in M-30 concrete to 3.96 % in M-30 HPC with micro silica (fig.15). Particularly for stilling basins considerable damage is caused by eddy currents coupled with high velocity flow and impact due to debris. Use of HPC with high strength and low abrasion loss was considered necessary.

Fig.15 Tala Hydroelectric Power Project

SELF COMPACTING CONCRETE (SCC)

Self compacting concrete can be compacted into every corner of a formwork, purely by means of its own weight and without the need for vibration. In spite of its high flowability the coarse aggregates are not segregated. SCC is produced from normal concreting materials and complies with the strength grades in the Code. The mix may incorporate steel and/or polypropylene fibers. SCC has been produced successfully with coarse aggregate up to 40mm in diameter. Sand can be finer than normal as the material less than 150 microns may help increase cohesion, thereby resisting segregation.

Cement and fillers are required for cohesion and stability in larger proportions than in traditional concrete. These fines can be inert or active materials such as GGBS or PFA. Admixtures are essential to ensure flow characteristics and workability retention. Properties of this concrete like flowability, workability etc cannot be determined by usual tests. Due to its properties SCC can be effectively used in heavily reinforced elements and underwater structures.

SCC was successfully designed and has been used in India at the Kaiga[8,11] nuclear power plant project for concreting the heavily reinforced columns (fig.16a and b).

Fig.16a- Use of SCC in Kaiga for heavily reinforced columns

Fig.16b – SCC, dense concrete without compaction

SCC is also used for some components of a cooling water pump house Tarapur Units 3 & 4[9,11]. Cement, fly ash, micro silica, superplasticisers and viscosity modifying agents were used. After successful mock up trials, some components walls which were heavily reinforced were concreted using SCC mix. The shuttering system was designed for a maximum pour height of 4.50 m. The SCC filled in the formwork and encapsulated the reinforced concrete without vibration and with an excellent surface finish. Approximately 230 m^3 of SCC were used in the pump house.

The second line of the Delhi Metro[10] has seen some of the most technically advanced construction systems to date including the use of Self Compacting Concrete. SCC was used in an effort to solve technically challenging concreting problems in a timely and cost effective manner. The mix design contains high amounts of fines, fly ash and chemical admixtures such as Glenium 51, Glenium Stream 2 and Pozzolith 300R. SCC helped in the following situations:

1. Casting of top down station columns and wall panels in a single 5 m lift, thus reducing pour time and formwork erection time by almost 50 %.
2. The complicated dome structure at Central Secretariat Station concreted without the need for vibration and improved finish quality.
3. Cross passage construction in metro tunnel was done with fast pouring rates, single pumping point and improved surface.

About 10-15000 m^3 of SCC is expected to be used in the Delhi Metro Project.

MUMBAI PUNE EXPRESSWAY

The Mumbai-Pune Expressway (fig.17) is the first international-standard expressway in India made of concrete. Nearly 70 km length of the expressway was constructed and opened to traffic in just 27 months. It has a dual three-lane carriageway, 12.45 m wide each with a separating median of 7.6 m. The expressway also involved construction of five twin tunnels having a total length of 6004 m. The construction of the expressway involved voluminous concrete construction activities. Nearly 1.1 million m^3 of concrete has been produced and used for pavement-quality concrete (PQC) and dry lean concrete (DLC). For the purpose of construction the entire expressway was divided into four sections.

Following parameters were specified by the Maharashtra State Road Development Corporation (MSRDC) for the PQC

Grade of concrete : M40
Flexural strength at 28 days : 4 MPa
Max w/c ratio : 0.45
Max size of aggregate : 20 mm
Max cement content : 425 kg/m^3
Min cement content : 350 kg/m^3
Slump at placing point : 30 ± 15 mm

Fig.17 Mumbai-Pune Expressway

Fig.18 Tunnel on Mumbai-Pune Expressway

The average standard deviation was around 3-4 MPa. This could be achieved due to the use of automated concrete batching plant and strict quality control. For resistance against abrasion and to limit segregation and bleeding the sum of the average values of flakiness and elongation indices were kept well below 25 % limit and varied between 20.8 to 23.98 for different sections.

Use of conventional support system inside the 6004 m long tunnels would have become heavy, uneconomical and also time consuming. Hence, shotcrete with fibre reinforcement and rock bolting system were adopted. 35 MPa grade shotcrete with 6.5 % micro silica and 5 % steel fibres were used for the lining in the tunnels. About 27000 m^3 of shotcrete was done (fig.18).

THE WAY AHEAD

Though India had a head start over others in the use of High Performance Concrete for a variety of situations, there have been constraints due to prevailing Codes and Specifications. The Bureau of Indian Standards have revised the concrete code in 2000 and provided for high strength concrete up to 80 MPa. However, the Indian Roads Congress whose Codes and Specifications govern the design and construction of bridges and flyovers have not yet upgraded the documents to facilitate use of high strength concrete. Some changes towards this end are in the draft stage.

Despite of such handicaps some enlightened users have already been adopting High Performance Concrete for various special structures and during the next decade there will certainly be a high rise in the use of HPC in the country.

REFERENCES

[1] Shama, Robert E. "Application of Prestressed Concrete in India," *Indian Concrete Journal*, Mumbai, **XVIII** (1944)

[2] D.Ghosh et al, "Adoption of Colcrete having special features in plugging the foundation No.2 of Second Hoogly bridge," *Journal of Indian Roads Congress*, New Delhi,.**47**[3], (1986)

[3] Amit Mittal, "Development of High Performance Concrete for containment dome at Kaiga Atomic Power Project," *Indian Concrete Journal*, Mumbai, Apr (1998)

[4] S.A.Reddi, "Design and construction of flyovers/urban viaducts: Recent trends," *Indian Concrete Journal*, Mumbai, Special Issue: Concrete Bridges,**77**[7] (2003)

[5] "Parliament Library building – Documentation," A CPWD Internal Publication, New Delhi, 2003.

[6] Personal Communication from Elkem Chemicals

[7] Rajbal Singh et al, "Design and Application of High Performance Concrete in Wangka Dam at Tala," *New Building Materials & Construction World*, New Delhi, **9**[11] (2004)

[8] S.G.Bapat, "Using Self Compacting Concrete in Nuclear Power Plants," Special Issue on Self Compacting Concrete, *Indian Concrete Journal*, Mumbai, **78** June (2004)

[9] Basu, "Use of SCC in a Pump House at TAPP 3 & 4, Tarapur," Special Issue on Self Compacting Concrete, *Indian Concrete Journal*, Mumbai, **78** June (2004)

[10] "Delhi Metro-Utilising the benefits of SCC," *The MBT Mirror*, Mumbai, **II**[3] July (2004)

[11] Gammon Bulletin, Mumbai, Apr-Jun 2004, Vol.134

FRACTURE CHARACTERISTICS OF HIGH STRENGTH HIGH PERFORMANCE CONCRETE

B. K. Raghuprasad
Professor
Civil Engineering Department
Indian Institute of Science
BANGALORE-560 012, India

B .H. Bharatkumar
Scientist
Structural Engineering Research Centre
CSIR Campus, Taramani
CHENNAI-600 113, India

ABSTRACT

Deterioration of concrete structures has become a major problem causing widespread concern about durability. Experiences shows that having low water/binder (w/b) ratio and chemical and mineral admixtures typically produces high performance concrete (HPC). High strength high performance concrete (HSHPC), having compressive strength more than 50 MPa, is now being used worldwide. However, such concretes exhibit a rather brittle mode of failure and it is necessary to evaluate the fracture characteristics. This paper presents the fracture characteristics of fly ash and slag based HSHPC.

INTRODUCTION

Concrete is the most widely and extensively used construction material in the world, due to its relatively low cost, availability of constituents, versatility and adaptability. In spite of the intrinsic technical and economic advantages of this material and the tremendous scientific advances that have been made in our understanding of its microstructure and engineering, deterioration of concrete has become a major global problem. It has been well established that, besides strength, there are other equally important criteria such as durability, workability and toughness, which have to be met. Such requirements have led to the development of high strength high performance concrete (HSHPC). It is reported that the concrete having low water/binder (*w/b*) ratio and containing chemical and mineral admixtures typically provide high performance concrete (HPC). Use of chemical admixtures helps to reduce the water content, thereby reducing the porosity within the hydrated cement paste. Mineral admixtures, such as fly ash, slag, silica fume, etc., act as pozzolanic materials as well as fine fillers; thereby the microstructure of the hardened concrete matrix becomes denser and stronger. It is also seen from literature that the high strength concrete mix (i.e. compressive strength > 50MPa) is known to exhibit a rather brittle mode of failure [1,2]. This may be due to the refinement in the pore structure of HPC, causing densification of the paste-aggregate transition zone, which, in turn may affect its fracture characteristics. Thus, it is necessary to evaluate the fracture properties of HSHPC, if the safety of structures built with such concrete and their durability are to be assured. This paper presents the results of the mechanical, durability and fracture characteristics evaluated for HSHPC. The influence of mineral admixtures on the fracture properties, such as fracture toughness (K_{IC}), based on linear elastic fracture mechanics approach, fracture energy (G_F) based on work-of fracture and fracture energy (G_f) and size of fracture process zone (c_f) based on size-effect are discussed. It is seen that the fracture characteristics of HSHPC is different than that of normal concrete.

EXPERIMENTAL INVESTIGATION

Ordinary portland cement of 53 grade was used. Crushed granite aggregates of maximum size 12.5 mm were used. The specific gravity, dry-rodded unit weight and water absorption of coarse aggregate were 2.71, 1550 kg/m^3 and 0.5% by weight of the aggregate. River sand passing 4.75 mm sieve was used. The specific gravity and the fineness modulus of sand were

2.62 and 2.48 respectively. A sulphonated napthalene formaldehyde type superplasticizer (SP) was used. Fly ash (Class F) from a thermal power plant near Chennai, India was used. The SiO_2, CaO and LOI contents of fly ash were 59%, 1.02% and 1.08% respectively. Ground granulated blast furnace slag supplied by a local manufacturer in India was used. The SiO_2, CaO and glass content of slag were 33.67, 32.45 and 92% respectively.

Totally 8 series comprising of 4 OPC based mixes, one having a water binder ratio (*w/b)* of 0.50 (without SP) and the other three having w/b ratios of 0.5, 0.4, 0.36 (with SP), 2 fly ash and 2 slag based mixes (with SP) having *w/b* ratio of 0.4, and 0.36 have been investigated. The mix proportions used are given in Table I. A pan type mixer machine was used for preparing the concrete mix. Slump test was carried out on fresh concrete to evaluate its workability.

Table I Details of concrete mix proportions

Sl. No.	Mix ID	Quantities, kg/m^3							Slump, mm
		Cement	Fly ash	Slag	Sand	Agg.	Water	SP	
1	O*-0.50	392	--	--	865	911	196	--	60
2	O-0.50	344	--	--	865	1018	172	0.70	60
3	O-0.40	430	--	--	793	1018	172	0.70	50
4	O-0.36	472	--	--	753	1018	172	0.70	55
5	F-0.40	323	107	--	754	1018	172	0.75	45
6	F-0.36	354	118	--	710	1018	172	0.75	45
7	S-0.40	215	--	215	793	1018	172	0.75	40
8	S-0.36	236	--	236	753	1018	172	0.75	40

O*- OPC without SP, O- OPC, F Fly ash, S –Slag mixes
0.50, 0.40, and 0.36 indicate the *w/b* ratio

The compressive strengths at the ages of 7, 28, and 56 days were evaluated using 100 mm cube specimens. Rapid chloride permeability test as per ASTM C 1202 (1990)[3] using 100 mm dia x 50 mm thick disc shaped specimens were carried out at the age of 56 days. The geometrically similar beam specimens (50 x 50 x 200mm, 50 x 100 x 400mm and 50 x 200 x 800 mm, notch depth = 0.3d) were tested using a servo-controlled UTM under constant displacement rate of 0.05 mm/min. A single point loading system was adopted in which the beam was simply supported over a span equal to four times the depth of beam. The load, deflection and crack mouth opening displacement were measured till failure. The output from the external LVDT, clip gauge and the load channel from UTM were connected to an ORION data logger to record displacement, CMOD and load and data were stored on a computer on-line at 2 second interval. The test was continued till the beam failed.

ANALYSIS OF RESULTS

Mechanical Properties

Mechanical and durability properties for the various mixes are given in Table II. It is seen from the results of mixes O*-0.50 and O-0.50, that the addition of superplasticiser reduces the water content from 196 to 172 kg/m^3 and does not affect the compressive strength at the age of 56 days. It is seen from the results that the use of fly ash (25% replacement) or slag (50% replacement) results in a decrease in compressive strength by 12 to 16% at 7 and 28 days and 7-10% at 56 days. Split tensile strength and flexural strength results also show similar trends. This is because of the slower pozzolanic reaction of the mineral admixture, which caused slow rate of setting and hardening. It is seen from literature that a higher rate of strength development similar to only OPC may be achieved by modifying the mixture proportion using an efficiency factor or application of different curing procedures[4-6]. The values of chloride permeability in terms of total charge passed through the specimen are reduced to lower than 600 Coulombs for various mixes with mineral admixtures.

Table II Mechanical and durability properties of various HSHPC mixes

Sl. No	Mix ID	Compressive strength (MPa) at			Split tensile strength (MPa) at 56 days	Flexural tensile strength (MPa) at 56 days	Chloride permeability, Coulombs
		7 days	28 days	56 days			
1	O*-0.50	30.67	44.69	49.36	3.17	5.55	5337
2	O-0.50	24.50	47.30	49.91	3.26	5.53	3967
3	O-0.40	41.93	63.49	65.99	4.38	6.20	3096
4	O-0.36	45.25	69.73	71.27	5.04	6.81	2604
5	F-0.40	35.65	54.73	59.35	4.14	5.61	556
6	F-0.36	39.76	60.17	66.46	4.86	6.26	500
7	S-0.40	37.07	55.55	58.76	4.16	5.50	601
8	S-0.36	38.06	58.60	65.23	4.71	6.32	550

Fracture Characteristics

Fracture characteristics, such as fracture toughness (K_{IC}) based on LEFM, fracture energy (G_F) based on work-of fracture and fracture energy (G_f) and size of fracture process zone (c_f) based on a size effect model for various HPC mixes are given in Table III. The details of evaluation of various fracture parameters are given by Bharatkumar et.al. (in press)[7].

Fracture toughness (K_{IC} or G_C) based on LEFM: Fracture toughness (K_{IC}) is the value of the critical stress intensity factor K, for which the crack starts growing. K_{IC} values for various mixes were obtained from a peak load based on LEFM approach and are given in Table III. It is seen that with a decrease in *w/b*, there is an increase in the fracture toughness (5-10%). It is also seen that there is a reduction in the fracture toughness (10-25%) due to addition of fly ash or slag. Taylor and Tait[8] have also observed that at initial ages, the fly ash was found to have increased fatigue resistance and toughness due to the spherical fly ash particles having a blunting effect. However with increase in age, the mixes containing fly ash were slightly less resistant to fatigue crack growth due to the poor bond between the gel and the large unhydrated fly ash particles, acting as flaws that were now larger in relation to other flaws in the matrix, which had reduced with continued hydration. It is also seen that the fracture toughness increases as the depth increases, consistent with evident from the literature that shows that the fracture toughness (K_{IC}) is dependent on the beam size. This variation of K_{IC} is also reflected in the corresponding toughness value G_C, since K_{IC} and G_C are directly related.

Table III Fracture characteristics of HSHPC mixes

Mix ID	Beam depth d (mm)	K_{IC}	G_{IC}	G_F	G_f	c_f	l_{ch}	S_e	$1/\beta$
	50	26.44	19.89	100			0.63	0.63	0.89
O*-0.50	100	32.18	29.48	122	44.03	10.95	0.38	0.38	1.78
	200	34.62	34.11	129			0.20	0.20	3.57
	50	26.82	20.36	67			0.41	0.41	1.28
O-0.50	100	33.90	32.54	98	40.49	7.61	0.30	0.30	2.57
	200	33.92	32.57	82			0.12	0.12	5.13
	50	32.76	26.43	88			0.40	0.40	1.39
O-0.40	100	35.19	34.08	125	43.50	7.04	0.29	0.29	277
	200	38.68	40.65	130			0.15	0.15	5.55
	50	36.48	31.53	114			0.45	0.45	1.78
O-0.36	100	37.93	34.08	127	46.37	5.48	0.25	0.25	3.56
	200	41.42	40.65	132			0.13	0.13	7.13
	50	26.79	18.63	76			0.37	0.37	0.73
F-0.40	100	33.22	28.65	93	44.08	13.4	0.22	0.22	1.46
	200	36.21	34.04	106			0.13	0.13	2.92
	50	26.99	17.87	70			0.29	0.29	0.74
F-0.36	100	33.26	27.15	96	44.84	13.17	0.20	0.20	1.48
	200	36.38	32.46	118			0.12	0.12	2.97
	50	26.15	17.85	81			0.39	0.39	0.63
S-0.40	100	29.00	21.94	100	40.72	15.55	0.24	0.24	1.26
	200	34.06	30.26	105			0.13	0.13	2.51
	50	27.85	19.21	76			0.33	0.33	0.87
S-0.36	100	32.53	26.20	100	41.09	11.21	0.21	0.21	1.74
	200	35.77	31.68	107			0.11	0.11	3.49

K_{IC} is in MPa $(mm)^{0.5}$, G_{IC}, G_F, G_f, are in J/m^2, c_f is in mm

O*- OPC without SP, O- OPC, F Fly ash, S –Slag mixes

Fracture energy (G_F) from work-of-fracture (W_F): Fracture energy G_F is the energy needed to create a crack of unit area and is also called as the specific fracture energy. The work-of fracture W_F was calculated by measuring the area under the load-deflection plot and the fracture energy was calculated and is given in Table III. For a classically brittle material $G_F = G_C$, however, for concrete, which is a quasi-brittle material, G_F is higher than G_C because stable crack growth before failure takes place in quasi-brittle materials. It is seen that G_F varies from 70 to 132 J/m^2, while G_C varies between 17 to 40 J/m^2, which are in the range reported in the literature[9]. Moreover, G_F is not constant, but varies with beam size; increases with an increase in depth of beam.

The post peak path of load-deflection plot seems to be steeper, when fly ash or slag is added due to densification and also there is a slight reduction in the peak load. Thus, the area under the load-deflection plot is decreased due to addition of fly ash or slag, which results in a decrease in G_F. Figure 1 shows the variation of fracture energy G_F with compressive strength for all series with and without fly ash or slag (maximum aggregate size used is 12.5 mm). It is seen that the fracture energy increases with increase in strength. The above relationship is similar to that suggested in CEB-FIP code.

Fracture energy (G_f) and fracture process zone (c_f) from size effect law: Fracture energy (G_f) and fracture process zone (c_f) obtained from size effect law[13], using geometrically similar specimens are given in Table III. It is seen that the addition of fly ash or slag decreases the fracture energy slightly. However, the fracture energy (G_f) increases with the increase in strength (Fig.1). It is also seen that the fracture energy based on size effect law (G_f) is less than fracture energy based on work-of fracture (G_F). The fracture energies G_F and G_f are two different material characteristics and are weakly related [(10)]. For the present investigation, the ratio of (G_F/G_f) for d=100 mm is around 2.5, consistent with that of Planas et.al. [(11)].

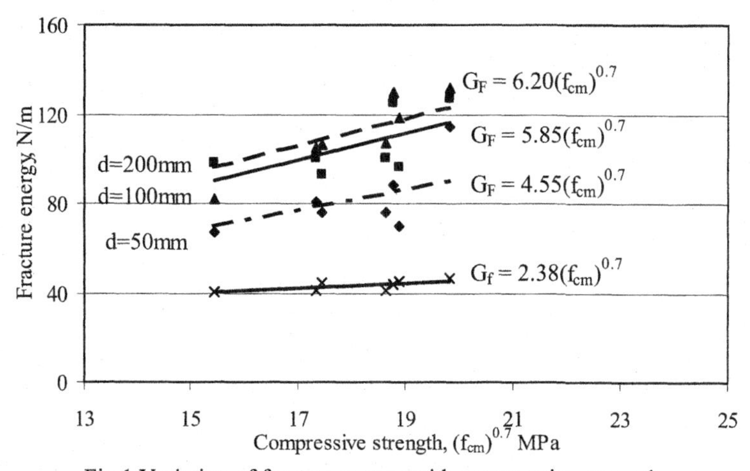

Fig.1 Variation of fracture energy with compressive strength

Intrinsic Brittleness

An increase in the fracture energy with strength should not be confused with an increase in its ductility. While discussion on the fracture properties of concrete, it is more convenient to use brittleness, inverse of ductility. The brittleness of concrete is characterized not only by the fracture energy or toughness, but also by a parameter related to it through other fracture and or elastic constants. According to the fictitious crack model (FCM), the intrinsic brittleness of concrete is quantified by its characteristic length:

$$l_{ch} = E\, G_F/f_t^2 \tag{1}$$

where E is the young's modulus, G_F is the fracture energy, f_t is the tensile strength. The intrinsic brittleness based on FCM (l_{ch}) for the various mixes are given in Table III. Figure 2 shows the variation of intrinsic brittleness of HPC as per FCM with the depth of specimen for various HPC mixes. The l_{ch} is found to vary between 120-450, which is in the range reported in the literature[9]. It is seen that as the concrete strength increases (i.e. decrease in *w/b* ratio) there is a reduction in the l_{ch} value indicating that the material is becoming brittle. This may be due to the densification of the HPC due to reduction in *w/b* ratio. The l_{ch} also seems to depend on the size of the specimen and as size increases there is an increase in l_{ch}. The relationship between the intrinsic brittleness of HPC (for 100 mm deep beams) and the compressive strength for the material investigated is:

$$l_{ch} = A\,(f_{cm})^{-0.3} \tag{2}$$

where A =781, is an empirical constant, f_{cm} is the cube strength in MPa and l_{ch} is the intrinsic brittleness in mm. This relationship is similar to that given by Hilsdorf and Brameshuber[12], indicating an increase of brittleness with increase in strength.

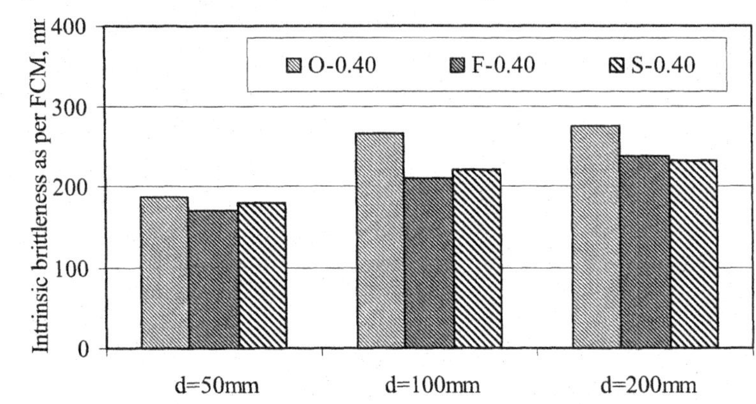

Fig.2 Variation of Intrinsic brittleness of HPC as per FCM (w/b =0.40)

According to this size effect model, the length of the fracture process zone c_f determined by the test method proposed by RILEM[13] is the measure of the brittleness. The value of c_f is found to be varying between 5.5 and 15.5, which is also in the range reported in the literature. It is seen from Table III that the intrinsic brittleness c_f reduces as the *w/b* ratio decreases (i.e. increase in strength) indicating that the concrete is becoming more brittle. Also the reduction in water content due to addition of SP results in a reduced c_f value. This indicates that the fracture process zone reduces as the strength increases due to densification of concrete resulting in a more homogenous mix.

The effect of fly ash or slag on c_f is different from its effect on l_{ch}, i.e c_f is found to increase due to addition of mineral admixtures indicating that the material is becoming less brittle or the length of fracture process zone is increased. The reason may be due to the fact that determination of c_f depends only on the peak loads of notched geometrically similar specimens, while l_{ch} depends on the complete load-deflection plot and on the material characteristics, such as elastic modulus and tensile strength, which are determined from un-notched specimens. However, it can be seen that the strength of specimens or structures made out of quasi-brittle materials such as concrete depend on their size and that the size effect is different for mixes with or without fly ash or slag. Thus, the addition of mineral admixture produces more micro-cracks ahead of the crack-tip, which causes the material to fail at lesser load, resulting in lower fracture strength with larger fracture process zone.

Brittleness of Concrete Structure

The brittle response of a concrete structure should not be confused with the intrinsic brittleness of concrete. The brittleness of concrete structural elements depends on their size. Small elements fail in a ductile or plastic manner, while large elements of the same material fail in a brittle and often catastrophic manner. Carpinteri[14] proposed a parameter proportional to $(l_{ch}/d)^{0.5}$ as the indicator of concrete structural brittleness, but later introduced the energy brittleness number s_e, defined as

$$s_e = G_F/f_t\, d \tag{3}$$

Bazant & Pfeiffer[15] proposed the structural brittleness number $1/\beta$, which reflects the geometry and initial notch depth of the specimen besides intrinsic brittleness of concrete and the structural size, as:

$$\beta = d\, g(\alpha_0)/c_f\, g'(\alpha_0) \tag{4}$$

where $g(\alpha_o)$ and $g'(\alpha_o)$ are functions of notch depth, d is the characteristic dimension of the structure and c_f is the intrinsic brittleness based on the size effect method.

The brittleness of a concrete structure based on Eq. 3 and Eq. 4 for the present investigation are given in Table III. It is seen that both values s_e and $1/\beta$ decrease with increase in the depth of beam. The effect of strength and mineral admixture is the same as that of intrinsic brittleness of HPC. Some suggest that β is capable of characterizing the type of failure (brittle or ductile) regardless of structural geometry. It quantifies how closely of the behaviour of a structure approximates LEFM and therefore is a convenient and effective measure of the brittleness of a structure or specimen. The brittleness number s_e (based on FCM) can compare only structures of similar geometry.

Size Effect on Nominal Strength

Size effect is defined through a comparison of geometrically similar structures of different sizes, and is conveniently characterized in terms of the nominal stress σ_N ($=P_{max}/td$). When σ_N values for geometrically similar structures of different sizes are the same, we can say that there is no size effect. A dependence of σ_N on the structure size is called "size effect". Bazant[16] assumed that the total potential energy released at fracture is proportional to the square of the crack length a, which scales proportionally to the specimen size (a/d = constant), while the energy dissipation is proportional to a, the width of the crack band being assumed constant and proportional to the maximum aggregate size g. The nominal stress as per this size effect law is:

$$\sigma_N = Bf_t / (1 + d/d_0)^{1/2} \tag{5}$$

Typical normalized plots in Fig. 2 show the variation of $\log(\sigma_N/Bft)$ with $\log(d/d_0)$. The strength limit as well as LEFM limit is also plotted. Experimental points seem to be following the size effect law and lie in between strength and LEFM failure criteria. It is seen that decrease in the w/b ratio for both OPC-based or mineral admixture-based concrete shifts the behaviour towards the applicability of LEFM (i.e $d/d_o > 10$). This indicates the depth required for application of LEFM is lower as there is increase in strength of concrete mix. Hence, high strength concrete is more brittle and LEFM can easily be applied to these fracture properties. However, for same the w/b ratio, it is seen that due to the addition of fly ash or slag, the depth required for application of LEFM is more, i.e. less brittle.

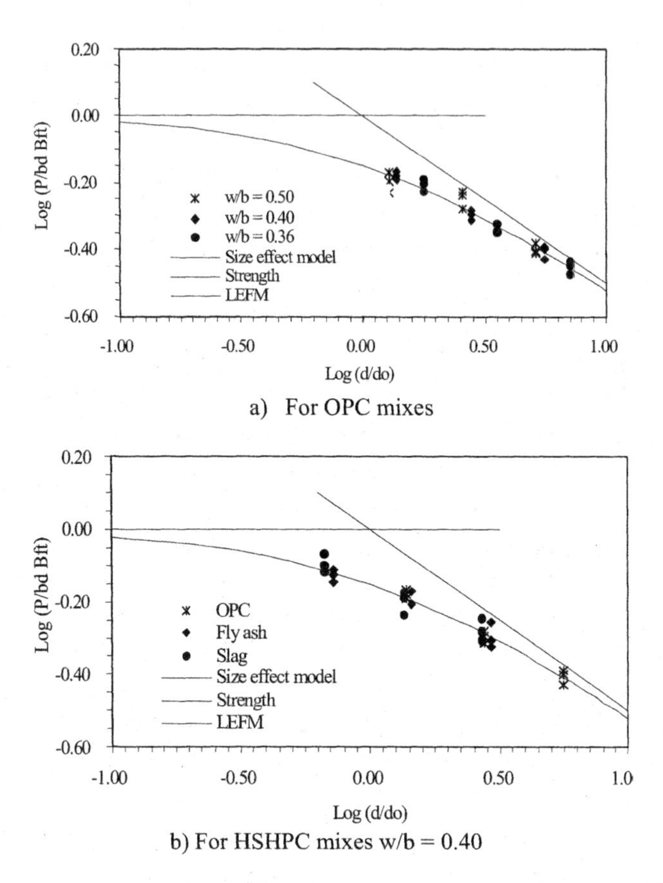

Fig. 2 Typical size effect plots for various HSHPS mixes

CONCLUSIONS

Based on the experimental investigations on the effect of mineral admixtures on the fracture characteristics of HSHPC, the following conclusions are made.

- There is an increase in the fracture toughness (K_{IC}) by 5-10% as *w/c* decreases, due to the densification of concrete leading to higher failure load. There is a reduction in the fracture toughness (K_{IC}) by 10-25% due to addition of fly ash or slag at the same *w/b* ratio and this may be due to the presence of larger particles of unhydrated fly ash or slag that act as flaw.
- It is also seen that the fracture energy based on size effect law (G_f) is less than fracture energy based on work-of fracture(G_F). As reported in the literature, the fracture energies G_F and G_f are two different material characteristics and are weakly related. For the present investigation, the ratio of (G_F/G_f) for d= 100 mm is around 2.5.
- The value of intrinsic brittleness of HSHPC (l_{ch} and c_f) is found to be in the range reported for high strength concrete in literature. The effect of fly ash or slag on the c_f is different from that on l_{ch}, i.e c_f is found to increase due to addition of fly ash or slag, indicating that the material tends to become less brittle, while l_{ch} is found to slightly

decrease for concrete mix containing fly ash or slag indicating that concrete becomes slightly more brittle than OPC. This contradicting behaviour may be due to the method of determination of c_f, which depends only on the peak loads of notched geometrically similar specimens, while l_{ch} depends on the complete load-deflection plot and on the material characteristics, such as elastic modulus and tensile strength. It may be said that c_f reflects structural brittleness, while l_{ch} reflects the intrinsic material brittleness. The intrinsic material brittleness is due to the material properties E, f_t, which are determined from unnotched beams. It is also influenced by G_F to a lesser extent. G_F is the only parameter, which reflects structural brittleness. Therefore, l_{ch} is a good combination of intrinsic material brittleness and structural brittleness and could be more reliable.

- HSHPC exhibits size effect and is depends on the constituents of the concrete mix. The depth required for application of LEFM is lower as there is increase in strength of concrete mix. However, for same the *w/b* ratio, it is seen that due to the addition of fly ash or slag, the depth required for application of LEFM is more, i.e. less brittle.

ACKNOWLEDGEMENT

One of the authors, Dr. B. H. Bharatkumar wishes to thank the Director, Structural Engineering Research Centre, Chennai, India for his kind permission to publish this paper.

REFERENCE

[1]G.A. Rao, and B.K. Raghu Prasad, "Size effect and Fracture Properties of HPC," pp.21-24 *Proc. of the 14th Engineering Mechanics Int. Conference (ASCE)*, Texas, Austin, 2000

[2]G.A. Rao, and B.K. Raghu Prasad, "Fracture Energy and Softening behaviour of High Strength Concrete," *Cement and Concrete Research*, **32**, 247-252 (2002)

[3]ASTM C1202, "Standard Test Method for Electrical Indication of Concrete's Ability to Resist Chloride Ion Penetration," *Annual book of ASTM Standard, 4.02, Concrete and Aggregates*, pp.624-629 (1990)

[4]R.N. Swamy, "Design for Durability and Strength Through the use of Fly ash and Slag in Concrete," *ACI SP-171*, pp. 1-72 (1997)

[5]P.K. Mehta, "Concrete Technology at the Cross-Roads-Problems and Opportunities," *ACI SP-144,* pp.1-30 (1994)

[6]B.H. Bharatkumar, R. Narayanan, B.K. Raghu Prasad, and D.S. Ramachandramurthy, "Mix Proportioning of High Performance Concrete," *Journal of Cement and Concrete Composites*, **23** 71-80, (2001)

[7]B.H. Bharatkumar, B.K., Raghuprasad, D.S. Ramachandramurthy, R. Narayanan, and S. Gopalakrihsnan, "Effect of Fly Ash and Slag on The Fracture Characteristics of High Performance Concrete," Accepted for publication in RILEM journal of Materials and Structures

[8]P.C. Taylor, and R.B. Tait, "Effect of Fly ash on Fatigue and Fracture Properteis of Hardened Cement Mortar," Cement and Concrete Composites, **21** [3] 223-232, (1999)

[9]S.P. Shah, S.E. Swartz, and C. Ouyang, "Fracture Mechanics of Concrete," John Wiley and Sons, New York, (1995)

[10]Z.P. Bazant, Yu Qiang and Zi Goangseup, "Choice of Standard Fracture Test for Concrete and its Statistical Evaluation," *International Journal of Fracture*, **118** 303-337, (2002)

[11]J. Planas, M. Elices, and G.V. Guinea, "Measurement of the Fracture Energy using Three-point Bend Tests: Part-2 Influence of Bulk Energy Dissipation," *Materials and Strcutres*, **25**, pp.305-312, (1992)

[12]H.K. Hilsdorf and W. Bramseshuber, Code-type Formulation of Fracture Mechanics Concepts for Concrete," *International Journal of Fracture*, **51** 61-72

[13]RILEM Committee FMT 89, "Size-effect Method for Determining Fracture Energy and Process Zone Size of Concrete," *Material and Structures*, **23** 461-465 (1990)

[14]A. Carpinteri, "Notch Sensitivity in Fracture Testing of Aggregative Materials," *Journal of engineering Fracture Mechanics*, **116** 467-481, (1982)

[15]Z.P. Bazant and P.A. Pfeiffer, "Determination Fracture Energy Properties from Size Effect and Britleness Number," *ACI Material Journal*, **84** 463-480, (1987)

[16]Z.P. Bazant, "Size-effect in Blunt Fracture: Concrete, Rock, Metal," *ASCE Journal of Engineering Mechanics*, **110** 518-535 (1984)

Fiber-Based Systems for High Performance

HIGH PERFORMANCE HYBRID COMPOSITES FOR THIN CEMENTITIOUS PRODUCTS: THE NEXT GENERATION

A.E. Naaman, T. Wongtanakitcharoen, and V. Likhitruangsilp
Department of Civil and Environmental Engineering
University of Michigan
2378 G.G. Brown Building
Ann Arbor, MI 48109-2125

ABSTRACT

Most generally a hybrid composite may imply the use of meshes (or textiles) of different materials (such as carbon and PVA), fibers of different materials or properties (length, diameter, modulus, strength), and/or a combination of them. Here the term "hybrid composite" is used to indicate a combination of continuous meshes such as kevlar or steel, with discontinuous fibers such as PVA or steel. The fibers are generally premixed with the mortar matrix. However, a prefabricated fiber mat used as core or spacer, or a fiber network made of dispersed fibers allowing infiltration by the mortar matrix, can be used. Thus, at one extreme, hybrid composites qualify as ferrocement or textile (FRP) reinforced composites and, at the other extreme, they qualify as fiber reinforced concrete. The study described in this paper explores different ideas to manufacture thin hybrid cement composites with the objective to facilitate constructability and reduce labor cost while maintaining or improving efficiency. The reinforcing systems tried include, respectively, 2D meshes and fibers, 2D meshes and a fiber mat, 3D meshes, and very high strength steel wires. Details of each reinforcing system are described. Bending tests are carried out on small beams or plates. The experimental results are presented and compared to demonstrate the effectiveness of each system. Moduli of rupture close to 100 MPa were achieved with less than 3% total volume fraction of reinforcement. These composites are equally strong under both positive and negative bending. It should be observed that the fibers, while contributing their share of resistance, also contribute to crack control, toughness and ductility, and help change the failure mode from interlaminar shear to either shear or bending.

INTRODUCTION

In a number of previous studies the case was made to develop better ferrocement and laminated cementitious composites by using meshes (or textiles or fabrics) made out of advanced fiber reinforced polymeric reinforcements [Refs. 12 to 16]. With a thickness ranging from about 5 to 50 mm, these composites are considered part of the family of thin concrete products. It was argued that while material cost is higher for FRP meshes than for steel meshes, FRP meshes can be tailored to exact requirements (that is, size, diameter, and opening) at little cost compared to steel wire mesh. Moreover, although FRP reinforcements are brittle, the mesh configuration and progressive failure of the separate yarns in the composite under loading help achieve a reasonably ductile composite response. These advantages may give FRP meshes (or textiles or fabrics) some advantages over meshes made from steel wires. Moreover, hybrid reinforcing systems comprising continuous meshes with discontinuous fibers may offer unique benefits [Refs. 1 to 18].

This paper focuses on exploring innovative ideas with hybrid compositions of fibers and 2D meshes, and three dimensional (3D) meshes or textiles. The main objectives are to improve composite performance, ease of construction and cost. The use of several layers of mesh in typical ferrocement structures significantly increases the cost of the composite not only because of material cost but also because of labor cost. Material and labor consume more than 90% of the cost of the composite and every attempt at reducing them is welcome.

Several ideas are described in this study with the objective of producing thin concrete products with improved performance and ease of construction. They include hybrid composites using 2D meshes and fibers, 2D meshes and a fiber mat, 3D meshes or textiles, and very high strength steel wires. Details of each reinforcing system are described below, followed by an experimental program where their effectiveness is demonstrated and compared.

ADVANTAGES OF HYBRID COMPOSITES WITH MESHES AND FIBERS

Most generally a hybrid composite may imply the use of meshes of different materials (such as carbon and PVA), fibers of different materials or properties (length, diameter, modulus, strength), and/or a combination of them. Here the term "hybrid composite" is used to indicate a combination of continuous meshes such as carbon or steel, with discontinuous fibers such as PVA or steel. The fibers are generally premixed with the mortar matrix. However, a prefabricated fiber mat used as core or spacer, or a fiber mat made of dispersed fibers allowing infiltration by the mortar matrix, can be used.

Several reasons have led to the idea of selecting hybrid composites for ferrocement structural components [15, 13]:

1. For bending applications, increasing the number of layers of mesh (or the volume fraction of reinforcement) does not lead to a proportional increase in bending resistance. This is because the farther the layers of mesh are placed from the extreme tensile fiber the less efficient they are. Indeed meshes close to the neutral axis contribute much less. This is confirmed not only analytically but also experimentally. Thus the layer of mesh that contributes the most is the extreme layer; optimizing its properties has a strong impact on composite properties and cost efficiency. Therefore, it seems that using only two layers of mesh in a thin ferrocement sheet is the most cost effective way to use the mesh. However, this can also limit the maximum strength that could be achieved otherwise.
When high strengths, high moduli meshes are used with normal strength mortar matrices, failure under bending load seems to be primarily due to either shear delamination of the extreme layer of mesh and its mortar cover, or to vertical shear, unless extensive ties are used. Numerous experimental tests with carbon meshes augmented by analysis of the composite confirm such behavior.

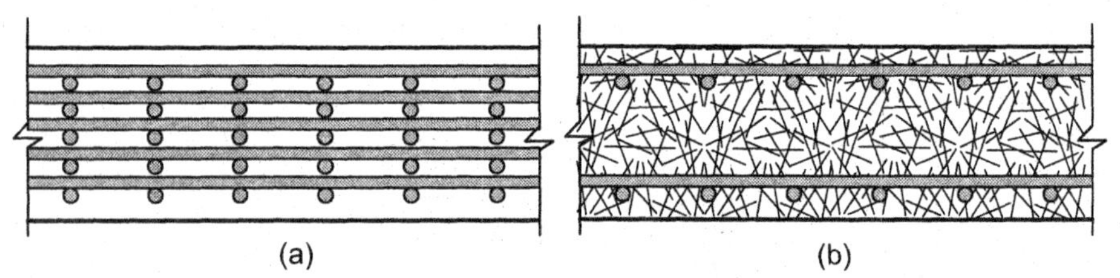

Figure 1 (a) Typical section of ferrocement with several layers of mesh. (b) Typical section of efficient hybrid composite with only two extreme layers of mesh and fibers

3. In numerous situations, ferrocement sheets or structural elements need to be drilled either to attach other structural or non-structural elements, or simply to provide a hole for piping or wiring. Drilling is easier when the mortar matrix is made of fine particles and when meshes

are FRP instead of steel. However, under best circumstances, drilling leads to numerous fracture sites in the matrix around the primary hole being drilled, leading to a poor bearing surface.

A remedy to the above three problems is the use of discontinuous fibers (or micro-fibers) in the matrix; essentially the conventional ferrocement composite of Fig. 1a is replaced by the hybrid composite shown in Fig. 1b. Appropriately selected fibers will add to the strength and toughness of the composite in bending, will help increase both the vertical shear resistance and interlaminar shear resistance at the interface of the extreme layer of mesh, and will help in allowing for cleanly drilled holes in the composite. Fibers also impart other properties such as reduced crack width and spacing, and improved impact resistance. As a result of using fibers in combination with meshes, a hybrid composite with optimized properties can be obtained.

ADVANTAGES OF THREE-DIMENSIONAL (3D) MESHES

As mentioned above, when the reinforcement used in ferrocement has a combined high strength and high elastic modulus (such as for steel or carbon meshes) failure of the composite is often controlled by shear or interlaminar shear (between layers of reinforcements) instead of bending. There is thus need to provide reinforcement, that is, shear reinforcement in the third direction. This can be achieved by providing reinforcement connecting the two extreme layers of 2D mesh together, such as shown in Fig. 2b to 2d. The resulting reinforcement system, when connected together, leads to a 3D type of mesh which has the advantage of providing an armature system that can be easily placed in a mold and infiltrated to its full thickness by a matrix. Thus the main advantage is that, while the reinforcement in the third dimension provides shear resistance, easier construction thus reduced cost ensues.

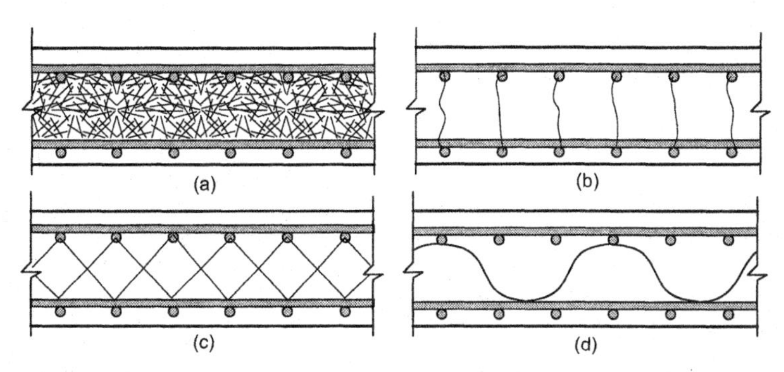

Figure 2 Examples of tri-dimensional reinforcements for thin concrete products: a) 3D fiber mat taken in sandwich between two extreme layers of 2D mesh; b) 3D mesh made from two extreme layer of 2D mesh and vertical wires or threads; c) 3D mesh made from two extreme layers of 2D mesh connected by zig-zag wires; d) 3D mesh made from two extreme layers of 2D mesh connected by sinusoidally shaped wires or 2D mesh.

The "Institut fur Textiltechnik – ITA" at the Technical University of Aachen, Germany, and the Institute of Construction at the Technical University of Dresden, Germany, have capability to produce 3D meshes (also described as textiles or fabrics) of different types [Refs. 6 and 8]. Two such meshes were produced by ITA in Aachen for this study and they were tested for comparison with other composites (Figure 3). Both are made of two extreme layers of 2D glass mesh with square openings connected by reinforcement in the third direction. For the case of mesh Type A the reinforcement in the third direction consists of strings made of polyethylene yarn; in the case of mesh Type B, the reinforcement is made of an undulating mesh connected (adhesively glued) at each contact surface with the top and bottom mesh layer. Approximate mesh properties are given below in Table 4. These meshes had different distances between the two extreme layers and were used as received thus leading to composites of different thickness. However, by comparing modulus of rupture for a given reinforcement volume, some conclusions can be drawn.

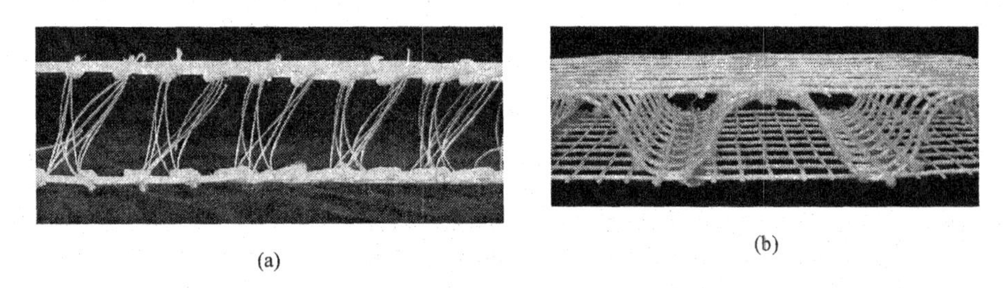

(a) (b)

Figure 3 Photos of 3D meshes or textiles (Produced by ITA, Aachen, Germany). (a) Type A and (b) Type B

ADVANTAGES OF THREE-DIMENSIONAL (3D) FIBER MAT

Another way to improve vertical and horizontal or interlaminar shear in a laminated composites, is to use fibers or micro-fibers in the matrix. Fibers can be premixed with the mortar matrix or used in the form of a fiber mat (Fig. 2a). A fiber mat is a tri-dimensional network of fibers connected in space and forming a volume with empty openings. Examples of mat using polypropylene fibers were used in prior studies [Refs. 9, 18]. When a fiber mat is used, the reinforcement network is simply infiltrated by a cement-slurry matrix. A fiber mat can be designed-tailored with a prescribed thickness, and with some flexibility in the equivalent volume fraction of fibers it adds to the composite. Volume fractions of less than 2% are desirable and possible for cement matrices. The fibers may be continuous or discontinuous. A fiber mat can be used by itself as reinforcement (adding strength, toughness and drill-ability to the composite), or it can be used as the core of a sandwich type construction. In the latter case, two extreme layers of mesh take in sandwich the fiber mat. The key advantage of this type of construction is to minimize construction labor due to placing since there is no need to provide spacers between the meshes; moreover, the composite obtained has a consistent thickness throughout. A commercialized fiber mat, produced by Kuraray Inc., Japan, is shown in Fig. 4 and was used in this study. It is made out of PVA fibers and is about 75 mm in thickness. When used in a ferrocement type plate, the equivalent volume fraction of fibers it provides is about 1%.

Figure 4 Examples of sandwich construction with 3D PVA fiber mat and various reinforcing meshes

ADVANTAGES OF HIGH STRENGTH REINFORCING MESH

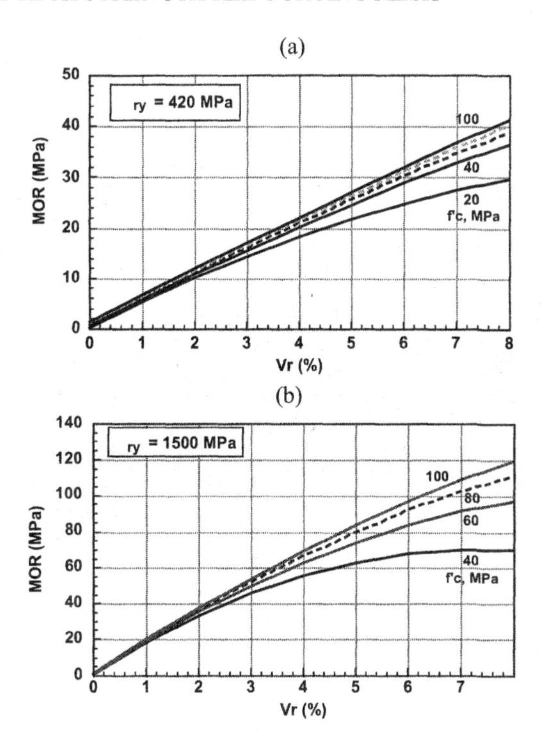

Figure 5 Theoretical values of modulus of rupture (MOR) for laminated cement composites at 2 different levels of yield strength of the reinforcing mesh.

Prior experimental investigations on standard ferrocement plates using conventional steel wire mesh reinforcement have concluded that better performance (tensile and bending resistance) can be achieved when high yield strength wires are used. This conclusion is also supported by analytical predictions and holds for FRP reinforcements as well [Refs. 13, 16]. Figure 5 illustrates the influence of increasing the yield strength of wires on the bending resistance (modulus of rupture) of ferrocement plates of rectangular cross-section. It can be observed that the modulus of rupture can be tripled if the yield strength of the wire mesh increases from 500 MPa to 1500 MPa. The chart assumes several layers of mesh are stacked in the ferrocement sheet, but it also applies when two extreme layers are used in thin sheets.

Note, however, that most commercially available steel meshes do not offer yield strengths above about 600 MPa. There is therefore a large gap between the yield strength of commercially available wire meshes used for ferrocement, and the yield strengths available for prestressing strands or piano wires, which can exceed 2400 MPa.

A product with trade name Hardwire© has been recently put on the market in the US. It is marketed as a substitute to adhesively bonded fiber reinforced polymeric (FRP) sheets or plates such as carbon or Kevlar, used for repair of reinforced concrete members. Hardwire© is similar to a 2D mesh. In the primary direction it comprises parallel steel strands spaced at approximately 6.25 mm (different spacing is also available); the strands are held in place (adhesively bonded) by a square mesh made from glass fibers. Thus the product looks like a wire mesh (Fig. 6). However the glass fiber mesh is not strong or significant and is used only as support to the steel strands. The strands are made each from five steel wires with approximate diameter of 0.3 mm each. The wires have very high yield strength of the order of 3150 MPa and are typically produced to fabricate tire cord for high performance tires. To simulate a two dimensional mesh, two layers of Hardwire© placed parallel to each other are needed. Although Hardwire© has been introduced as an alternative to adhesively bonded FRP sheets, it is also suitable as reinforcement for ferrocement and thin concrete products, as demonstrated below.

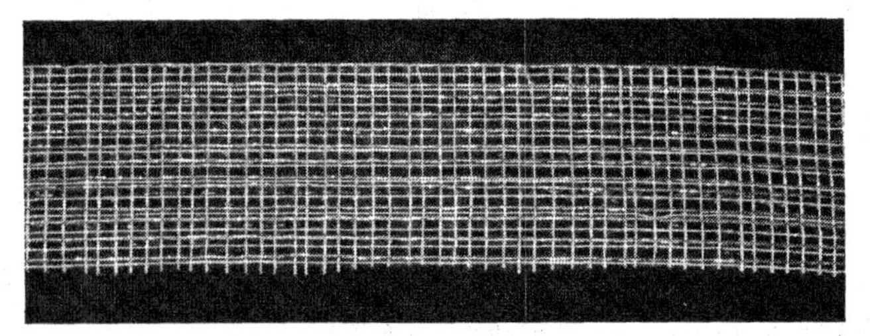

Figure 6 Hardwire© reinforcement with a 6.25 mm strand spacing held by a square glass fiber mesh with 6.25 mm opening.

EXPERIMENTAL PROGRAM

An exploratory experimental program was carried out on the bending response of ferrocement (or reinforced concrete thin sheets) using the main ideas described above, namely the use of hybrid composites with a fiber mat, the use of 3D mesh, and the use of very high strength wires. Twelve series of tests were carried out. Because each product (such as a 3D mesh or mat) had specific dimensions, the dimensions of specimens were not the same for all series of tests. Details of the experimental parameters are given in Table 1. All specimens were tested in four point bending. Typical bending test set-up is shown in Fig. 7. Since the main objective was to reduce labor cost by using only two extreme layers of reinforcement, no attempt was made to match the volume fraction of reinforcement for the different series of tests. Thus performance comparison is made primarily based on the modulus of rupture, cracking characteristics and the failure mode, for a given total volume of reinforcement. For each series, at least 3 specimens were prepared and tested. The average response was obtained either using a specially designed averaging program, or selecting a specimen representative of the average response of each series.

Note that three matrix mixtures were used depending on the fabrication process. For infiltration of fiber mat, a slurry matrix containing cement and fine sand was used. For premixed fibers, a high strength mortar (HSM) matrix was used. Mix proportions are given in Table 2. Fiber properties are given in Table 3 and mesh properties, when available, are given in Table 4. Note that the Kevlar mesh had the following properties: 10 yarn per inch in the longitudinal (L) direction and 6 yarns per inch in the transverse (T) direction; leno weave; 1500 denier per yarn; 2800 MPa tensile strength.

Table 1 Experimental Program

Series no.	**Matrix**	**Reinforcement**						**Fiber**		**Specimen Size (mm)**
		Mat		**Mesh**						
		Type	**Vf (%)**	**Type**	**Vr (%)**	**VrL**	**VrT**	**Type**	**Vf(%)**	
1	Slurry 2	PVA mat	1	-	-	-	-			75x75x540
2				Square steel mesh Gauge 16	0.831	0.416	0.416			
3				Square steel mesh Gauge 19	0.352	0.176	0.176			
4				Kevlar mesh	0.033	0.020	0.012			
5				Hardwire© (unidirectional)	0.294	0.294	-			
6	HSM	None	0	3-D Glass Type A	1.368	0.684	0.684	-	-	75x12.5x300
7	HSM			Hardwire© 1 tensile layer	0.882	0.882	-	PVA	1	
8	HSM			Hardwire© 2 extreme layers	1.763	1.763	-	PVA	1	
9	HSM			3-D Glass Type B	0.684	0.342	0.342	-	-	75x25x300
10	Slurry1			Hardwire© 2 extreme layers	0.294	0.294	-	Torex	3.50	75x37.5x540
11	HSM			Hardwire© 2 extreme layers	0.294	0.294	-	Torex	1.50	
12	HSM	None	0	3-D Glass Type A	1.368	0.684	0.684	PVA	1	75x12.5x300

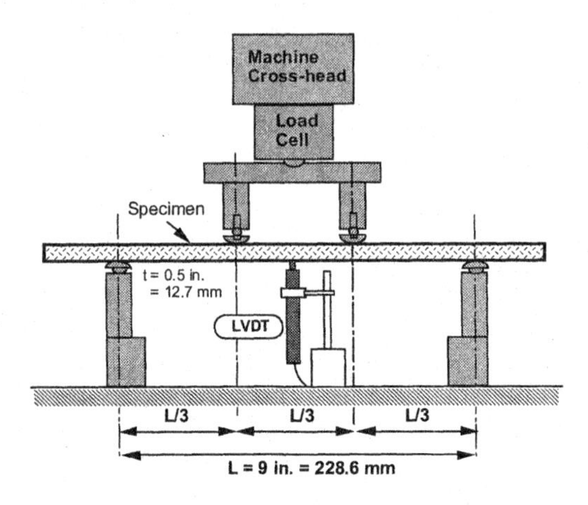

Figure 7 Bending test set-up for Series 6 to 8

Table 2 Mix proportions by weight

Matrix ID	Cement	Sand	Fly ash	Silica Fume	Superplastizer	Water	f_c' (Mpa)
6 (HSM)	0.8	1.0	0.2	0.07	0.04	0.26	84
Slurry1	0.8	1	0.2	-	0.01	0.40	43.4
Slurry 2	1	0.6	0.2	-	0.02	0.54	N/A

Table 3 Fiber properties

Fiber	Diameter (mm)	Length (mm)	Specific Gravity	Tensile Strength (Mpa)	Elastic Modulus (Gpa)
PVA	0.014	1.61	1.30	1585	38.1
Torex	0.3	30	7.85	2618	200

Table 4 Mesh properties

Mesh no.	Wire or Yarn Equivalent Diameter (mm)	Spacing (mm)	Elastic Modulus (Gpa)	Tensile strength (Mpa)
3D Type A	0.7	7.5	-	-
3D Type B	0.7	7.5	-	-
Gauge 16	1.60	12.5	200	420
Gauge 19	1.03	12.5	200	420
Kevlar	-	-	124	2800
Hardwire©	0.33	6.25	-	3150

RESULTS AND ANALYSIS

Since several specimens have different dimensions, the main results are presented, for a better comparison, in terms of the equivalent elastic bending stress,. The equivalent elastic bending stress is obtained from the following equation:

$$\sigma_e = \frac{6M}{bh^2} \tag{1}$$

where M is the bending moment, b is the width of the plate specimen, and h is its depth. The bending moment is obtained from the load applied to the specimen. For a specimen tested in bending at third point, as shown in Fig. 7, the bending moment is calculated from:

$$M = \frac{PL}{6} \tag{2}$$

where P is the applied load and L is the span length. The maximum value of bending stress obtained from Eq. (1) for the maximum load is defined as the modulus of rupture (MOR) of the composite. The equivalent elastic flexural stress accommodates automatically different geometric properties and test conditions, providing a good basis for comparison.

Table 5 Summary of test results of high performance hybrid cementitious composite materials.

Series no.	**First Cracking Stress [Mpa]**	**Deflection at First Cracking Stress [mm]**	**Max. Equivalent Bending Stress [Mpa]**	**Deflection at Max. Equivalent Stress [mm]**	**Toughness at Max Equivalent Stress**	**Mode of Failure**	**Average Crack Spacing [mm]**
1	1.61	0.10	2.77	1.31	15.52	Flexural Failure, Tension	NA
2	2.87	0.17	7.96	3.30	129.09	Flexural Failure, Concrete	NA
3	2.45	0.20	4.45	2.10	43.20	Flexural Failure, Concrete	NA
4	0.97	0.06	6.76	7.40	200.98	Shear Failure	NA
5	1.85	0.12	14.20	3.75	182.11	Shear Failure	NA
6	2.99	0.10	14.22	5.58	299.30	Flexural Failure, Concrete	25
7	9.80	0.38	94.08	11.25	3495.52	Hardwire Failure	NA
8	9.45	0.28	108.36	9.63	3353.65	Hardwire Failure	NA
9	4.17	0.11	13.13	4.40	231.91	Shear Failure	30
10	6.94	0.63	61.49	11.98	2575.79	Hardwire Failure	3
11	6.03	0.25	56.63	10.95	2291.05	Hardwire Failure	NA
12	9.30	0.28	13.2	5.69	33.62	Flexural Failure, concrete	NA

Series 1 to 5 of Table 1

The equivalent elastic bending stress is plotted versus the midspan deflection for each specimen tested. Typical results for Series 1 to 5 (Table 1) are illustrated in Fig.8. Series 1 uses only a 3D

PVA fiber mat, while Series 2 to 5 are hybrid series using a PVA fiber mat taken in sandwich between two extreme layers of reinforcement. Small beams 75 mm deep were tested in four point bending.

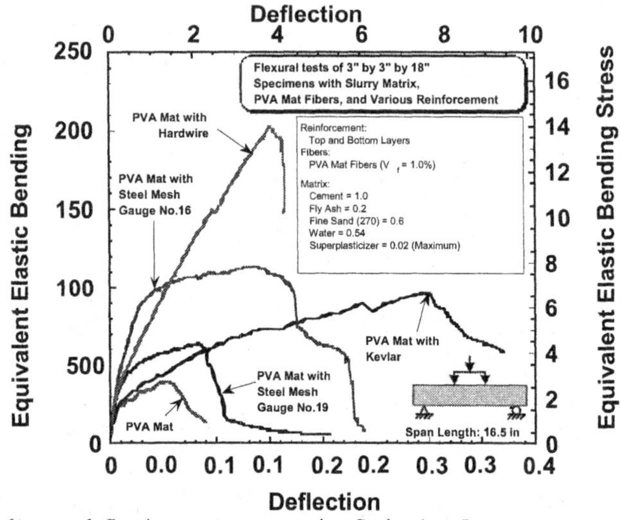

Figure 8 Typical stress-deflection curves comparing Series 1 to 5.

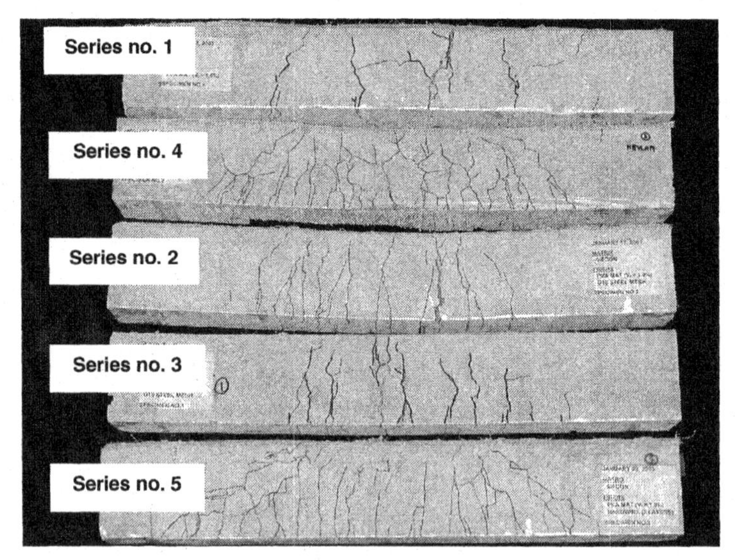

Figure 9 Crack pattern and failure specimen of series 1 to 5

It can be observed that a specimen reinforced with the PVA mat and with no other reinforcement, achieves a modulus of rupture of the order of 3 MPa. Adding two extreme layers of square welded steel steel mesh as commonly used for ferrocement leads to increasing the modulus of rupture to about 4.5 MPa when the volume fraction of reinforcement in the longitudinal direction, $V_{rL} = 0.17\%$ for the gauge 19 mesh (Series 3), and to about 8 MPa when $V_{rL} = 0.416\%$ for the gauge 16 mesh (Series 2). The highest strength (about 14 MPa) was achieved with the Hardwire© reinforcement with $V_{rL} = 0.294\%$, and the best best ductility (deflection capacity) was achieved with the Kevlar mesh; this is essentially due to the modulus of elasticity of Kevlar which is lower than that of steel. In all cases, good ductility and good multiple cracking developed (Fig. 9).

Series 6 to 8 of Table 1

Specimens of Series 6 to 8 (Table 1) had a depth of 12.5 mm. Series 6 used a Type A 3D glass mesh (Fig. 3), while series 7 and 8 used either one or two layers of Hardwire© (Fig. 6). The average response of these series is compared in **Fig. 10** where the equivalent elastic bending stress is plotted against the deflection. Series 6 leads to a modulus of rupture of the order of 14 MPa, while a modulus of rupture close to 100 MPa is achieved when Hardwire© reinforcement is used (Series 8). Series 8 which had two layers of reinforcement, one near each extreme fiber, did only a little better than Series 7 which had only one layer of reinforcement in the tensile zone. In both cases failure occurred by failure of the reinforcement in tension and good multiple cracking was observed (Figs. 10b and 10c).

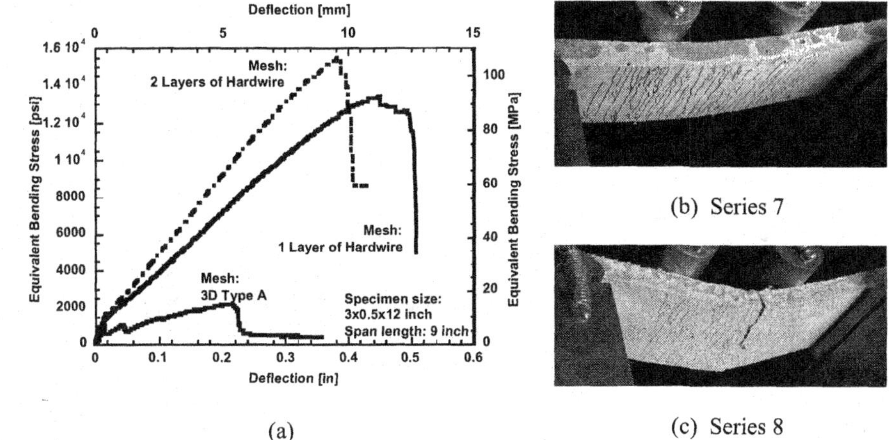

(b) Series 7

(a) (c) Series 8

Figure 10 (a) Typical stress-deflection curves comparing Series 6 to 8. (b) Typical cracking in specimens of Series 7. (c) Typical Cracking in specimens of Series 8

Series 10 and 11 of Table 1

Because the specimens with Hardwire© reinforcement lead to very good results, similar tests were carried out using thicker hybrid composite specimens (h = 37.5 mm) with Hardwire and steel fibers, leading to Series 10 and 11. For Series 10 a slurry infiltration process was used while for Series 11 a mortar premixed with steel fibers was used. Figure 11a illustrates the results. It can be observed that both series lead to very similar response suggesting that the additional amount of fibers used in Series 10, in comparison to Series 11, was not needed for bending resistance. In both series failure occurred by failure of the wires in the tensile zone implying that the fiber reinforced cement matrix provided sufficiently high ductility in compression (Fig. 11b).

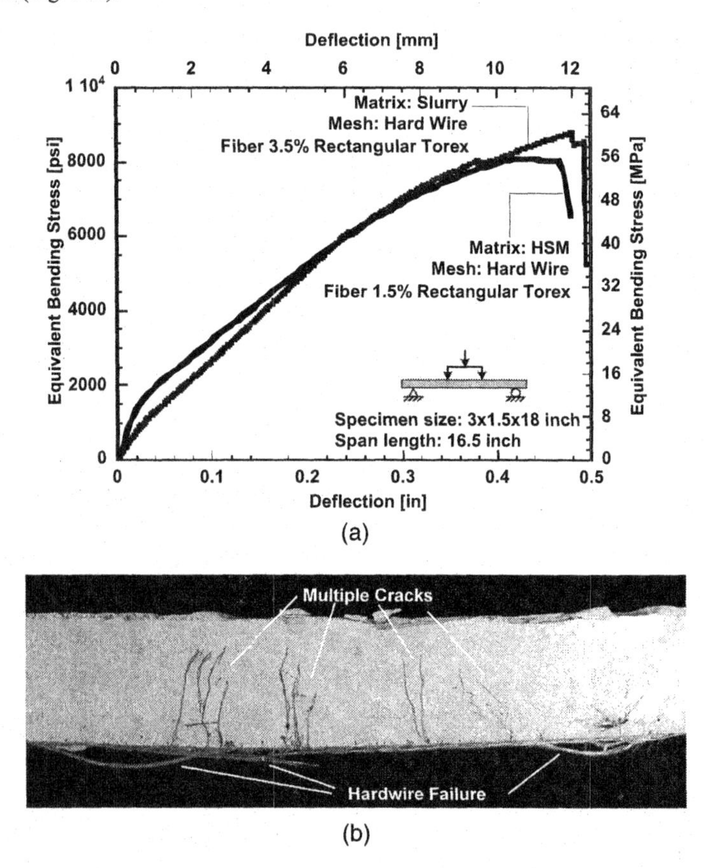

Figure 11 (a) Typical stress-deflection curves comparing Series 10 and 11. (b) Crack pattern and failure of Hardwire© reinforcement in specimen of series 10

Comparison of 3D Glass Meshes

The two 3D glass meshes used (Type A and Type B, Fig. 3) are compared in Fig. 12 to 15. The stress-deflection curves generally show an initial linear portion up to first cracking followed by a second portion with a linear trend interrupted by jagged lines corresponding to the formation of cracks (Figs. 12 and 14). In the case of mesh Type A (with vertical strings connecting the two extreme layers of mesh) failure occurred by failure of the mesh in tension

(Fig. 13). In the case of mesh Type B (sinusoidal interlaminar element connecting the two extreme layers of mesh) failure occurred when a crack formed along one portion of the sinusoidal mesh (Fig. 15). A quick analysis of this failure indicates that the failure crack was along a plane of high principal tension not crossed by transverse reinforcement; in fact the crack followed the shape of the sinusoidal mesh (Fig. 15). As a result from this observation, it was concluded that Type A mesh would be more desirable than Type B since the vertical strings act as better shear reinforcement than the sinusoidal elements.

Since the modulus of the mesh system is relatively small compared to steel meshes, cracking leads instantly to large crack widths; this is because the neutral axis of bending in the elastic state moves up along the section to very near the extreme compression fiber. The load-deflection curve experiences a jolt and a sudden drop such as shown in Figs. 12a.

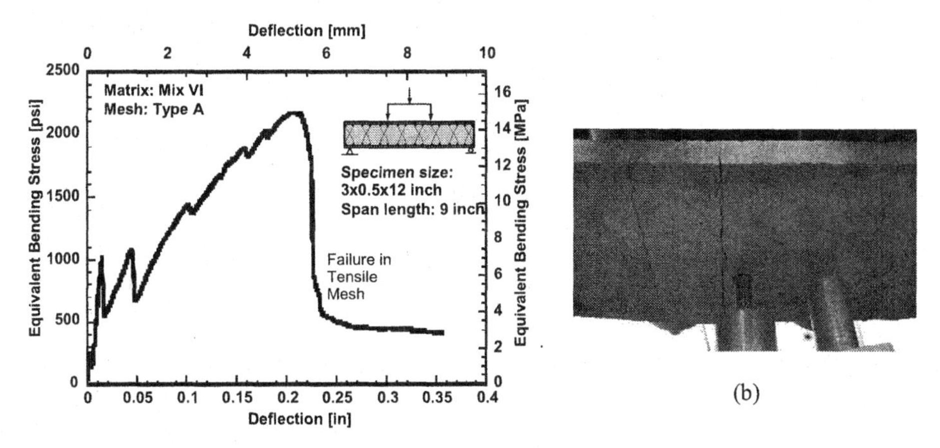

(a)

Figure 12 (a) Stress-deflection curve of specimen with 3D Type A fiber glass mesh. (b) Cracking development.

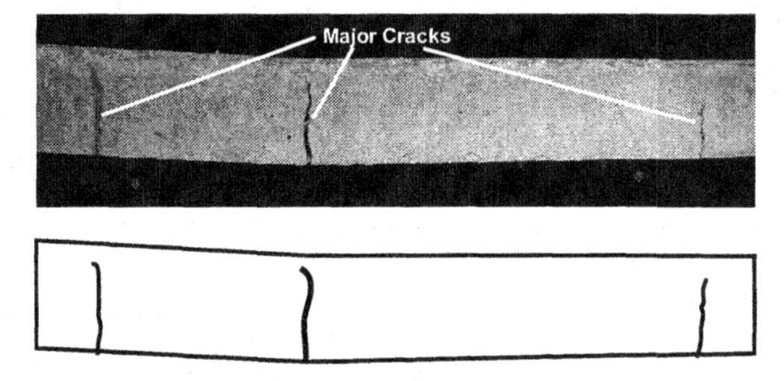

Figure 13 Flexural crack pattern in specimen using 3D Type A fiber glass mesh

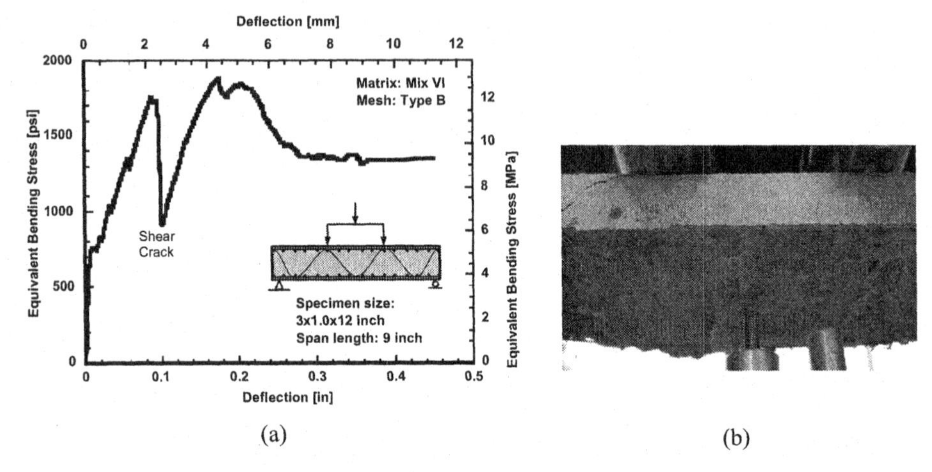

Figure 14 (a) Stress-deflection curve of specimen with 3D Type B fiber glass mesh (b) Cracking development.

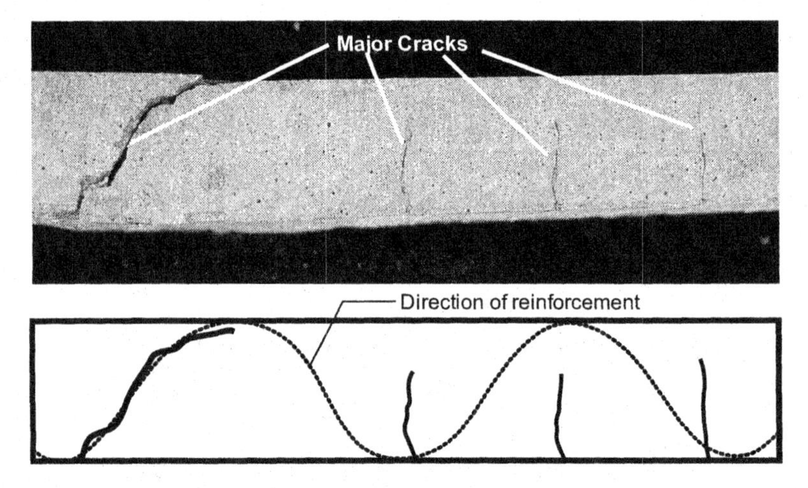

Figure 15 Crack pattern and failure shear crack that developed in specimen using 3D Type B glass mesh

Adding micro-fibers to the cement matrix help smoothen the response. Indeed, series No. 12 has the same reinforcement (3D Glass mesh, Type A) as series 6, but also 1% by volume short and fine PVA fibers were added. The stress at first cracking increases, and cracks move up along the cracked sections much more slowly leading to a less jagged load-deflection response. Also better ductility is observed prior to failure. A comparison of results is illustrated in Fig. 16. Referring to Table 5, the toughness index at peak load for Series 12 is smaller than that of Series

6, although the absolute toughness is higher. This is because the denominator for the toughness index is the area up to first cracking and that value is higher for Series 12 than for Series 6. Indeed, the area under curves at the maximum equivalent bending stress is 282.42 lb/in for Series 6 and 355.44 lb/in for Series 12, while the area up to first cracking is 0.94 lb/in for Series 6 and 10.57 lb/in for Series 12.

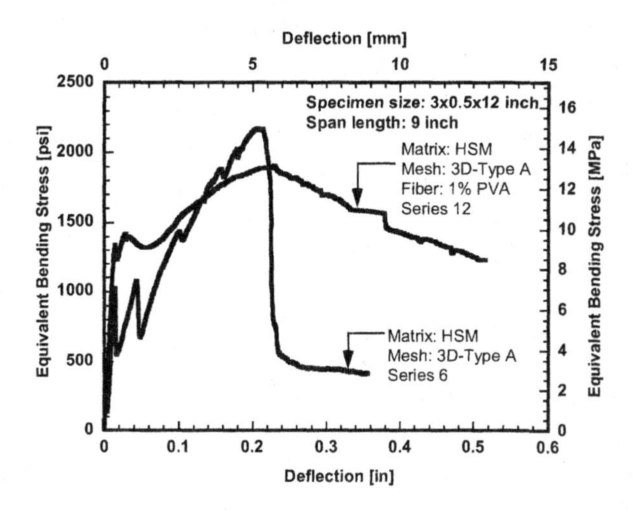

Figure 16 Effect of adding micro-fibers on bending stress versus deflection curve of specimens with 3D-Type A glass mesh

Because the 3D glass meshes have an elastic modulus lower than that of steel, the crack width was generally larger in these series than for the series with steel meshes. Failure occurred by failure of the mesh layer in the tensile zone. Because of the enormous potential of 3D meshes (or textiles) as reinforcement in thin cement products, their evaluation using different types of fiber materials should be actively pursued.

CONCLUDING REMARKS

This study attempted to explore different methods of building high performance thin cement based composites where labor cost is minimized. Three manufacturing configurations were considered: 1) a sandwich type system using a fiber mat taken in sandwich between two layers of reinforcing mesh and infiltrated by a cement matrix; 2) 3D meshes (or textiles or fabrics) infiltrated by a cement matrix; and 3) steel meshes of very high strength used in extreme layers only in combination with a fiber reinforced cement matrix. The following conclusions can be drawn:

1. The combination of discontinuous fibers and meshes (hybrid composite) leads to significantly improved flexural performance, in terms of strength, toughness, increased shear resistance, and cracking. It is also cost effective in comparison to the use of several layers of mesh without fibers.

2. The combined use of only two layers of mesh reinforcement, placed near the extreme surfaces, with discontinuous fibers (premixed or in the form of fiber mat) offered an optimum combination for bending behavior and should be further explored in future studies.
3. When using a combination of discontinuous fibers and meshes, particular attention should be placed to insure good penetration of the mesh system by the matrix. A poor matrix penetration of the mesh will affect the performance of the composite and may negate the benefits of using fibers. Superplasticizers and other additives are recommended to insure proper fiber dispersion and matrix penetration of the mesh system.
4. The use of only two layers of very high strength steel mesh such as Hardwire©, in combination with fibers, in 12.7 mm thick ferrocement plates, leads to equivalent elastic bending strengths of about 100 MPa. Such strengths should cover most common applications of ferrocement and thin concrete products today. Note that these values apply for positive and negative bending and were the highest observed in this study for 12.7 mm thick products. The total volume fraction of reinforcement was 2.76% including 1% fibers.
5. Sandwich construction with two layers of mesh separated by a core fiber mat, showed overall excellent behavior and should be further explored especially with high strength reinforcements.
6. The use of FRP meshes (or textiles or fabrics) in ferrocement and thin concrete products seems logical in applications where corrosion and weight of the structure are of concern, where non-magnetic properties are desired, and where the production process can be simplified, such as with 3D meshes. There is need to pursue the development of 3D meshes or textiles to be specifically used as reinforcement of cement matrices.
7. The combination of 3D meshes and fibers offers a number of advantages in improving composite response, such as increased cracking stress, finer cracks, and better ductility prior to failure.

As observed from the above results, hybrid composites do achieve excellent mechanical performance. The discontinuous fibers fulfilled their function by improving the shear resistance of the composite (vertical and interlaminar shear) and by taking a fair share of the resistance otherwise provided by intermediate meshes. Other benefits in using fibers (or fiber mats) in a hybrid configuration, include increased cracking strength, finer microcracks or shrinkage cracks, smaller crack widths under load, and improved "drill-ability" of the composite.

There is real need to develop FRP meshes (or textiles or fabrics), both 2D and 3D, with parameters especially optimized for thin concrete products. These include mechanical parameters such as strength, modulus, and ultimate elongation; geometric parameters such as yarn diameter, yarn spacing, specific surface; and practical parameters such as rigidity of the mesh for handling and placing, thickness and width of the mesh. 3D meshes (or textiles) with reinforcement for shear, and/or micro-fibers, can make a real breakthrough in the future.

ACKNOWLEDGEMENTS

The research work of the first author was supported in the past by grants from the US National Science Foundation (CMS-9908308 and CMS-0096700) and by the University of Michigan. Their support is gratefully acknowledged. Professor Thomas Gries and Andreas Roye from the Institut fur Textiltechnik – ITA at the Technical University of Aachen, Germany, kindly provided the 3D glass meshes (textiles) used in this study. The authors also wish to thank Professor Brameshuber from the University of Aachen in Germany for helping to obtain the 3D textiles used in this study, and Dr. Frank Jesse from the Technical University of Dresden for providing valuable suggestions about the 3D meshes used.

REFERENCES

[1]ACI Committee 549, "State-of-the-Art Report on Ferrocement," ACI-549-R97, in Manual of Concrete Practice, American Concrete Institute, Farmington Hills, Michigan, 1997, 26 pages

[2]P. Balaguru, A.E. Naaman, and W. Weiss, Co-Editors, "Concrete: Material Science to Application – A Tribute to Surendra P. Shah," American Concrete Institute, SP 206, Farmington Hills, Michigan, April 2002, 580 pages

[3]El-Debs, M. K. and Ekane, E. B., "Tension Tests of Mortar Reinforced with Steel Meshes and Polymeric Fibers," in *"Ferrocement 6 - Lambot Symposium,"* Proceedings of Sixth International Symposium on Ferrocement, A.E. Naaman, Editor, University of Michigan, CEE Department, June, 1998

[4]El-Debs, M. K. and Naaman, A. E., "Bending Behavior of Ferrocement Reinforced with Steel Meshes and Polymeric Fibers," *Journal of Cement and Concrete Composites*, Vol. 17, No. 4, December, 1995, pp. 327-328

[5]Guerrero, P. and Naaman, A. E., "Bending Behavior of Hybrid Ferrocement Composites Reinforced with PVA Meshes and PVA Fibers," in *Ferrocement 6 - Lambot Symposium* Proceedings of Sixth International Symposium on Ferrocement, A.E. Naaman, Editor, University of Michigan, CEE Department, June, 1998

[6]Hegger, J., Editor: "Fachkolloquium der Sonderforschungsbereiche 528 und 532," AACHEN, 2001

[7]Hussin, M. W., and Swamy, R. N., "Flexural Behaviour of Ferrocement Sections with Steel Fiber," in *Ferrocement: Proceedings of the Fifth International Symposium*, P.J. Nedwell and R.N. Swamy, Editors, E & FN Spon, London, 1994, pp. 416-34

[8]Kurbach, M., Editor, Proceeding of Second "Colloquium on Textile Reinforced Structures – CTRC2," Editor, Dresden, Germany, Sept. 2003

[9]Naaman, A.E., S.P. Shah and J.L. Throne, "Some Developments in Polypropylene Fibers for Concrete," in Proceedings of the ACI International Symposium on Fiber Reinforced Concrete, Special Publication SP-81, American Concrete Institute, Detroit, 1984, pp. 375-396

[10]Naaman, A. E. and Al-Shannag, J., "Ferrocement with Fiber Reinforced Plastic Meshes: Preliminary Investigation," in *Ferrocement: Proceedings of theFifth International Symposium*, P.J. Nedwell and R.N. Swamy, Editors, E & FN Spon, London, 1994, pp. 435-445

[11]Naaman, A.E., and Guerrero, P., "Bending Behavior of Thin Cement Composites Reinforced with FRP Meshes," Proceedings of First International Conference on Fiber Composites in Infrastructures, ICCI 96, Edited by H. Saadatmanesh and M. Ehsani, University of Arizona, Tucson, January 1996, pp. 178-189

[12]Naaman, A.E., and Chandrangsu, K., "Bending Behavior of Laminated Cementitious Composites Reinforced with FRP Meshes," ACI Symposium on High Performance Fiber-Reinforced Concrete Thin Sheet Products, Edited by A. Peled, S.P. Shah and N. Banthia, American Concrete Institute, Farmington Hills, ACI SP 190, 2000, pp. 97-116

[13]Naaman, A.E., *Ferrocement and Laminated Cementitious Composites*, Techno Press 3000, Ann Arbor, Michigan, USA, 2000, ISBN 0-9674939-0-0, (www.technopress3000.com), 370 pages

[14]Naaman, A.E., and Reinhardt, H.W., Editors, "High Performance Fiber Reinforced Cement Composites (HPFRCC4)," RILEM Proceedings 30, Part 4: Hybrid and Textile Reinforcements, Rilem Publications SARL, Bagneux, France, 2003

[15]Naaman, A.E., "Progress in Ferrocement and Hybrid Textile Composites," Proceeding of Colloquium on Textile Reinforced Structures – CTRC2, M. Curbach, Editor, Dresden, Germany, Sept. 2003, pp. 325-346

[16]Paramasivam, P., Mansur, A., et al., Editors, Proceedings of the "Seventh International Symposium on Ferrocement and Thin Reinforced Cement Composites," National University of Singapore, June 2001

[17]Shirai, A., and Ohama, Y., "Flexural Behavior and Applications of Polymer-Ferrocements with Steel and Carbon Fibers," in *Fiber Reinforced Concrete Modern Developments,* N. Banthia and S. Mindess, Editors, The University of British Columbia, 1995, pp. 201-212

[18]Swamy, R. N., Hussin, M. W., "Continuous Woven Polypropylene Mat Reinforced Cement Composites for Applications in Building Construction," in *Textile Composites in Building Construction*, P. Hamelin and G. Verchery (Eds.), Part 1, pp. 57-67, 1990

FIBRE BASED SYSTEMS FOR HIGH PERFORMANCE

S. K. Kaushik
Professor and Head
Department of Civil Engineering
Indian Institute of Technology, Roorkee,
Roorkee, India- 247 667

ABSTRACT

Fibre reinforcement is generally used to enhance tensile properties and energy absorption capacity of matrices erstwhile strong under compression. Fibre addition has also been found to aid in attaining better impact and abrasion properties in pavement, confinement in concrete compression members, reduction of plastic shrinkage of concrete, etc.

This paper reviews developments made in recent years in high performance cement based composites, which includes hybrid fibre systems, fibre- reinforced self compacting concrete, and alkali resistant glass fibre reinforced concrete. The research and development work being carried out at IIT Roorkee presently in the related areas is also reported.

INTRODUCTION

Fibre reinforced material systems offer an enhanced performance and a high cost-benefit ratio over traditional materials. The advantages may be in the form of low density coupled with high mechanical properties or high durability as the designed composite materials are resistant to most common environments. The composites are commonly formed by intimate mixing of two distinct phases, a fibrous reinforcement and a continuous medium termed the matrix. The matrix serves three important functions: (1) it holds the fibre in place, (2) it transfers loads to the high stiffness fibre, and (3) it protects the fibre.

The composites may be grouped broadly into two categories. In the first category, fibres with high stiffness and strength, but brittle in nature are embedded in a ductile matrix. Fibre volume fraction of up to 40 percent can be embedded in such types of composites. Since the surface area of the fibres is large, it is possible to obtain perfect bond between the fibres and the matrix. The bond is strong enough to make the composite stronger and stiffer than the matrix and more ductile compared to the fibre behaviour. In fiberglass, a polymeric matrix such as polyester resin has been used as the matrix. Similarly, composites made of high strength carbon (graphite) fibres and a resin matrix, provide high performance for aerospace applications. The failure strain of the matrix, which is larger than the failure strain of the fibre, permits the use of the full potential strength of the fibres[1]. Fibre Reinforced Polymer (FRP) composites developed in recent years with glass fibres or carbon fibres and polymer matrix are about four times lighter than steel with an equal strength[2]. The strength/stiffness of FRPs is almost entirely attributed to fibres, since a typical polymeric matrix has negligible strength/stiffness in comparison to the fibres.

The second category of composites is comprised of composites with strong fibres in a relatively weak (in tension) brittle matrix. The thrust is on improving the ductility of the matrix. The fibres contribute to the increase in strength as well. In these composites since the ultimate strain capacity of the matrix is lower than the strain capacity of the fibres, the matrix fails before the full potential capacity of the fibre is achieved. The fibres that bridge the cracks formed in the matrix contribute to the energy dissipation through processes of de-bonding and pull out. A portland cement based matrix is a typical example. The fibre can be metallic, mineral, polymeric, or naturally occurring.

This paper discusses some insight into FRPs along-with some typical experimental test results obtained by the author. Then, cement based fibre composites are discussed with emphasis on three types: namely glass fibre- reinforced concrete composites, steel fibre reinforced self-compacting concrete and mechanical property enhancement in structural members through conventional steel fibre- reinforced concrete. Some of the test results from ongoing investigations under the supervision of the author are highlighted in this section.

FIBRE- REINFORCED PLASTICS

In recent years, FRP's originally developed for aerospace industry, are being considered for applications in the repair of buildings due to their low weight, ease of handling and rapid implementation. Their use for seismic resistant structures, which can absorb shock and vibrations through tailored microstructures has found applications. FRPs are also in use for the retrofitting of seismic deficient structural elements such as piers of bridges, building columns and repair of masonry walls to enhance or restore the flexural or shear capacity in high intensity earthquake zones by providing additional reinforcements in the form of overlays (wrapping). FRP bars are gaining acceptance as alternative reinforcing bars in reinforced concrete members due to their high corrosion resistance. A wide variety of FRP rebars are commercially produced and these ranges from simple smooth rebars to rebars treated to improve bond characteristics. To exploit the frictional contribution to bond, the bar's surface can be coated with sand before curing of the thermosetting resin, using of excessive resin that lead to irregular humps, forming deformation pattern mimicking those on the surface of the steel bars, braiding of fibres and a combination of some of them. Most of the surface treatments improve the bond strength of FRP bars to the same order, as that of the steel.

FRP pre-stressing tendons are found suitable for pre-stressing concrete girders and deck slabs; these are manufactured in the form of round or flat bars, ropes, cables, or strands. In these also, a variety of surface textures are possible, including smooth, braided, rough, sand coated, dimpled and indented. Field applications of concrete structures pre-stressed with FRP reinforcements have been consistently increasing since 1980s. More than fifty structures using FRP composites have been built around the world., mostly either pedestrian bridges or vehicle bridges. These structures were built mostly as demonstration projects to validate the technology, to gain experience and to study long- term durability and performance. The first GFRP pre-stressed bridge was the Lunensche Gasse Bridge, built in Dusseldorf Germany in 1980. The single -span slab bridge had a span of 6.55 m and 100 Polystall GFRP un-bonded, tensioned tendons were used in the construction. One of the interesting application of the CFRP tendons was in the construction of the Headingley Bridge in Canada in 1993 where in the construction of the five spans (each 32.5 m), four out of the forty pre- cast, pre- tensioned AASHTOI – girders contained FRP tendons. The first bridge pre-stressed with CFRP was completed in Japan in 1988. In Canada, the Beddington Trail Bridge, represents the first Highway Bridge in the world to use CFRP and fibre optic sensing technology. In the USA, the first demonstration using CFRP and GFRP cables was built in Rapid City, South Dakota, in 1992. In Japan, the use of AFRP tendons for pre- tensioned and post- tensioned concrete structures is on the increase since 1990[3-4].

FRP is rapidly gaining preference in retrofitting applications over the conventional methods like externally bonded steel plates hitherto used for strengthening flexural members, concrete jacketing of R. C. columns, etc. Bonding one layer of GFRP, the load carrying capacity of flexural members can be enhanced in the range of 20 % to 70 %. The increase in strength is lowered with increasing strength of the concrete. Fig.1 shows typical test results obtained by the author in strengthening concrete beams with GFRP (ultimate tensile strength of 575 MPa) wrap/s. The compressive strength of the concrete at 28 days age was 35 MPa[5].

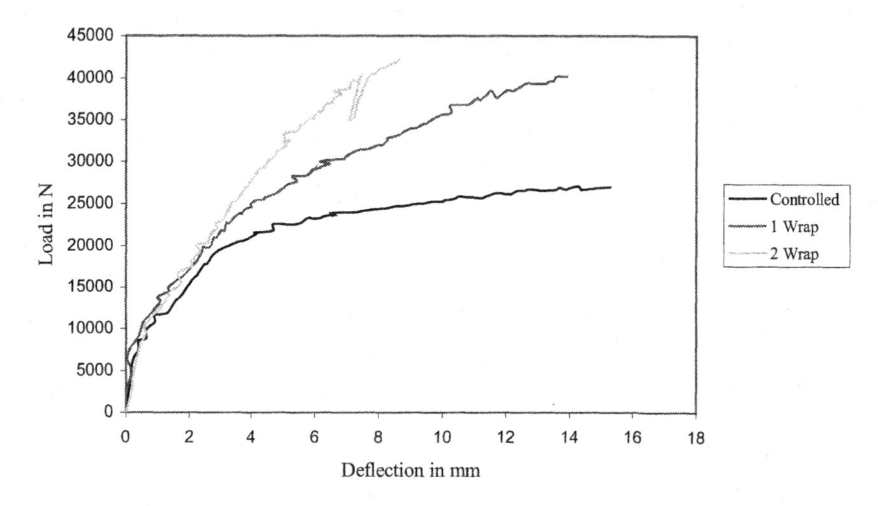

Fig. 1 Typical Load- Deflection Curve of GFRP Strengthened Reinforced Concrete Beam[5]

As evident from Fig.1, the deflection ductility of the structural member reduces with increasing the number of FRP wraps. The steel strain ductility, however, increases as can be seen from Fig. 2[4].

FRP is ideally suited as the confinement material for cement concrete columns[6-7]. Confinement is applied to members in compression, with the aim of enhancing their load carrying capacity, or, in case seismic up-gradation to increase their ductility. From Fig. 3, it can be seen that unlike steel, that has a plastic zone, material for FRP is elastic up to failure. Therefore, the confining action on concrete due to FRP wrap increases until failure. The amount of this action depends on the lateral dilatation of concrete, which in turn is affected by the confining pressure[7]. Comparative material characteristics of steel and FRP are provided in Table I. The detailed research about the behaviour of FRP wrapped RC columns is being carried out in several parts of the world, and also in IIT Bombay, India.

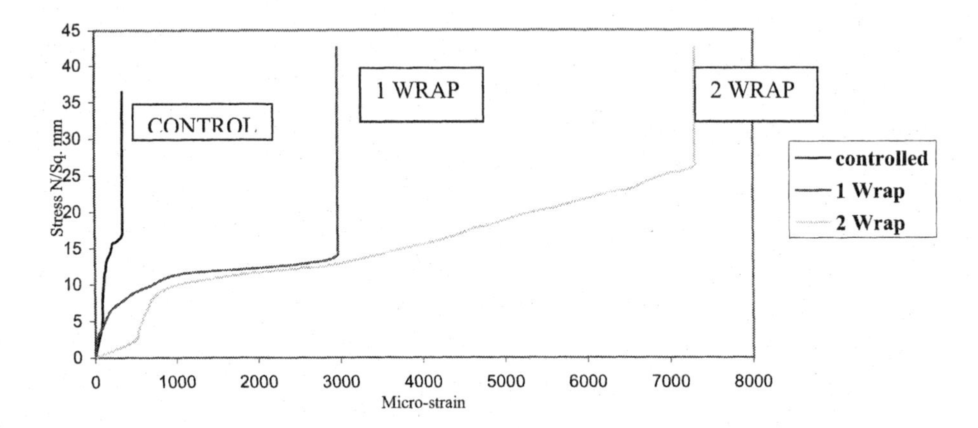

Fig. 2 Increase in Strain Ductility With Incorporation of FRP[5]

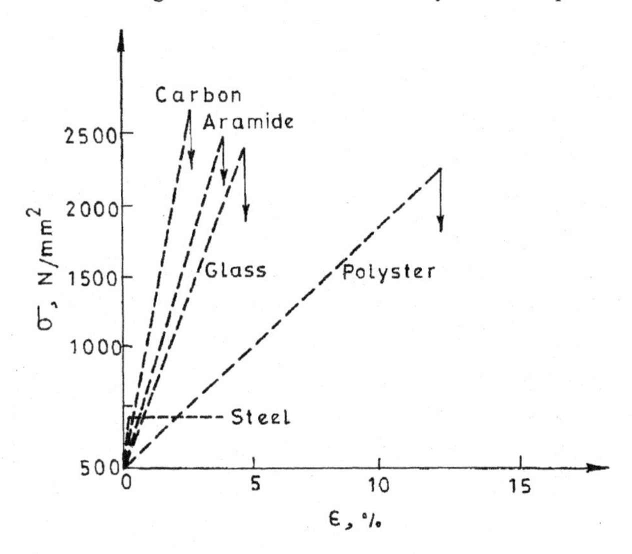

Fig. 3 Stress- Strain Characteristics of FRP Material and Steel

Table I Comparative material characteristics of steel and FRP

Type of Fiber	Modulus of Elasticity (GPa)	Tensile Strength (GPa)
Carbon	240-640	2.5-4.0
Aramid	124	3.0-4.0
Glass	65-70	1.7-3.0
Polyester	12-15	2.0-3.0
Steel	210	0.25-0.55

CEMENT BASED COMPOSITES

Fibres such as steel, glass, etc. are added to cement mortar or concrete to increase the tensile strength and toughness. The volume fraction of such fibres is generally limited to one percent. The last two to three decades have seen development of high volume fibre based composites like SIFCON and more recently RPC. SIFCON offers increase in toughness to the order of a magnitude higher than the conventional fibre reinforced concrete. RPC can exhibit tensile and flexural strength close to steel, with a lower density. Polymeric fibres such as made with polypropylene etc. are used in very small quantities (even less than 30 kg/m^3) to prevent plastic shrinkage cracking.

Alkali Resistant Glass Fibre – Reinforced Concrete

Commercially available Cem FIL AR glass fibres possess a tensile strength of the order of 1700 MPa, which is more than three times that of steel. Incorporation of such fibres in concrete greatly increases its tensile strength and energy absorption capacity[8]. Durability of conventional E- glass fibres is very low as the embrittlement of fibres occurs due to prolonged exposure in alkaline environment of concrete. However, studies made on CEM FIL AR glass fibres indicate lesser reduction in strength and ductility of the composite than with the E- glass fibre composites. Therefore, in temporary structures, where high tensile strength or energy absorption capacity may be the requirement, GFRC may be used with advantage. Fig. 4 shows that even with 0.5 % fibre content, the peak strain obtained in GFRC can be in the range of 1 % and the ultimate strain is of the range of 3 %, which are substantially higher than the corresponding values of 0.2 % and 0.35 % in plain concrete[8].

Recent Advances in Applications of Steel Fibre- Reinforced Concrete

The research in recent years on steel fibre reinforced concrete has been more directed towards its use as hybrid fibres, selectively using them to increase the specific requirements in structural members such as fatigue, beam-column joint efficiency, or as confinement of concrete compression members. An upcoming field is fibre-reinforced self-compacting concrete. This area is finding immediate applications in construction of high performance road deck panels and pavement construction in addition to the aesthetic applications.

Hybrid Fibres

Until recent past, most fibre reinforcement applications have employed macro steel fibres that average 0.5 to 1.0 mm in diameter and 25 to 60 mm in length. In last few years, advancements have taken place in micro- fibres that are less than 10 mm long and less than 100 microns in diameter. While macro- fibres reinforce at the level of large aggregates, micro-fibres reinforce mortar and paste phases and thus complement the reinforcement provided by the macro-fibres. Another advantage of micro-fibres is a fibre to fibre spacing that does not exceed a few microns. Thus a micro-crack between two fibres can stabilize, which, in turn, impedes a coalescence of micro-cracks, resulting in increased tensile strength. Combining macro-fibres and micro-fibres provides a hybrid high performance cement based material useful for a variety of applications, including thin repairs and patching. Banthia and Bindiganavile have reported the use of 30 mm long, 0.7 mm diameter. Together with 4.5 mm long with 80 μm diameter steel fibres in the repair of a severely damaged parking garage in Vancouver. Such applications are expected in India too now[9].

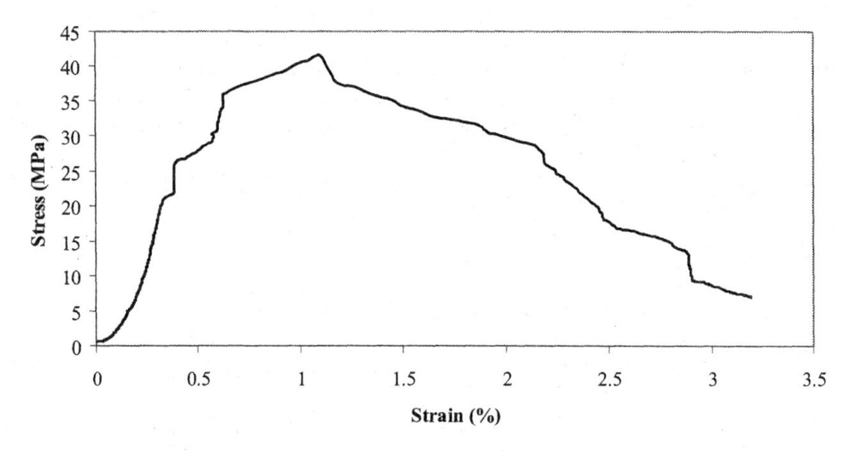

Fig. 4 Typical Stress-Strain Curve in Compression Obtained For GFRC at a 0.5 % Cem–FIL AR Glass Fibre Content at 180 days Age[8]

Confinement of Compression Members with Steel Fibres

Inelastic deformability of reinforced concrete columns is essential for overall strength and stability of structures. High strength concrete, when subjected to short term or sustained loads, tends to be brittle when loaded to failure, lacking the plastic deformation typical of normal strength concrete. This type of behavior raises questions on the suitability of HSC to structures in seismic regions. In recent years, research in many parts of the world has spurned up in the area of possible confinement of high strength columns with suitable means, one of which has been steel fibres too. An ongoing comprehensive experimental and analytical work at Indian Institute of Technology, Roorkee, India under the supervision of the author has indicated that suitable fibres coupled with moderate amount of lateral ties could be a good alternative to the high amount of lateral confining steel required for high strength concrete columns to attain desired ductility. The study has shown that only nominal amount of hooked end steel fibres (weight fraction =1%) are required to enhance the ductility ratio from 3.2 to 5.8 of M80 reinforced concrete column designed for the same ultimate load. Fig. 5 shows a typical set up for the study. Fig. 6 illustrates the influence of adding 1 % hooked end steel fibres to a 80 MPa concrete column. It suggests that the steel fibres could be a good option to improve the ductility performance of HSC columns. Confinement provisions of IS 13920-1993 and ACI 318-2002 codes are observed not to be sufficient for columns with high concrete strengths and high axial load levels[10].

Fig. 5 Test Set Up For Confinement Study

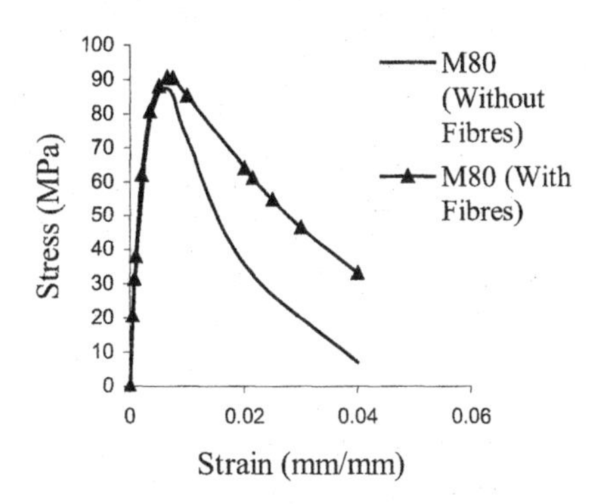

Fig. 6 Stress-Strain Curves for RCC Columns with & With out Fibres

Fibre- Reinforced Self- Compacting Concrete

Self- compacting concrete (SCC) was developed in 1988 in Japan to lead to a rheological behaviour that supersedes vibrating compaction. As aggregate content is lower in SCC as compared to the normal concrete, shrinkage is expected to be more in SCC than the conventional concrete. That in turn supports crack formation due to drying and may lead to a decline in durability. The addition of fibres is expected to have positive influence on crack formation. Fibre reinforcement of self- compacting concrete is expected to be a step towards more durable concrete and high performance concrete. Several investigators have focused on developing the mixes with addition of fibres without forfeiting self- compactability[11-13].

An ongoing study is under progress at IIT Roorkee to investigate shear behaviour of fiber reinforced self-compacting concrete (FRSCC) using steel fibers up to 1 %. The test results indicate significant improvement in ultimate shear strength and cracking behaviour in the FRSCC as compared to ordinary SCC. An increase in ultimate shear strength up-to 50% as compared to ordinary SCC, has been observed in FRSCC. A significant decrease in the rate of increase of crack width up-to 70 % in FRSCC is also observed which signifies the utility of steel fibers in SCC as crack arresting elements. Beams with FRSCC have been observed to fail less vigorously as compared to beams of ordinary SCC. Crack width for the same load condition has been observed to be significantly lower in FRSCC as compared to that in the beams made with SCC of similar composition and strength (but without fibbers). Fig. 7 shows crack pattern as observed with SCC and Fig. 8 shows crack pattern as observed in FRSCC.

Fig. 7 Crack Pattern in Conventional SCC Beams in Shear

Fig.8 Crack Pattern in Fibre Reinforced SCC under Shear

MISCELLANEOUS APPLICATIONS OF FIBRE BASED COMPOSITES IN INDIA

One of the earliest applications of FRC material in India was the use of corrugated roofing sheets made out of coir fibre reinforced concrete in a major leprosy settlement in a village near Titalagarh in Orissa under the technical guidance of Swiss Centre for Appropriate Technology, Switzerland. These roofing sheets have successfully withstood many monsoon seasons since their production in 1980. Such FRC roofing sheets are also now being used in various villages in Andhra Pradesh. The production of these FRC sheets is very simple and does not require any special skills. Applications of glass fibre composites reported in India includes those in ITC Hotels, Mumbai, Hotel, Leela, Bangalore and renovation of 166 year old Handloom House, Mumbai. Steel fibre reinforced concrete has also been used by the International Airport Authority of India in the construction of a certain stretch of airfield pavement at New Delhi. The thickness of the pavement was 300 mm as compared to 400 mm required in the case of concrete without fibres. This reduction in thickness was made possible by the higher flexural strength of SFRC. The pavement is now in service for more than fifteen years and no cracking, spalling and damage of any kind has been observed. SFRC has been used in a big way in the tunnel lining in the Uri hydroelectric project in Jammu and Kashmir. Fibre shotcreting technique with hooked steel fibres has been used in this prestigious project. EE fibres (18.0*0.6/0.4*0.3 mm; l/d 37.6 and 46.0) from Norway have been extensively used[14].

The Indian railways are the fourth biggest networks in the world. It has about more than one lac bridges of various categories with wooden sleepers[15]. Trial testing of FRP sleepers on the track has been undertaken recently. The use of FRP for components of rail coaches was initiated in the early 1980s for toilet floor inlays, bathroom fittings, seat and backrest for AC chair cars, toilet doors, roof sealing sheets etc [3,15].

As FRP exhibit excellent resistance to the marine environment, so the FRPs can be used for the construction of deckhouses, hatch houses, hatch covers, kings posts etc. In India beginning has already taken place in high speed boats, naval vessels, fishing boats, high capacity trawlers and other components of the ship[4].

CONCLUDING REMARKS

Fibre based systems are now looked for tailor made applications, be those in for applications involving higher strength to cost ratio, corrosion free reinforcing rebars, strengthening and rehabilitation of structures, or attaining high abrasion and impact resistance together with easy placement in pavement construction through adoption of the fibre reinforced self compacting concrete etc.

REFERENCES

[1]Balaguru, P.N. and Shah, S.P., "Fiber-Reinforced Composites," pp.530. McGraw –Hill, 1992

[2]IncWu, Hwai-Chung, "Design Flexibility of Composites For Construction," pp.421-432, Proceedings of ICFRC International Conference On Fibre Composites, High Performance Concretes and Smart Materials, 8-10 January, 2004, Chennai, India

[3]Sharma, S.K., Ali, M.S.M., Rao, M.V.B. and Sikdar, P.K. pp. 366- 384. "Applications of FRP As Modern Construction Material," Proceedings INAE National Seminar on Engineered Building Materials and Their Applications, Indian Institute of Technology, Bombay, January, 17-18, 2003

[4]Soudki, K.A., "FRP Reinforcements For Pre-stressed Concrete Structures," *Progress in Structural Engineering and Materials*,**1**[2] 135-142 (1998)

[5]Admane, Niranjan, Behaviour of RC Beams Strengthened With GFRP Laminates," pp. 110 M. Tech. Dissertation, I. I. T. Roorkee, 2004,.

[6]Desai, Yogesh, M, "FRP Composites- From Theory to Field Applications," pp. 256- 279. *Proceedings INAE National Seminar On Engineered Building Materials and Their Applications*, Indian Institute of Technology, Bombay, January, 17-18, 2003

[7]Mukherjee, Abhijit, "Non – Metallic Fiber Composites For Structures," pp. 314- 326. *Proceedings INAE National Seminar On Engineered Building Materials and Their Applications*, Indian Institute of Technology, Bombay, January, 17-18, 2003

[8]Goyal, Amit, "Long Term Property Evaluation of Glass Fibre Reinforced Concrete," M. Tech. Thesis, Department of Civil Engineering, Indian Institute of Technology, Roorkee, ROORKEE, India, June, 2004

[9]Banthia, Nemkumar and Bindiganavile, Vivek, "Repairing With Hybrid- Fibre Reinforced Concrete," *Concrete International*, June 29-32 (2001)

[10]Sharma, Umesh, Bhargava, Pradeep and Kaushik, S. K., " Post- Peak Behaviour of Confined Fibre- Reinforced Concrete Columns," *Indian Concrete Journal*, **78**[5] 47-55 (2004)

[11]Grunewald, Steffe and Walraven, J. C., "Maximum Content of Steel Fibres In Self Compacting Concrete," pp. 137-147. *Proceedings of the Second International Symposium On Self Compacting Concrete*, October, 2001, Tokyo, Japan,

[12]Bauml, M. F. and Wittmann, Folker, H., "Improved Durability of Self- Compacting Concrete By Addition of Fibres," pp. 527-536. *Proceedings of the Second International Symposium On Self Compacting Concrete,* October, 2001, Tokyo, Japan

[13]Hwang, Chao- Lung and Tsai, Chih- Ta, "Durability Design Consideration and Application of Fibre- Reinforced Concrete Composites in Taiwan", pp.3- 17. Proceedings of ICFRC International Conference On Fibre Composites, High Performance Concretes and Smart Materials, 8-10 January, 2004, Chennai, India

[14]Kaushik, S.K., Kumar, Praveen and Sharma, Umesh, "Fibre- Reinforced Composites: Indian Scenario", Proceedings of National Seminar On Reinforcement Today and Tomorrow, India Chapter of American Concrete Institute, Mumbai, June, 13-14, 2003,

[15]Subramanian, P. S., " Development of a Railway Sleepers In A new Generation Material For Indian Railways", pp. 99-107. Proceedings of the International Conference and Exhibition on Reinforced Plastics (ICERP), Feb., 7-9, 2002, IIT Madras

GUIDELINES FOR DESIGN OF REINFORCED CONCRETE STRUCTURAL ELEMENTS WITH HIGH STRENGTH STEEL FIBRES IN CONCRETE MATRIX

N. Lakshmanan
Director
Structural Engineering Research Centre
CSIR Campus, Taramani
Chennai 600 113

T. S. Krishnamoorthy
Deputy Director
Structural Engineering Research Centre
CSIR Campus, Taramani
Chennai 600 113

ABSTRACT

Fibre reinforcement in concrete matrix is known to improve several properties, such as delayed crack initiation, energy absorption, resistance to impact, increased fatigue strength and resistance to abrasion. Use of high strength steel fibres in concrete matrix has been investigated by researchers all over the world. Structural Engineering Research Centre at Chennai has been undertaking R&D activities in the area of steel fibre reinforced concrete over the last three decades. This paper presents the results of all these investigations with a view to derive design recommendations for use of steel fibre reinforcement in reinforced concrete structural elements.

INTRODUCTION

Random oriented fibre reinforced concrete is one of the most promising composites used in the construction. Generally, for structural applications, steel fibres should be used in a role supplementary to reinforcing bars. Steel fibres relatively inhibit cracking and improve resistance to material deterioration as a result of fatigue, impact, and shrinkage or thermal stresses. In applications where the presence of continuous reinforcement is not essential to the safety and integrity of the structure (e.g., slabs on grade, pavements overlays and shotcrete linings), the improvements in flexural strength, impact resistance, and fatigue performance associated with the fibres can be used to reduce section and to enhance performance or both. Some full scale tests have shown that steel fibres are effective in supplementing or replacing the stirrups in beams.

The mechanical properties of fibre reinforced concrete (FRC) are influenced by: the type of fibre; fibre length-to diameter ratio (aspect ratio); the amount of fibre; strength of matrix; the size, shape and method of preparation of the specimen; and the size of the aggregate. Fibres influence the mechanical properties of concrete and mortar in all failure modes. The strengthening mechanism of the fibres involves transfer of stress from the matrix to the fibre by interfacial shear or by interlock between the fibre and matrix, if the fibre surface is deformed. Besides the matrix itself, the most important variables governing the properties of FRC are the orientation factor and the fibre content. Fibre efficiency is controlled by the resistance of the fibres to pullout, which in turn depends on the bond strength at the fibre matrix interface. Also, since pullout resistance is proportional to interfacial area, non-round fibres offer more pullout resistance per unit volume than larger diameter fibres. Therefore, for a given fibre length, higher aspect ratio is more beneficial. Most mixes used in practice employ fibres with an aspect ratio less than 100, and failure of composites, therefore is, due primarily to fibre pullout. However, increased resistance to pullout without increasing the aspect ratio is achieved in fibres with deformed surface or end anchorage; failure may involve fracture of some of the fibres, but it is still usually governed by pullout.

STRESS-STRAIN CHARACTERISTICS OF STEEL FIBRE REINFORCED CONCRETE

Fibre reinforced concrete cylindrical specimens with 150mm diameter and 300 mm length were tested for their stress strain characteristics under compression. Different fibre volume fractions (V_f) (0, 0.75, 1.0, 1.25 and 1.5 percent by volume) were investigated. Trough shaped steel fibres with an aspect ratio of 80 were used in the investigations. The mix was designed for a characteristic strength of 25 MPa. Fig. 1 shows the typical stress –strain plot sharing the effects of inclusion of steel fibres in concrete matrix[1]. As the volume of fibre reinforcement is increased between 0 to 1.5%, there is an increase in strength as well as strain at peak-stress. For the purpose of design, it is proposed not to assume any increase in strength, and take the value of strain at peak stress to be equal to 0.006. This also agrees with the value suggested by Fanella and Naaman[2]. The failure strain is considered as the post-peak strain corresponding to a stress level of 0.75 times the peak stress. The failure strain increases with increase in volume fraction of steel fibres and varies in the range of 0.009 to 0.015. For design purposes, the following relationship between failure strain and volume fraction of fibres is proposed.

$$\varepsilon_u = 0.006 + 0.003\, V_f \qquad \text{for } 0.5\% < V_f \leq 2.0\% \tag{1}$$

$$f_u = 0.75\, f_{peak} \tag{2}$$

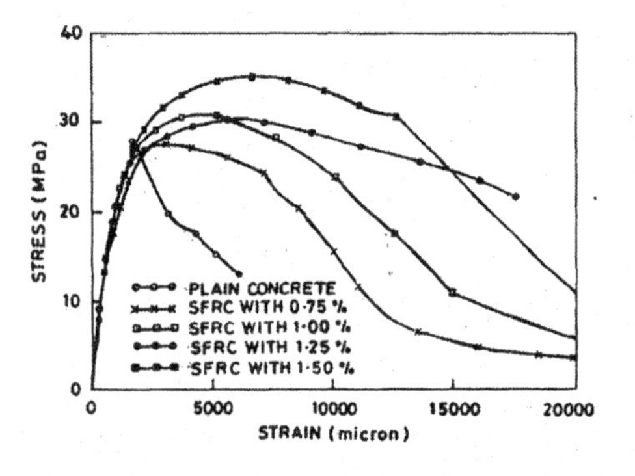

Fig.1 Stress-strain plot for SFRC specimen

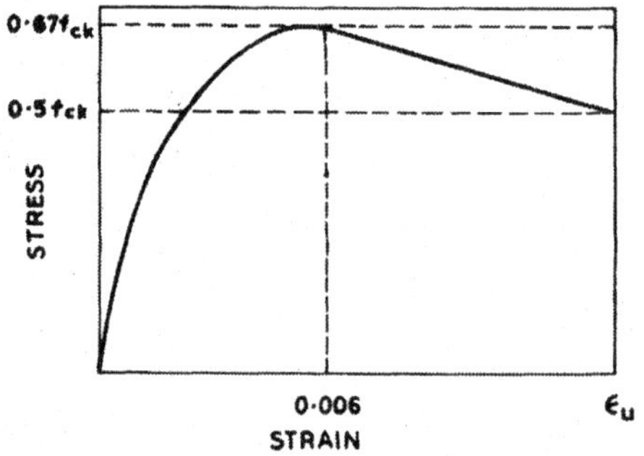

Fig.2 Idealised stress-strain curve

Table I shows the variation of energy absorption in fibre reinforced concrete as compared to plain concrete using the theoretical stress-strain model proposed. The values of the stress block parameters for use in flexural design, namely the average stress and the depth to centre of compression as a fraction of neutral axis depth are also given in Table I.

Fig 2 shows the idealized stress- strain curve used for deriving Table II. Typical stress-strain variation of SFRC specimens under cyclic loading are given in Figs. 3 and 4. Even when cycling at peak load, the envelope curve reasonably matches with the monotonic counterpart suggesting the suitability of SFRC in seismic resistant design of structures.

Table I Characteristics of SFRC

Sl. No.	V_f (%)	Failure strain	Energy absorption f_{ck} x 10^{-3}	Stress parameter x f_{ck}	Centre of gravity of stress block
1	Plain	0.0035	1.90	0.54	0.42
2	0.5	0.0075	3.52	0.47	0.40
3	1.0	0.0090	4.41	0.49	0.42
4	1.5	0.0105	5.25	0.50	0.44
5	2.0	0.0120	6.12	0.51	0.45

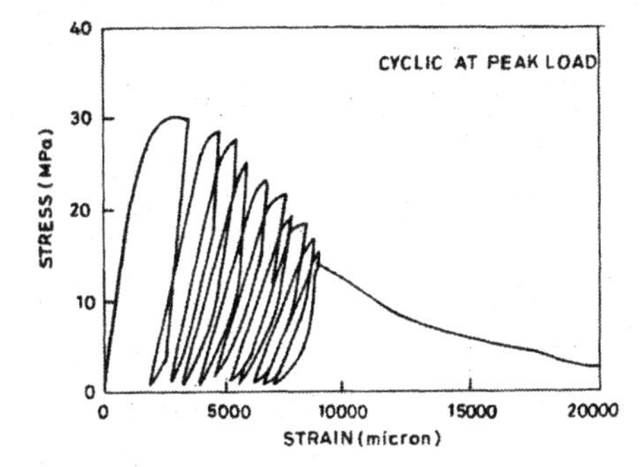

Fig.3 Stress-strain plot for SFRC specimen under cyclic compression loading (V_f = 0.75%)

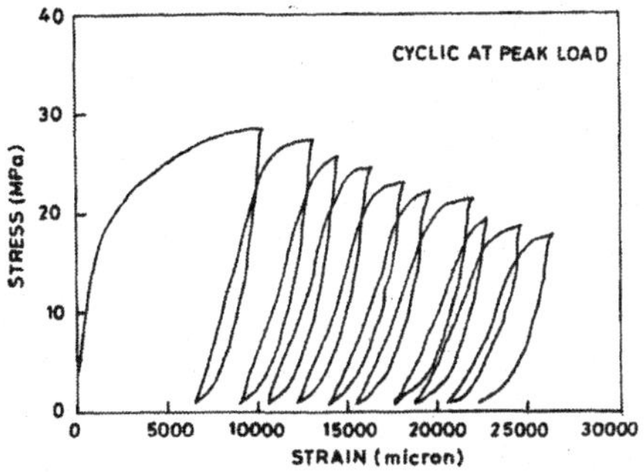

Fig.4 Stress-strain plot for SFRC specimen under cyclic compression loading (V_f = 1.5%)

BEHAVIOUR OF SFRC BEAMS UNDER FLEXURE

A number of singly and doubly reinforced concrete beams with varying volume percentage of steel fibres have been tested[3]. Table II gives the results of the investigations. The RC beams in serial number 1 to 8 (100 x 150 mm) were designed for a characteristic strength of 30 MPa, while the RC beams from serial 9 to 12 (100 mm x 200 mm) had a characteristic strength of 20 MPa. While the first series had 2 nos. of 10 mm dia as tension reinforcement, the second series had equal top and bottom reinforcement of 2 nos. of 10 mm diameter. The SFRC beams in the S series were tested for their shear strength and the beams given in Table II failed in flexure. While calculating the theoretical capacity of SFRC beams, the tensile stress developed across the crack in the tension zone was represented by a rectangular stress block across the entire tensile zone as suggested by Craig[4]. As can be seen from Table I, the stress block parameters derived are nearly the same for RC and SFRC beams. The following simplified expression for flexural capacity is proposed.

$$M_{u,SFRC} = M_{u,RC}(1+10V_f) \qquad (3)$$

Where, $M_{u,SFRC}$ is the ultimate moment capacity of SFRC beams, $Mu_{,RC}$ is ultimate moment capacity of reinforced concrete beam with same % of longitudinal steel, and V_f is the volume fraction of steel fibres.

Table II gives the comparison of experimental results with values obtained using Craig's model, and the simplified approach. The above simplified approach was also used to predict ultimate capacities of beams reinforced equally in tension and compression faces. A very good agreement is seen validating the use of simplified expression.

The maximum deviation obtained using the simplified theory is within 10% of the experimental value, and always gives a conservative estimate, and hence can be used with confidence in the design of flexural members.

Table II Ultimate load capacities of SFRC beams [3,5,6]

Sl.No	Identification	Type	V_f (%)	Ultimate capacity, kN		
				Test	Theory	Simplified
1	Com	-	0	44.2	44.1	44.2
2	TA	Trough	1.0	50.0	51.1	48.3
3	TB	Trough	1.5	55.8	54.8	50.6
4	TC	Trough	2.0	53.5	55.9	52.8
5	SA	Straight	1.0	48.0	49.2	48.3
6	SB	Straight	1.5	50.0	50.0	50.6
7	CA	Crimped	1.0	52.0	51.5	48.3
8	CB	Crimped	1.5	52.5	53.5	50.6
9	B6	RC	0	44.0	-	44.8
10	B1	Straight	0.50	50.0	-	47.1
11	B2	Straight	0.75	51.0	-	48.2
12	B3	Straight	1.00	52.5	-	49.3
13	S5	Straight	0	132.0	-	132.0
14	S6	Straight	0.5	150.0	-	140.0
15	S7	Straight	1.0	154.0	-	146.0
16	S8	Straight	1.5	152.0	-	150.0

Steel fibre reinforced concrete beams with equal tension and compression reinforcement are extremely ductile and have very large failure deformations. A simplified model was proposed to compute the failure rotation at a plastic hinge[7].

$$\theta_u = \phi_u(2l_p) \tag{4}$$

$$l_p \approx d \tag{5}$$

$$\phi_u = (\varepsilon_{ce} + \varepsilon_{se})/d \tag{6}$$

$$\varepsilon_{ce} = 0.006 + 0.02(b/z) + (\rho_s f_y/138)^2 \tag{7}$$

$$\varepsilon_{se} = ((2\,l_p - l_g)\,\varepsilon_{s1} + l_g.\,\varepsilon_{su})/2 \tag{8}$$

where, θ_u is failure rotation at plastic hinge, l_p is length of plastic hinge, d is effective depth ϕ_u is failure curvature, ε_{ce} is equivalent failure compressive strain in concrete, b is width of beam, z is distance of plastic hinge from point of contra flexure, ρ_s is volume percentage of steel, f_y is yield strength of longitudinal steel, l_g is gauge length over which percentage elongation is measured, ε_{su} is failure strain of steel, ε_{s1} is average strain in steel outside the gauge length, and ε_{se} is equivalent average strain in longitudinal steel.

Using the above relations, the failure rotation in reinforced concrete beams equally reinforced in compression and tension were computed, and compared with experimental results. Since comparisons involve very large strains, scatter is inevitable. The methodology was justified by comparing the limiting values with experimental results (Table III).

$$\theta_{u,max} = (\varepsilon_{se} + \varepsilon_{ce} + 0.01)\ 2l_p/d_c \quad (9)$$
$$\theta_{u,min} = (\varepsilon_{se} + \varepsilon_{ce} - 0.01)\ 2l_p/d_c \quad (10)$$

Table III Failure rotations of SFRC beams reinforced equally on the tension and compression faces

Sl.No	Beam	$\theta_{u,min}$	$\theta_{u,obs}$	$\theta_{u,max}$	ε_{ce} %	ε_{se} %
1	B1	5.91	6.76	8.01	1.46	5.14
2	B2	5.91	6.65	8.01	1.46	5.14
3	B3	5.91	8.09	8.01	1.46	5.14
4	B4	6.38	7.76	8.51	1.78	5.19
5	B5	10.90	12.52	13.07	3.17	7.86

The enormous rotation capacities available in SFRC beams with equal tension and compression reinforcement makes them extremely suitable for use in blast and earthquake resistant design of structures. SFRC beams were tested under cyclic loading by loading the beams to various preloading levels, releasing them and again loading them till failure. The reloading stiffness at various preloading levels were obtained as,

$$\begin{aligned} C &= C_u(1-0.75\ V_f) \leq 1.00 && \text{at } M/M_{cr} \leq 1.0 \\ &= 0.42\ V_f + C_u(1-0.42\ V_f) && \text{at } M/M_u = 0.8 \\ &= C_u && \text{at } M/M_u = 1.0 \end{aligned} \quad (11)$$

where, C is the ratio of effective value of flexural rigidity at a particular load stage to the flexural rigidity of the gross cross section of the SFRC Beam. The value of C for the RCC beam is given by:

$$C = [f^3 + (1-f^3)\ EI_{cr}/EI_g] \quad \text{for } M_{cr}/M_u < f \leq 1.0 \quad (12)$$

where, $f = M_{cr}/M$, M is moment at any loading stage, EI_{cr} is= flexural rigidity of the cracked section, and EI_g= flexural rigidity of gross section. The schematic variation of effective flexural rigidity coefficient C for SFRC beams is given in Fig. 5. The above expressions derived based on the results of six beams were later used to derive the monotonic values of EI_{eff} of three beams TA, SA and CA, which had 1% fibre reinforcement at various load stages.

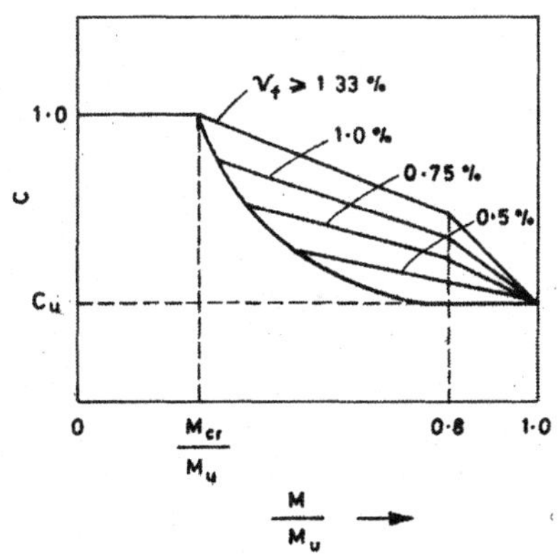

Fig.5 Schematic variation of effective flexural rigidity co-efficient for SFRC beams

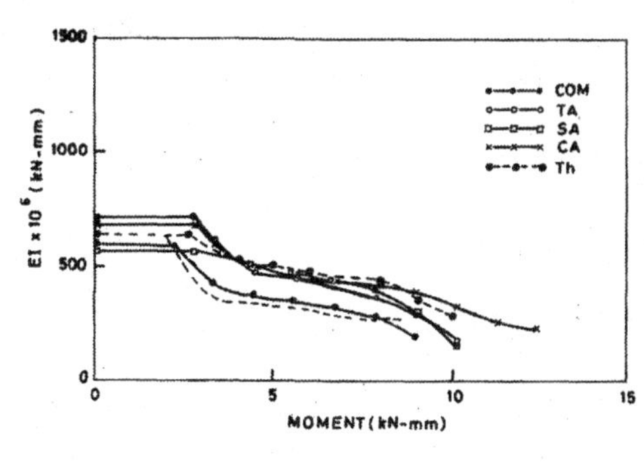

Fig.6 Typical variation of EI (average) with moment under static load (V_f=1.0%)

Fig. 6 shows the variation of EI_{eff} obtained in experiments with those obtained analytically. An excellent correlation is seen validating the use of the approach suggested for obtaining EI at various bending moment levels, which can be used in deformation computations.

SHEAR STRENGTH FOR SFRC BEAMS

A series of 32 SFRC tee beams (2 in each series S1 to S16) were tested under shear span to depth ratio of 2.0. The flange was introduced to avoid pre-mature flexure failure. The fibre reinforcement used was 0, 0.5, 1.0, and 1.25% by volume. Two legged stirrups at spacing of 0.75d, 0.5d and 40 mm were used. Control RC beams were tested only with longitudinal steel rods. The beams with web steel failed by diagonal tension, whose shear capacity could be predicted using the relation:

$$V_{uc} = f_t \left(1 + 20\, V_f\right) bd \tag{13}$$

where, $f_t = 0.3\sqrt{f_{ck}}$, N/mm^2

The beams with close stirrup spacing of 40 mm failed in flexure whose capacities wee well predicted using equation (3). All other beams failed by a combined mechanism of failure in bond, crack at an inclination of less than 45 (horizontal axis to crack) and compression hinge. The idealized mode of failure is shown in Fig. 7. The equation of equilibrium of moment gives:

$$\left(V_c + V_S\right) a = f.A_s.f_y d + A_{sv}.f_y \left(\frac{d}{s}\right)\left(\frac{\cot^2\theta}{2}\right) \tag{14}$$

where V_c is concrete contribution to shear, V_s is steel contribution to shear, $f = \left(\frac{L - d\cot\theta}{L_{df}}\right)$ is a factor less than 1.0, L is as shown in Fig 7., L_{df} is modified developed length, A_s is area of longitudinal steel, A_{sv} is area of stirrup steel, and $V_c = 0.1\left(1 + 300 V_f\right)\sqrt{f_{ck}}\, bd\left(1 - f^2\right) \leq V_{uc}$

$$V_u = V_c + A_{sv} f_y \frac{d}{s} \cot\theta \tag{15}$$

Figs 8 and 9 give the comparison of theoretical and experimental values suggesting the validity of the approach suggested.

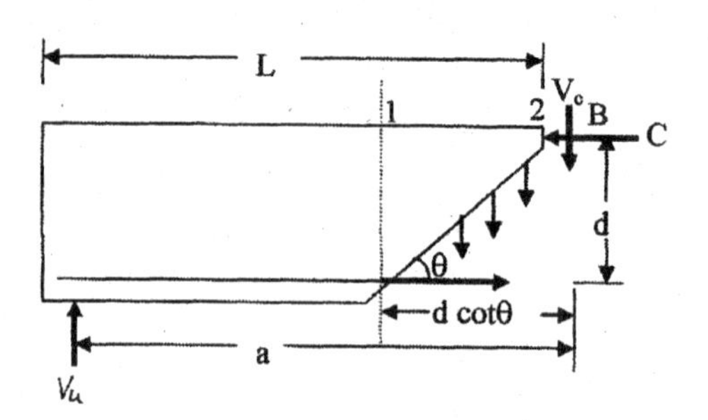

Fig.7 Idealised mode of failure

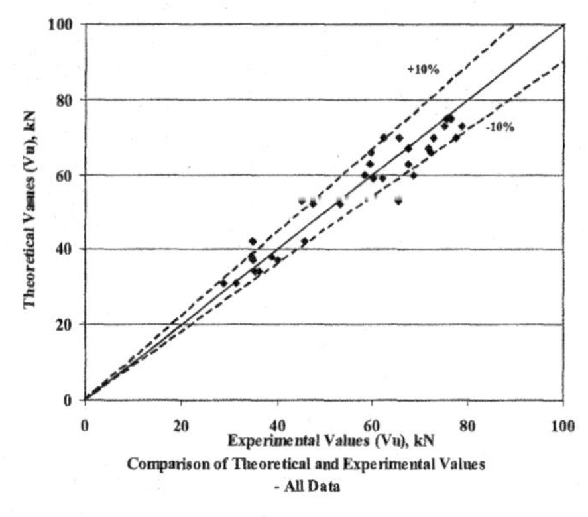

Fig.8 Comparison of Theoretical and Experimental Values - All Data

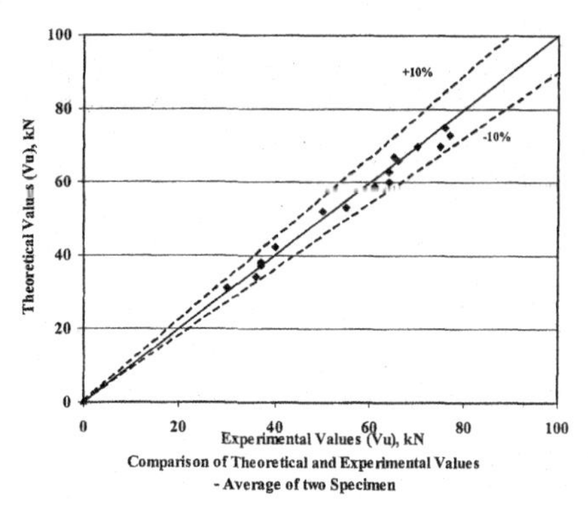

Fig.9 Comparison of Theoretical and Experimental Values Average of two Specimens

SFRC UNDER FATIGUE LOADING

The most significant influence of incorporation of fibres in concrete is to delay and control the tensile cracking of concrete. Thus, inherently unstable tensile crack propagation in concrete is transformed into a slow and controlled crack growth. The fibre provides a ductile constituent in a brittle matrix, and the resultant composite has ductile properties which are significantly different from plain concrete. This inherent feature of fibre reinforced concrete improves the dynamic properties like energy absorption, and behaviour under fatigue loading over the plain concrete.

The fatigue performance of 100 x 150 x 1500mm steel fibre reinforced RC beams was investigated[8]. The maximum loads for the fatigue test were varied as a percentage of the static ultimate load of RC beams without fibre. The minimum load was kept constant as 10kN for all the test specimens. The frequency of loading was kept as 5 Hz. At the end of 5,000, 25,000, 50,000, 1,00,000 cycles and so on, deflections, strains and crack width were monitored. It was found that the addition of fibres to RC beams improved their performance both under static and fatigue loadings. It was seen that among the three types of fibres tested, the fatigue behaviour of beams containing trough shaped fibres was better when compared to beams with other types of fibres.

Estimation of Deflection

For the purpose of practical calculation of long term cyclic deflection of reinforced concrete members sustaining a larger number of repeated cycles of load, Lovegrove and Salah El Din[9] have developed the equation :

$$Y_n = AY_1 \ln(N) \qquad (16)$$

Where Y_n is deflection at nth cycle; Y_1 is the deflection at first cycle; N is the number of cycles ; and $A = 0.225$.

The constant A of equation (16) was evaluated by regression analysis from the deflection data obtained from the present investigations. It was found that there was considerable scatter in the calculated deflection values as compared to the present test results. According to Lovegrove and Salah El Din (9), equation (16) does not cater to lower number of cycles. At lower number of cycles, the difference between the calculated values of deflection and the values obtained in the tests was found to be very high. A plot has been made between the non-dimensional variable (Y_n/Y_1) and the log number of cycles. It was seen that a linear equation can be defined to evaluate (Y_n/Y_1) with ln (N). The expression of Y_n has been derived and given in equation (17).

$$Y_n = Y_1[1 + A\ln(N)] \tag{17}$$

Here Y_1 is the deflection at first cycle of the FRC beam; and $A = 0.0274$ (from regression analysis). The equation (17) was checked for all the deflection data points of the present investigation. Fig.10 shows the variation of theoretical value of deflection (using equation 17) with the values obtained from the tests. It is found that there is good agreement in the prediction of deflection using the empirical relationship derived given by equation (17).

Estimation of Fatigue Life

An empirical relationship was also obtained to predict the number of cycles to failure using the proposed deflection equation. Equation (17) was differentiated with respect to number of cycles to obtain the equation :

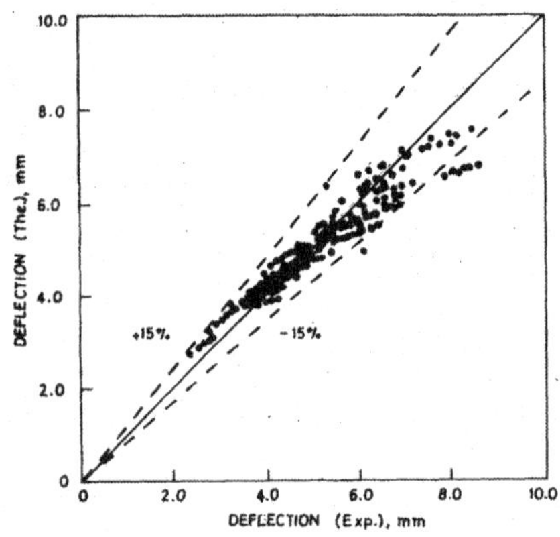

Fig.10 Variation of theoretical value of deflection with the experimental values

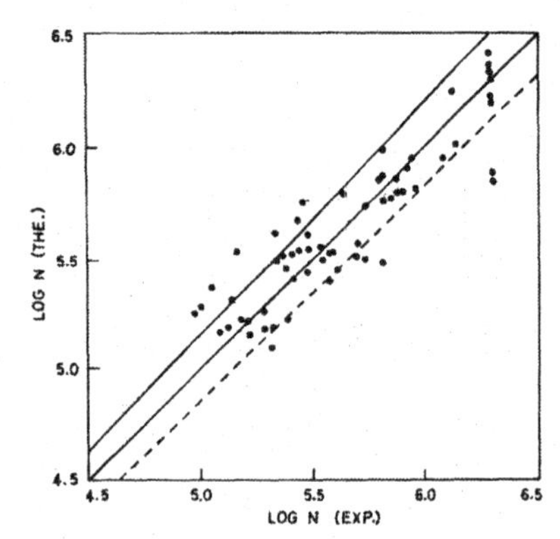

Fig.11 Variation of the predicted and experimental number of cycles to failure

$$(dY_n / dN) = (AY_1 / N) \tag{18}$$

$$N = AY_1 / C \tag{19}$$

Where C is the slope representing the variation of deflection with increased number of load cycles given by:

$$C = dY_n / dN \tag{20}$$

At failure,

$$N_f = (AY_1 / C_f) \tag{21}$$

As stated earlier, the number of cycles to failure depends on the fibre type, percentage of fibre volume fraction, and the load range. The effect of all these variables is included in the proposed equation for evaluating the value of C_f as

$$C_f = C_1[(P_{max} - P_{min})/(P_u - P_{min})]^{C_2} /(1 - BV_f) \quad (22)$$

Where P_{max} is the maximum load applied on the beam; P_{min}, minimum load applied on the beam ; P_u, ultimate load ; and V_f, the volume fraction of the fibre.

The effect of load range has been included in the non-dimensional form by the parameter $[(P_{max} - P_{min}) / (P_u - P_{min})]$, and the effect of volume percentage and type of fibre is taken into account directly through the expression $(1\text{-}BV_f)$ and indirectly through the ultimate strength (P_u). The constants B, C_1 and C_2 were evaluated based on regression analysis and the following values were obtained for the three types of fibres used in the present investigation.

B = 0.29 for trough-shaped fibres $\qquad C_1 = 5.0 \times 10^{-6}$
= 0.28 for straight fibres $\qquad C_2 = 5.0$
= 0.09 for crimped fibres

Hence, the number of cycles to failure is given by

$$N_f = (AY_1)/\{C_1[(P_{max} - P_{min})/(P_u - P_{min})]^{C_2} /(1 - BV_f)\} \quad (23)$$

It is important to note that Y_1 depends on a number of parameters such as the peak load, type of fibre, and percentage of volume fraction of fibre and is based on the behavior of SFRC beams under monotonic loading. Based on equation (23), the number of cycles to failure was predicted by knowing the deflection at first cycle, ultimate static load, the volume fraction of the fibre, fibre type, and maximum and minimum loads applied during fatigue loading. Fig.11 shows the variation of the predicted and experimental number of cycles to failure on a log scale. It is seen that the scatter is limited to ± 3% on log (N) for majority of beams tested. The following equations are useful for fatigue design.

$$Y_n = Y_1 [1 + A \ln (N)] \quad (24)$$
$$N_f = (AY_1/C_f) \quad (25)$$

Where $C_f = C_1 [(P_{max} - P_{min}) / (P_u - P_{min})]^{C_2} / (1\text{-}BV_f)$

CONCLUSIONS

The results of various investigations conducted at SERC, Chennai are presented in this paper. The investigations include the flexural and fatigue behaviour of steel fibre reinforced concrete.

Based on the experimental investigations on the SFRC cylindrical specimens, for the purposes of design, a relationship between failure strain and volume fraction of fibres is proposed.

A simplified approach was developed to predict the ultimate capacity of SFRC beams based the experimental results of the investigations on the flexural behaviour of SFRC. The experimental values was compared with values obtained using Craig's model, and the simplified approach. The above simplified approach was also used to predict ultimate capacities of beams reinforced equally in tension and compression faces. A very good agreement is seen validating the use of simplified expression.

A simplified model was proposed to compute the failure rotation at a plastic hinge of SFRC beams with equal tension and compression reinforcement and compared with the experimental values. The enormous rotation capacities available in SFRC beams with equal tension and compression reinforcement makes them extremely suitable for use in blast and earthquake resistant design of structures. An approach to evaluate the shear capacity of SFRC beams has also been suggested

Empirical equations have been proposed to evaluate the deflection of SFRC beams and to obtain the number of cycles to failure of SFRC beams, subjected to fatigue loading.

REFERENCES

[1]B.H. Bharatkumar, T.S. Krishnamoorthy, K. Balasubramanian, and S. Gopalakrishnan, "Behaviour of Steel Fibre Reinforced Concrete subjected to Cyclic Loading," *International Symposium on Innovative World of Concrete-98,* Calcutta, India, pp 7.31-7.38 (1998)

[2]D.A. Fanella, and A.E. Naaman, "Stress- Strain Properties of Fibre Reinforced Mortar in Compression," *ACI-Journal,* **82** [4] 475-483 (1985)

[3]K. Balasubramanian, B.H. Bharatkumar, S. Gopalakrishnan and V.S. Parameswaran, "Flexural Behaviour of Steel Fibre Reinforced Concrete Beams under Static Load," *Journal of Structural Engineering* , **25** [3] 167-172 (1998)

[4]R. Craig, "Flexural Behaviour and Design of Reinforced Fibre Concrete Members," in *Fibre Reinforced Concrete Properties and Applications, edited by S.P. Shah and Batson, ACI SP-105,* ACI Detriot, pp 517-563 (1987)

[5]V.S. Parameswaran, T.S. Krishnamoorthy, K. Balasubramanian and N. Lakshmanan, "Behaviour of Fibre Reinforced Concrete Beams with Equal Tension and Compression and Reinforcement," *International Symposium of fibre Reinforced Concrete, Edited by V.S. Parameswaran and T.S. Krishnamoothy*, Madras, pp 1.57-1.67 (1987)

[6]N. Lakshmanan, T.S. Krishnamoorthy, K. Balasubramanian, B.H. Bharatkunar and S. Gopalakrishnan, "Shear Behaviour of Steel Fibre Reinforced Concrete Beams with Low Shear span to Depth Ratios," Paper under communication.

[7]N. Lakshmanan, V.S. Parameswaran, T.S. Krishnamoorthy, and K. Balasubramanian, "Ductility of flexural members Reinforced symmetrically on the Tension and Compression Faces", *The Indian Concrete Journal,* **65** [8] 381-389 (1991)

[8]K. Balasubramanian, B.H. Bharatkumar, S. Gopalakrishnan and V.S. Parameswaran, "Fatigue Behaviour of Steel Fibre Reinforced Concrete Beams," *Journal of Institution of Engineers, India,* **25** [3] 167-172 (2000)

[9]M. Lovegrove, and Salah El Din, "Deflection and Cracking of Reinforced Concrete under Repeated Loading and Fatigue," in *Fatigue of Concrete Structures, SP-75, Edited by S.P. Shah,* ACI Detroit, pp 133-152 (1982)

STEEL FIBER REINFORCED CONCRETE-APPLICATIONS IN INDIA

V. S. Parameswaran
Chief Executive, International Centre for FRC and President & CEO
Design Technology Consultants
Chennai-600 020, India

K. Balasubramanian
Scientist
Structural Engineering Research Centre
Chennai 600 113, India

S. Gopalakrishnan
Director-Grade Scientist & Advisor(Management)
Structural Engineering Research Centre
Chennai 600 113, India

ABSTRACT

During the last two decades, significant developments have taken place to improve the mechanical and durability properties of cement concrete. The sustained efforts of researchers to have unmatched excellence in construction materials have led to development of a number of new concrete composites having very high compressive strength and durability properties with potential for civil engineering applications. One such versatile concrete composite that was developed in early 1960s and which had since then undergone many modifications in design and production aspects is Steel Fiber Reinforced Concrete (SFRC). This paper briefly presents the developments and applications in India of SFRC and its related high-end derivatives such as Slurry Infiltrated Fibrous Concrete (SIFCON), Slurry Infiltrated Mat Concrete (SIMCON) and Compact Reinforced Concrete (CRC).

INTRODUCTION

The idea of combining two or more materials to obtain a composite is not new to engineers. Perhaps, the earliest application of the now well known composite theory relates to the making of clay bricks using natural fibers such as straw, the latter vastly improving the properties of the otherwise brittle matrix. When two different kinds of materials with contrasting properties of strength and elasticity are combined together, they realize a great portion of the theoretical strength of the "stronger" component. These combined materials are called "two-phase composites". In an ideal two-phase composite, the strength of the weak phase is thus improved by the strong phase.

Conventional reinforced concrete is a two-phase composite in which the matrix when cracked, is held by the reinforcing bars. On the other hand, when concrete is reinforced by short, closely spaced fibers, an ideal two-phase composite is produced, so much so that even the cracking strength is increased by the closely dispersed fibers acting as "crack arresters". Asbestos cement is an example of a cement composite in which the fibers play an important role by contributing to the strength and ductility of the composite.

Fibers are produced from steel, carbon, glass, plastic, polypropylene, nylon, rayon, asbestos and basalt. Natural fibers such as cotton, coir, sisal, etc. in various shapes and sizes are also abundantly available. It is now well established that low-modulus fibers such as nylon and polypropylene are unlikely to lead to usable strength improvement of the composites; their elongation, enabling the composites to absorb large amounts of energy and resist impact and shock loading. However, for structural applications using concrete, steel and glass fibers are generally preferred, since they possess very high modulus of elasticity and lead to strong, stiff and durable composites.

Application of SFRC and its other special versions depend mostly on the ingenuity of the designer and builder in taking advantage of their relatively superior static and dynamic strength properties, ductility, energy-absorbing characteristics, abrasion resistance and fatigue strength as compared to conventional concrete. In recent years, the type and scope of application of these materials have increased and expanded in line with their overall use. Growing experience and confidence of engineers, designers and contractors have led to new areas of use. Traditional application where SFRC was initially used as pavements has now gained wide acceptance in heavy-duty and container yard floors due to comparable cost and assurance of better performance and durability.

The advantages of SFRC have now been well recognised and utilised in many precast concrete applications as well, where designers are looking for thinner sections and more complex shapes. Applications include building panels, sea-defense walls and blocks, piles, blast-resistant storage cabins, coffins, pipes, kerbs, prefabricated tanks, domes, swimming pools, composite panels and ducts. Recent advancements in concrete technology have also helped in developing fiber shotcreting techniques, which have become popular and competitive in the construction of tunnels and in repair and rehabilitation of distressed concrete structures. Major applications of SFRC in India are briefly outlined in this paper.

PAVEMENTS AND INDUSTRIAL FLOORS

Early use of SFRC has been in the construction of heavy-duty pavements and industrial floors where its high strength and excellent impact and abrasion resistance were prerequisites to resist high traffic loads and general wear and tear from heavy moving vehicles. In addition to the serviceability aspects, SFRC has also been found to be economically competitive with conventional materials for such applications. Placing and compaction of SFRC inside the forms is easier and faster, since it eliminates the time needed for placement of the mesh reinforcement. Also, a typical 30-40% reduction in slab thickness that is obtainable with SFRC means smaller volume of concrete for a given area resulting in further savings in placement time.

There are several pavement applications of SFRC in India. One of the early applications relates to the construction of an airfield pavement at the New Delhi airport by the International Airport Authority of India[1]. The thickness of the pavement was reduced by 25% (from 400 mm to 300 mm) which was made possible by taking into account the higher flexural strength of SFRC in the design which was found to be 80 to 100 kg/cm^2 for the pertinent FRC mix. Conventional methods of handling and placing were adopted with a slight modification to the conventional concrete mixer.

A heavy duty industrial floor measuring about 10,000 square meters has been built with SFRC in Ankleshwar, Gujarat during 1994-96 for a paint manufacturing facility which has thus far successfully withstood the effects of impact and abrasion caused by heavy moving vehicles with practically no maintenance expenses till now (Fig. 1).

Figure 1. SFRC Industrial Floor

ROADS AND BRIDGE DECKS

Because of its excellent abrasion and impact resistance, SFRC has found ready acceptance in the construction of roads and bridge decks. The mining industry, particularly the coal and ore mining industry has used significant volumes of SFRC in haulage roads, wash down pads and hard standing areas. A large stretch of haulage road has been recently laid in the mines area in Bilaspur by the South Eastern Coal Fields using SFRC produced with indigenously available round, crimped steel fibers. The fiber dosage has been as much as 60-80 kg per cubic meter of concrete.

SFRC has been successfully used in the construction of a 2.2 km haulage road at Mathura by the Indian Oil Corporation (Fig. 2). A significant reduction in thickness was possible by the higher flexural strength of SFRC which was found to be 80 to 100% more than the ordinary mix. Conventional methods of handling and placing were adopted with a slight modification to the conventional concrete mixer.

A few applications of SFRC in bridge related construction has also been reported. Two bridges across river Narmada near Allahabad and a fly-over in Bangalore made use of SFRC to strengthen the end-anchorage zones of precast bridge girders used in the superstructure.

Figure 2. SFRC Road under Construction in Mathura

OVERLAYS AND TOPPINGS

Unbonded overlays of SFRC have been used in many countries to rejuvenate old concrete floors in various stages of distress. Overlays are normally limited to a minimum thickness of 75mm although 100mm is more typical. Thin toppings of SFRC have also been used on timber floors in old warehouses.

SFRC overlay consisting of round, crimped steel fibers of 0.45mm diameter and 36 mm length has been cast on an experimental basis over an area of about 100 sq.m in a textile processing unit in Tirupur, Tamilnadu. The overlay was laid over a plain concrete sub-base and was 65 mm thick with a fiber content of 50 kg per cubic meter. Another application of SFRC has been in the overlay over a concrete sub-base in a highway under pass in Tirutani near Chennai where the authorities have preferred fiber concrete for its very high abrasion, cracking and impact resistant properties.

MARINE AND HYDRAULIC STRUCTURES

SFRC has also been successfully used for constructing waterfront marine structures which have to resist deterioration at the air water interface and impact loadings. These include jetty armors, breakwaters, floating pontoons and caissons. Other potential applications are under-water storage structures, water-front warehouse floors and wharf decking. It has been successfully used in hydraulic structures too, such as, spillway channels, tunnels, dams and their abutments, and stilling basins. The ease of molding fiber concrete to compound curves makes it attractive for ship hull construction either alone or in conjunction with ferrocement. In India, however, SFRC has been used only to a limited extent for repair of some offshore jetties and irrigation structures.

SFRC MANHOLE COVERS AND FRAMES

It has been estimated that every kilometer of urban road in India may require 15 to 20 manhole chambers, which, in turn, require an equal number of covers and frames. The demand for manhole covers is, therefore, phenomenal. Presently, grey cast iron is being used for the manufacture of these covers which are very expensive, brittle and are also liable to be pilfered for their scrap value. Several efforts have been made in the country in the past with little success in finding a suitable alternate material which meets the stipulated loading and functional requirements.

On the basis of extensive laboratory investigations carried out at the Structural Engineering Research Center (SERC), Chennai, technology for the design and production of SFRC manhole covers and frames for heavy, light and medium-duty applications has been developed. The covers and frames are produced using the well known pressure-cum-vibration technique to achieve dense and strong concrete with very low water content in the mix. The technology has been well received by the industry and as many as 35 firms are already manufacturing the units in various parts of the country. The investigations at SERC, Chennai covered optimization of production technology and included tests on several tens of covers to evaluate their static, impact, and fatigue resistance, even though the tests covering the latter two loadings are not specified in the current Indian codes. A view of SFRC manhole covers being produced in a factory near Hubli is shown in Fig. 3.

Figure 3. SFRC Manhole Covers and Frames

STEEL FIBER REINFORCED SHOTCRETE (SFRS)

Recent years have seen increasing application of steel fiber reinforced shotcreting technology for a variety of structures. The new technique is more economical and involves less site-oriented jobs and makes use of small diameter, short steel fibers which are gunited over the surface along with wet concrete as compared to conventional mesh reinforcement tied to the surface prior to shotcreting operation in traditional plain concrete shotcreting. The major advantages of using SFRS are less surface cracking, easier finishing, better quality control, elimination of about 90% of meshing and anchoring, and improved mechanical properties. The resulting shotcreted surface exhibits substantially superior toughness index and impact strength compared to the one obtained by plain concrete shotcreting. Fiber shotcreting is currently being employed successfully for shaft and tunnel linings, rock slope stabilization, storage structures, domes & shells and repair of deteriorated concrete structures[2,3].

In India, SFRS has been extensively used primarily for tunnel lining and repair works. In the recently completed Uri Hydro Electric project near Jammu, the pen stock tunnels were fiber shotcreted using remote controlled robots (Fig. 4). SFRS was also used in the tunnel lining work for the Baglihar Hydro Electric Project in the district of Doda in Jammu and Kashmir. The shotcreted panels from the site were tested for their energy absorption characteristics at SERC, Chennai as per EFNARC specifications. The Konkan Railways authorities have also built a number of tunnels passing through difficult mountain terrains using SFRS technology. One advantage found in rail tunnel work is that the scaffolding required for mesh installation can be totally eliminated and traffic interruption is minimized.

Figure 4. Robot Controlled Fiber Shotcreting

SLURRY INFILTRATED FIBROUS CONCRETE (SIFCON)

Slurry infiltrated fibrous concrete is a relatively new material and can be considered as a special type of SFRC[4]. In SFRC, the fiber content usually varies from 1 to 3% by volume whereas in SIFCON, it varies between 5 and 20%. The matrix of SIFCON consists of cement paste or flowing cement mortar as opposed to regular concrete in SFRC. The process of making SIFCON is also different because of the high fiber content. In SFRC, the fibers are added to the wet or dry mix of the concrete during mixing, but SIFCON is prepared by infiltrating cement slurry into a bed of pre-placed and pre-packed fibers. SIFCON has been successfully used for refractory application, pavements and overlays, and structures subjected to blast and dynamic loading. Because of high ductility and impact resistance, the composite has excellent potential for constructing structural components which need to resist high impact force and exhibit high ductility, such as explosive- storage cabinets, blast-resistant doors, high-security vaults, repair of concrete bridge decks, test tracks for heavy vehicles, missile silo structures and precast shapes where standard modes of reinforcement are ineffective. SIFCON has not yet been used for major applications in India, even though extensive research has been carried out establishing its suitability for very special structures.

SLURRY INFILTRATED MAT CONCRETE (SIMCON)

SIMCON uses non-woven steel fiber mats to reinforce the concrete matrix, thereby producing concrete components with extremely high flexural strength[5]. The steel fiber is directly cast from molten metal using a chilled wheel concept and then interlaid into a 1.2 to 5 cm thick mat. This mat is then rolled and coiled into sizes and weights convenient to a customer's application and can range up to 48 in. wide and 500 lb weight. Since the mat is already in a preformed shape, handling problems are minimised and balling is not a factor.

SIMCON can also be considered as a preplaced fiber concrete, the only difference between SIMCON and SIFCON being that the fiber is placed in a mat rather than as discrete fibers. The advantage of steel fiber mats over a large volume of discrete fibers is that the mat configuration provides inherent strength and utilizes fibers with an aspect ratio as high as 500. The fiber volume is less than half that required for SIFCON, while achieving similar flexural strength and energy absorbing toughness. As in the case of SIFCON, there has not been any major application of SIMCON in India till now.

COMPACT REINFORCED CONCRETE (CRC)

CRC is a new type of innovative composite made up of a very strong cementitious matrix toughened with a high concentration of fine steel fibers along with an equally large proportion of evenly distributed conventional steel reinforcing bars[6]. CRC has structural similarities with reinforced concrete, but is much more heavily reinforced and exhibits mechanical behaviour more like that of structural steel, having almost the same strength and exceptionally high degree of ductility. Components made out of CRC remain substantially uncracked till the yield limit of the main reinforcement (about 3mm/m), whereas conventional reinforced concrete typically cracks at about 0.1 - 0.2 mm/m. CRC was developed in 1986 by Aalborg Portland as a ductile version of the Densified Small Particles (DSP) materials.

CRC specimens are normally produced using 10-20% volume of main reinforcement (in the form of steel bars of diameter from about 5mm to perhaps 40 to 50mm) and 5-10% by volume of fine steel fibers. Compared with conventional concrete, production of CRC structures is incredibly complex and requires special techniques.

CRC has potential applications in off-shore and defense structures, high- rise buildings, earthquake-resistant structures, industrial floors, wear-protection surfaces, and large bridges, besides for production of security products such as currency safes/vaults, ammunition-storage vaults etc. While extensive R&D work has been carried out in numerous institutions in the country on the properties and production of CRC, there is not yet a single major application of this exotic material to any important structure in the country. Development of CRC tracks for movement of armored vehicles in remote areas as well as missile-resistant CRC aircraft hangers and other defense-related structural components is under way in a few defense R&D laboratories.

OTHER APPLICATIONS OF SFRC

Application of SFRC to biological shielding of atomic reactors and also to waterfront-marine structures which have to resist deterioration at the air-water interface and impact loadings has made some head way in the country with the availability of indigenously made steel fibers of various shapes and dimensions. The latter category includes jetty armor, floating pontoons, and caissons.

Laboratory investigations have indicated that steel fibers can be used in lieu of stirrups in shear reinforcing of frames, beams, and flat slabs and also as supplementary shear reinforcement in precast, thin-webbed beams. Steel fiber reinforcement can be added to critical local zones of precast prestressed concrete (e.g. end zones) and cast-in-place concrete to eliminate much of the secondary reinforcement. Fiber reinforced concrete may also be an improved means of providing ductility to blast-resistant and seismic-resistant structures especially at their joints, owing to the ability of fibers to resist deformation and undergo large rotations. The ductility of SFRC would permit the development of plastic hinges from over load conditions.

Both steel and vegetable fibers have been used in the development and production of several FRC building components by the Central Building Research Institute, Roorkee. Some of the components developed by the Institute are precast doubly curved roofing tiles, lintels and floor planks with thicknesses varying from 20mm to 75 mm.

Vast potentials exist for the use of SFRC as a repair material for retrofitting of distressed concrete structures particularly, jetties and bridges (Fig. 5). SFRC has been used for repair of a few off-shore jetties in Gujarat and for repair and strengthening of a few highway bridges and dams in Tamilnadu.

Figure 5. Repair of a Bridge using SFRC

REFERENCES

[1] Raghavendra, N., Kulshrestha, H.K., and Rattan Lal, "Fiber Reinforced Concrete for Airfield Pavements," *Indian Concrete Journal,* pp.64-67 (1985)

[2] Skatun, O., and Christiania, S., "Applications of Wet Process Steel Fiber Reinforced Shotcrete in Scandinavia," *Proceedings of Developments in Fiber Reinforced Cement and Concrete, RILEM Symposium, FRC-86, Vol. 2, Sheffield, UK.*

[3] Colin D. Johnston and Paul D. Carter, pp. 7-16 "Fiber Reinforced Concrete and Shotcrete for Repair and Restoration of Highway Bridges in Alberta," *International Symposium on Recent Developments in Concrete Fiber Composites, Transportation Research Board, Washington, D.C., (*1989)

[4] Lankard, D.R., "Slurry infiltrated fiber concrete (SIFCON)," *Concrete International,* **6**[12] 44-47 (1984)

[5] Lloyd, E. Hackman, Mark, B. Farrell and Orville O. Dunham, *Slurry Infiltrated Mat Concrete (SIMCON), Concrete International,* **14**[12] 53-56 (1992)

[6] Bache, H.H.,"Compact Reinforced Composite, Basic Principles", AALBORG PORTLAND, *Cement-og Betonlaboratoriet,* CBL Report No. 41 (1987)

FLEXURAL CRACKS IN BEAMS WITH GLASS FRP REBAR AND FIBER REINFORCED CONCRETE

D.C. Jansen
Assistant Professor
Dept. of Civil and Env. Engineering
California Polytechnic State University
San Luis Obispo, California 93407

W.K. Lee
Graduate Research Assistant
Dept. of Civil and Env. Engineering
Stanford University
Stanford, California 94305

ABSTRACT

Fiber reinforced polymer (FRP) rebar have attracted considerable attention for applications where corrosion of conventional steel reinforcement is problematic. The most common FRP rebar contain glass fibers due to their relatively low cost. In comparison to steel, glass FRP rebar have lower elastic moduli, higher ultimate tensile strength, and no ductility. With their low modulus of elasticity, use of FRP rebar results in decreased flexural rigidity of members and larger crack widths under service loads. Ten plain concrete or fiber-reinforced concrete beams reinforced with standard steel rebar and glass FRP rebar were tested to failure under third-point flexural load. The work presented includes the experimental results from the beams with steel rebar or glass FRP rebar. An image analysis system was used to monitor flexural cracks in the specimens. A modified Gergely-Lutz model was applied to the measured crack widths. In the beams with the steel rebar, the standard Gergely-Lutz expression predicted crack widths well, and the incorporation of fibers had little influence on crack widths. In the beams with FRP rebar, the standard Gergely-Lutz expression under-predicted crack widths, and use of synthetic fibers in the concrete significantly reduced the size of flexural cracks.

INTRODUCTION

Corrosion of reinforcing steel occurs in reinforced concrete structures in the presence of moisture, and the process is exacerbated by extreme environmental conditions (high humidity regions or marine atmospheres) or aggressive chemical exposure (such as to de-icing salts).[1] The majority of structures susceptible to corrosion are highway bridges. The cost of repair of the deteriorating highway infrastructure exceeds several billion dollars annually and is a world-wide problem.[2] Current methods of combating corrosion in reinforced concrete attempt to improve either the reinforcing bar (e.g. epoxy coatings, galvanized or stainless steel) or the concrete (e.g. latex modified concrete, corrosion inhibitors, or use of silica fume). However, use of these methods is inhibited by such factors as cost and questions of effectiveness. In addition, while these methods slow corrosion, they do not prevent it completely.

Recently, advanced composite materials have been applied to combat the problem of corrosion. Within the last few decades, civil engineers have realized the numerous benefits of composite materials, and much research has been performed in order to produce sufficient information so that structures utilizing composite technology can be safely and properly designed.[1, 2, 3, 4, 5] One form of composite that is being studied is the composite reinforcing bar that is similar to the traditional steel reinforcing bar. These composites, known commonly as fiber-reinforced polymers (FRPs), have high tensile strengths in comparison to steel, and more importantly are resistant to corrosion. Additional benefits to the FRP bars are their electromagnetic and radio transparency and their low weight compared to steel. However, FRP bars also have several negative aspects, as most types of FRP bars have low elastic moduli and relatively poor bond to concrete compared to steel. The low values of elastic modulus, in combination with poor bond characteristics, lead to excessive cracking and deflections of FRP-reinforced members under service loads.

To ameliorate the problems associated with the use of FRP bars in reinforced concrete, it is being proposed to use fiber-reinforced concrete (FRC) in place of plain concrete in FRP-reinforced members. The purpose of this study was to determine whether the use of FRC would improve the cracking response of beams reinforced with FRP bars. In this study, plain and fiber-reinforced concrete beams with varying FRP reinforcing bar types and reinforcing ratios were tested in bending, and the cracking response was measured in the constant-moment region. The cracking responses of the specimens are studied in order to determine any improvements in the crack widths that may come as a result of the fiber reinforcement. In addition, the observed cracking responses are used to determine corrective bond coefficients for use in the modified Gergely-Lutz equation for predicting crack widths in FRP-reinforced beams.

EXPERIMENTAL PROGRAM

Material Properties

Reinforcing Bars: The stress-strain behavior of the glass fiber reinforced polymer (GFRP) rebar, typical of most types of FRP bar, was linear elastic until failure, with no ductility or yielding. The GFRP bars had a helically-wound fiber on the outside to produce surface deformations as well as to provide lateral confinement. The surface of the GFRP bars was coated with a coarse silica sand to improve its bond to concrete. The control beams contained deformed steel reinforcing bars. The steel and GFRP reinforcing bars used can be seen in Figure 1, and the mechanical properties of the reinforcing bars are provided in Table I.

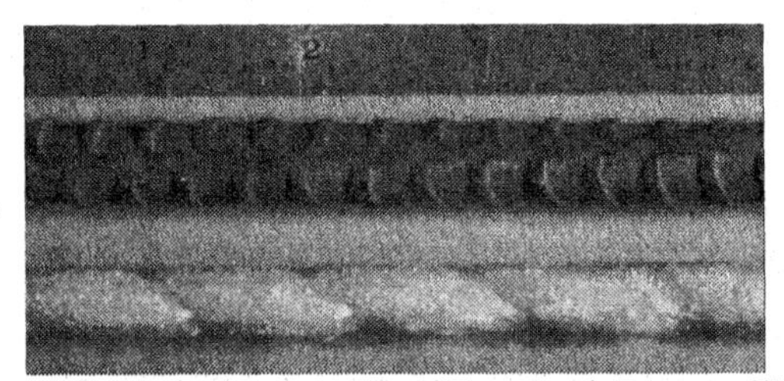

Figure 1. Reinforcing Bars Used (Top: Steel, Bottom: GFRP)

Table I. Mechanical Properties of the Rebar

Bar Type	Bar Size (U.S.)	Nominal Diameter (mm)	Elastic Modulus (GPa)	Yield Strength (MPa)	Tensile Strength (MPa)
GFRP	#2	6.4	37.8	N/A	507
GFRP	#3	9.5	43.3	N/A	769
GFRP	#4	12.7	45.6	N/A	690*
Steel	#3	9.5	183	361	
Steel	#4	12.7	162	448	

* Data provided by manufacturer.

Concrete: Normal strength concrete was used in this investigation. The concrete contained a Type II Portland cement and a Grade 100 ground granulated blast-furnace slag. The maximum nominal size of the coarse aggregate was 9.5 mm. The concrete had a water-to-cementitious materials ratio of 0.46. The ratio of cement:slag:fine aggregate:coarse aggregate:water was 1:1:0.27:0.27:0.91. An air-entraining admixture was used at a dosage of 82 ml/100 kg of cementitious material for the plain concrete. The air content for plain concrete was 7 percent.

The fiber-reinforced concrete used in this investigation contained polypropylene fibers at a fiber volume fraction of 1 percent. The polypropylene fibers had a tensile strength of 540 MPa and an elastic modulus of 9.5 GPa. The fibers were 40 mm in length with an aspect ratio of 90. The FRC had the same mixture proportions as the plain concrete, with the exception that an ASTM C94 Type F high range water-reducer (superplasticizer) was used to provide adequate workability. Superplasticizer was used at a dosage of 240 ml/100 kg of cementitious material. The compressive strength of the FRC was about 23 percent lower than the plain concrete.

Reinforced Beam Specimens: Beams were cast using a standard laboratory drum mixer. Two beams and nine 200 mm by 100 mm compression cylinders were cast at a time. Each beam was filled in three layers, and compaction was performed with an internal vibrator. The beams and cylinders were cured at room temperature (16° to 27° C) for 24 hours. The beams were wrapped in wet burlap and covered in plastic and allowed to cure for 14 days at room temperature. Three cylinders, known as companion cylinders, were placed with the beam specimens to be cured under the same conditions. After 14 days, the plastic and wet burlap were removed and the specimens were allowed to cure in the ambient environment until testing. Specimens were tested at an age between 27 and 31 days. The companion cylinders were tested at the same time as the reinforced beams specimens to provide a better estimate of the mechanical properties of the concrete in the beams as opposed to using properties of the moist-cured cylinders. The remaining six cylinders were placed in a 100 percent relative humidity environment at 23° C (73° F) after de-molding for curing until testing. The moist-cured specimens were tested at an age of 28 days as per ASTM standards. Table II lists the beams tested along with the associated compressive strengths of the concrete.

Specimen Geometry and Testing Configuration

The beams tested were rectangular in cross-section and prismatic. Specimen geometry is shown in Figure 2. The beams contained no compressive reinforcing and no internal shear reinforcing. Beams were tested in a model F3-300 Riehle screw-driven testing machine with a 1.33 MN load capacity. A reinforced beam specimen in the testing machine is shown in Figure 3. Shear reinforcement consisted of nine external steel stirrups evenly spaced on each of the two outer spans to prevent shear failure. The beams were tested with a cross-head displacement rate of 0.5 mm/min.

Digital Image Analysis for Crack Width Determination

A digital image analysis system was used as a method of data acquisition to measure the formation and growth of cracks in a specimen during testing. The digital image analysis system is a non-destructive method of data acquisition that does not require the direct attachment of any instrumentation to the specimen. Digital image analysis, or computer vision as it is often called, is a method of measuring relative displacements across the surface of a specimen. The principle upon which computer vision is based is the matching of image subsets between two different images. A full description of the computer vision system used can be found in Lee.[6]

Table II. Test Specimens and Concrete Compressive Strengths

Specimen	Reinforcing Bar Type	Bar Size	Fiber Type	Companion Compressive Strength (MPa)	Moist-Cured Compressive Strength (MPa)
G2N0	GFRP	#6 (#2 U.S.)	None	42.7	37.8
G3N0	GFRP	#10 (#3 U.S.)	None	39.9	38.6
G4N0	GFRP	#13 (#4 U.S.)	None	38.7	35.1
G2P1	GFRP	#6 (#2 U.S.)	Polypropylene	32.6	30.5
G3P1	GFRP	#10 (#3 U.S.)	Polypropylene	32.0	24.8
G4P1	GFRP	#13 (#4 U.S.)	Polypropylene	31.1	25.3
S3N0	Steel	#10 (#3 U.S.)	None	39.3	35.1
S4N0	Steel	#13 (#4 U.S.)	None	49.0	42.6
S3P1	Steel	#10 (#3 U.S.)	Polypropylene	28.9	22.9
S4P1	Steel	#13 (#4 U.S.)	Polypropylene	37.2	34.8

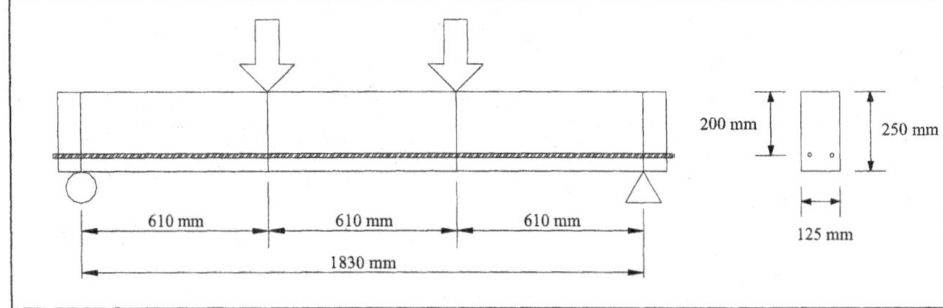

Figure 2. Specimen Geometry

Figure 3. Photograph of Beam Being Tested

RESULTS

The variables in this study that affected crack width were the presence of fiber reinforcing and the properties of the reinforcing bars (size, elastic modulus, and bond characteristics). To determine the effect only of the fibers and to attempt to eliminate the other factors in making comparisons, the Gergely-Lutz equation was used and will be discussed in following sections.

To compare the cracking behavior between the different beam specimens, plots were made of the moment versus the crack width. The crack widths were determined within the constant-moment region of the beam, as that was the region that was analyzed using the digital image analysis system. The computer vision system that was used allowed for the determination of the crack widths at almost any height of interest in the specimen. The crack widths were taken at the height of the reinforcing bars rather than calculating an average crack width value over the height of the beam since the crack width increases proportionally with the distance from the neutral axis (this statement will be justified in following sections). In addition, rather than taking an average value between the multiple cracks that formed, the individual cracks are all shown in the plots.

Crack Widths in GFRP Reinforced Beams

For beams reinforced with GFRP bars, the fibers improved the cracking behavior of the beams, with respect to both the size and number of cracks observed. Figure 4a shows the moment versus crack width plot comparing the specimens reinforced with #2 GFRP bars. The number of cracks present in both beams was the same, as two cracks had formed in both specimens. However, the widths of the cracks in the beam with fiber reinforcing were significantly lower than in the beam without fiber reinforcing. Comparing the two beams at a moment of 5 kN-m, the maximum crack width in specimen G2N0 is approximately 0.75 mm, while the maximum crack width in specimen G2P1 is only approximately 0.30 mm.

The plots of moment versus crack width for specimens G3N0 and G3P1 (#3 GFRP bars) are shown in Figure 4b. The crack spacing in the case of these two beams was reduced, as the beam with fiber reinforcing had three cracks in the constant-moment region in comparison with the two cracks seen in the beam without fiber reinforcing. Again, the crack widths were also reduced by the fibers. At any given applied moment, the maximum crack width seen is lower for the FRC beam than the plain concrete beam. Not only is the maximum crack width lower when fiber reinforcing is used, but the total crack width (the sum of the widths of all cracks) is also reduced. At a moment of 6 kN-m, the total crack width for specimen G3N0 is approximately 0.87 (found by adding the crack widths for the two cracks, 0.40 mm plus 0.47 mm). The total crack width for specimen G3P1 is only 0.58 mm. The fibers clearly help to improve the cracking both in terms of maximum crack width as well as total crack width.

For the two over-reinforced beams, specimen G4N0 and G4P1 (#4 GFRP bars), the fiber reinforcing greatly improved the cracking response that was observed as shown in Figure 4c. In specimen G4N0, only three cracks formed, all within a short time span of the others. As loading continued, these three cracks continued to grow without the formation of additional cracks. In contrast, specimen G4P1 saw the formation of several additional intermediate cracks following the formation of the first cracks. The formation of new cracks is made possible by the presence of the fibers which transferred stress across the cracks in addition to the reinforcement allowing formation of successive, intermediate cracks in the specimen. This effect was pronounced in the two over-reinforced specimens, as the number of cracks observed in specimen G4P1 was twice the number observed in specimen G4N0.

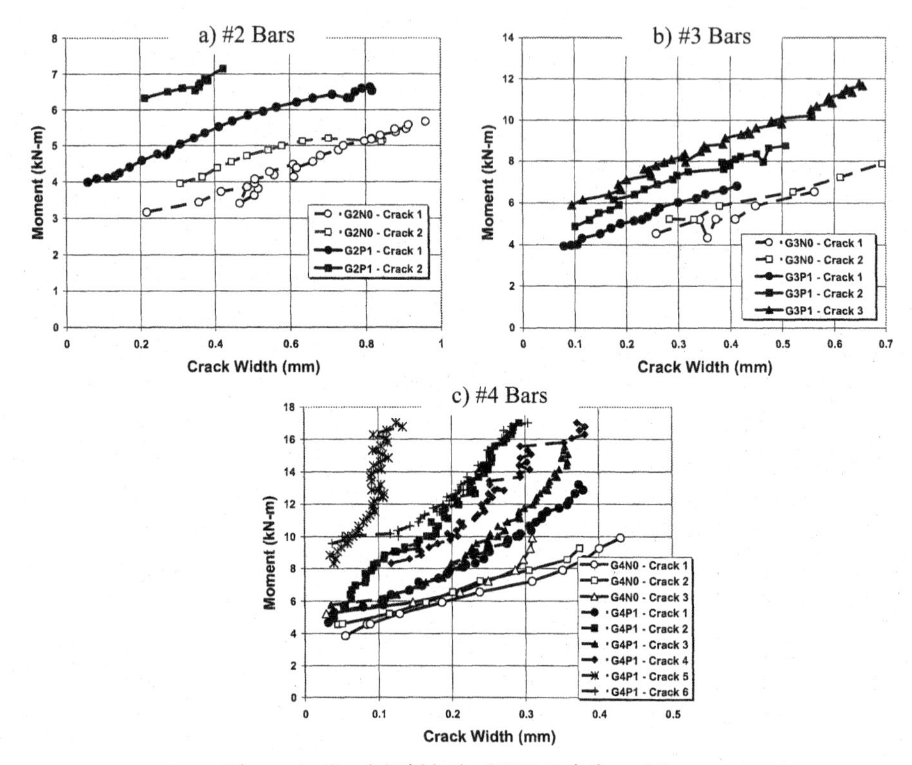

Figure 4. Crack Widths in GFRP Reinforced Beams

Crack Widths in Steel Reinforced Beams

The presence of fiber reinforcing did not seem to affect greatly the cracking response of beams reinforced with steel bars. The number and width of cracks observed in the steel-reinforced specimens did not seem to vary between beams with plain concrete and those with FRC. The plots of moment versus crack width for specimens S3N0 and S3P1 and specimens S4N0 and S3P1 can be seen below in Figures 5a and 5b, respectively. While it appears there are two cracks in specimen S3P1 and only one crack in specimen S3N0, in actuality, second crack had formed in specimen S3N0 but was outside the range of the viewing area of the digital image analysis thus could not be measured. The fibers have no effect on the crack widths while the steel remains linear elastic, which is the region of importance in design considerations. While the crack widths are decreased with the use of fibers after the steel begins to yield, the actual width of the cracks after that point are of little concern, as the member will be considered beyond service load conditions. In addition, the decrease in crack widths is not significant.

The cracking behavior between specimens S4N0 and S4P1 appear to be similar. There is little difference in the observed cracking response in these two beams. The moment versus crack width curves are rounded rather than linear due to the rounded stress-strain behavior of the reinforcing bars used.

The fact that the fibers did not greatly affect the cracking response of the specimens is likely due to the high elastic modulus of the steel reinforcing bars and the low elastic modulus of the fibers. Higher fibers contents or use of stiffer fibers (e.g. steel fibers) may lead to a more pronounced effect on the cracking response of the steel-reinforced beams.

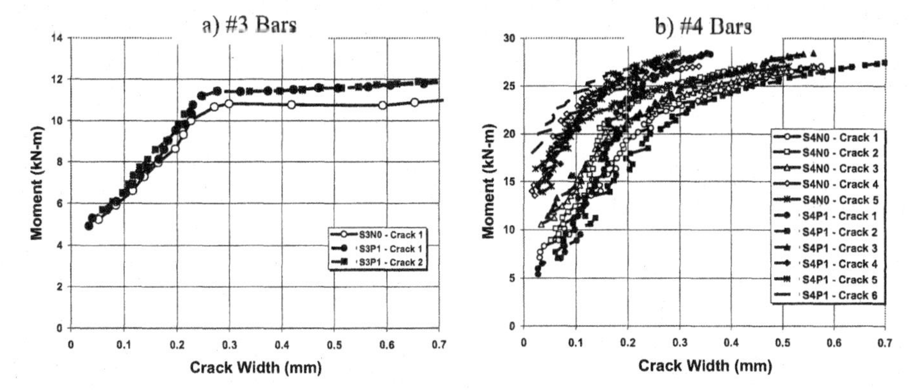

Figure 5. Crack Widths in Steel Reinforced Beams

CRACK WIDTH MODELING

Use of Modified Gergely-Lutz Equation

ACI Committee 440[7] has modified the well-known Gergely-Lutz equation for predicting crack widths for use with FRP-reinforced concrete members. The original Gergely-Lutz equation was an empirical equation developed based on data from numerous steel-reinforced concrete specimens.[8] The modification allows for the different elastic moduli and bond characteristics of FRP rebar.[9, 10, 11] The modified Gergely-Lutz equation is shown in Equation 1:

$$w = \frac{2.2}{E_f}\gamma \cdot k_b \cdot f_f \cdot \sqrt[3]{d_c A_r} \qquad (1)$$

where w is the crack width, γ is the ratio of distances from the tension face and from the centroid of the steel to the neutral axis, f_s is the stress in the tensile reinforcement, d_c is the thickness of concrete cover measured from the tension face to the center of the bar closest to that face, E_f is the elastic modulus of the FRP reinforcing, A_r is the concrete area surrounding one bar, equal to the total effective tension area of concrete surrounding the reinforcement and having the same centroid, divided by the number of bars, and k_b is a corrective bond coefficient that accounts for the differences in bond characteristics between steel reinforcing and FRP reinforcing.

To calculate the crack width from the applied moment using the Gergely-Lutz equation, it is necessary to express the stress in the reinforcement in terms of the moment. The strain in the FRP reinforcing can be expressed by:

$$\varepsilon_f = \frac{M(d - \bar{y}_{cr})}{E_c I_{cr}} \qquad (2)$$

where ε_f is the strain in the FRP reinforcing, M is the moment, d is the depth to the centroid of the tensile reinforcing, $\bar{y}_{cr}$ is the depth to the centroid of the elastic cracked concrete section, E_c is the elastic modulus of the concrete, and I_{cr} is the moment of inertia of the cracked section.

The stress in the reinforcing can be calculated by multiplying the strain determined in Equation 2 by the elastic modulus of the FRP reinforcing as shown in the following equation:

$$f_f = E_f \varepsilon_f = \frac{E_f}{E_c} \cdot \frac{d - \bar{y}_{cr}}{I_{cr}} \cdot M \quad (3)$$

where f_f is the stress in the FRP reinforcing and E_f is the elastic modulus of the FRP reinforcing. Substitution of Equation 3 into Equation 1 yields the modified Gergely-Lutz equation expressed as a function of the applied moment:

$$w = \frac{2.2}{E_c} \cdot \frac{d - \bar{y}_{cr}}{I_{cr}} \gamma \cdot k_b \cdot M \cdot \sqrt[3]{d_c A_r} \quad (4)$$

Calculating Crack Width at Height of Reinforcing

The Gergely-Lutz equation is used for calculating the maximum crack width at the extreme tensile face of the specimen. The value of γ is equal to the ratio of the distance between the distance from the neutral axis to the tensile face of the specimen (where the crack width is to be calculated) and the distance from the neutral axis to the centroid of the tensile reinforcing. Therefore, if the crack widths vary linearly throughout the height of the cross-section, the crack width at the height of the reinforcing bars can be calculated with a γ value of 1. For the purpose of applying the Gergely-Lutz expression to the data in this study, the crack widths are determined at the level of the rebar and γ is set to 1.

Use of the Gergely-Lutz equation for calculating the crack width at the height of the reinforcing bars is dependent on the assumption that the crack widths vary linearly with height. This assumption was found to be valid based on the results from the digital image analysis. The digital image analysis allowed for the calculation of the crack width at multiple heights in the specimen; therefore the variation of crack width with height could be found as shown in Figure 6 for specimen C3N0. At the time this image was taken, two fairly large cracks had formed in the specimen and indeed vary linearly with height. This behavior was typical of all specimens and was independent of the load or the number of cracks present thus confirming the validity of $\gamma = 1$ to calculate the crack widths at the height of the reinforcement with the Gergely-Lutz equation.

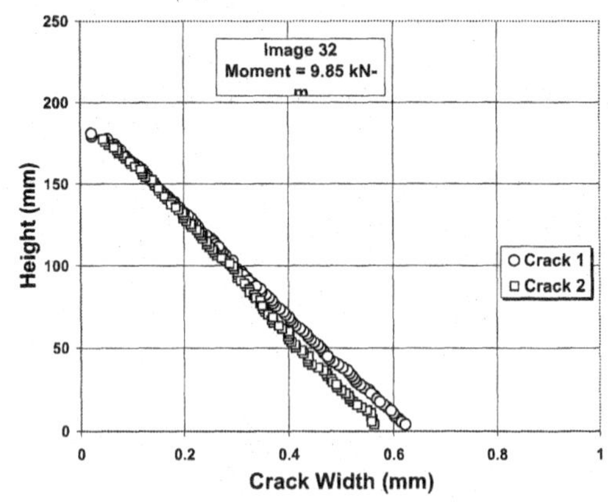

Figure 6. Crack Width at various Heights (Specimen C3N0, M=9.85 kN-m)

Comparison Between Specimens

The modified Gergely-Lutz equation was compared with the experimentally observed crack width behavior in the beams. However, the observed behavior was fundamentally different than what is predicted by the Gergely-Lutz equation. The Gergely-Lutz equation for predicting crack widths is a linear equation that passes through the origin. However, the observed response was that the plot of moment (which is a function of the stress in the reinforcing) versus the crack width will not, if extrapolated backwards, pass through the origin. This behavior was seen in all specimens tested, as the plots all intercepted the ordinate at a point that was not the origin. This behavior has also been seen in similar research on GFRP reinforced beams.[4] The observed behavior made using the Gergely-Lutz equation problematic. If a linear regression was performed in which the line was forced through the origin as is expected, a poor fit was obtained. However, to simply perform a linear regression on the data without regard to the location where the plot crossed the y-axis would not have allowed for a comparison between the data from different specimens.

To account for the differences seen in the cracking behavior between specimens and with respect to the Gergely-Lutz equation, an attempt was made to find a consistent method for comparing the data. Linear regressions were performed on the plots of moment versus the maximum crack width for all specimens. The best-fit lines were found to all intersect the y-axis at a moment equal to approximately 70 percent of the cracking moment (the average value for the y-intersect was found to be 0.703 of the cracking moment with a standard deviation of 0.124). A proposed modification to the Gergely-Lutz equation that is used is to allow the equation for maximum crack width to pass through a point on the y-axis equal to 70 percent of the cracking moment and is given in Equation 5:

$$w = \frac{2.2}{E_c} \cdot \frac{d - \bar{y}_{cr}}{I_{cr}} \gamma \cdot k_b \cdot (M - 0.7M_{cr}) \cdot \sqrt[3]{d_c A_r} \tag{5}$$

Linear regressions were then performed with the lines forced through the y-axis at a moment equal to 70 percent of the cracking moment. The moment versus maximum crack width and the best-fit linear regressions for two specimens are shown in Figure 7.

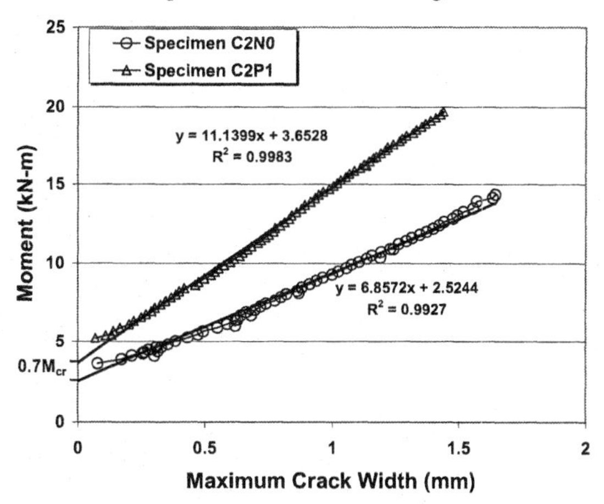

Figure 7. Moment versus Maximum Crack Width

Corrective Bond Coefficients

The slope of the best-fit linear regressions can be used to back-calculate the corrective bond coefficient, k_b, using Equation 5. The bond coefficients were calculated for all specimens tested, and a summary of the results are provided in Table III.

For a given reinforcing bar type, there does not appear to be any correlation between the bond coefficient and the bar size. There is a fair amount of scatter within the values for a given bar type; however, the values of the bond coefficient for FRC beams is consistently smaller than the value for plain concrete beams. While it is difficult to come to any conclusions regarding the effect of bar size on the bond coefficient, it is clear that the addition of fibers to the concrete will consistently lead to reduced values when FRP rebar are used.

The average bond coefficient found for steel bars with plain concrete was approximately equal to 1, which is expected, as the original Gergely-Lutz equation was based on the use of steel-reinforced concrete. In the original Gergely-Lutz equation, a reduction in the bond coefficient from a value of 1 means improved bond characteristics of the reinforcing bar in comparison to steel. In a more general sense, it means there is a decrease in the crack widths observed. Therefore it would be expected that while the addition of fiber reinforcing to the concrete would not actually improve the bond characteristics of the reinforcing bars, a reduction in crack widths would be observed that would show a decrease in the bond coefficient.

Table III. Calculated Bond Coefficients.

		Corrective Bond Coefficient, k_b		Average Corrective Bond Coefficient, k_b	
Rebar Type	Rebar Size	Plain Concrete	FRC	Plain Concrete	FRC
Steel	3	1.08	0.99	0.96	0.85
	4	0.84	0.71		
GFRP	2	0.87	0.60	1.08	0.64
	3	1.46	0.73		
	4	0.90	0.60		

The expected reduction in the value of the corrective bond coefficient with the use of FRC is seen in the steel-reinforced specimens, as the value of k_b is reduced from 0.96 to 0.85. However, this is a very slight reduction, and is consistent with the observed cracking response, as the crack widths were seen to be almost unaffected by the presence of the fibers.

An average bond coefficient of 1.08 was found for GFRP-reinforced members with plain concrete. This value is close to a value of 1, meaning that the GFRP bars have bond characteristics similar to that of steel. The bars used in this study had lugs and a coarse sand-epoxy coating meant to improve the bond characteristics of the bars. Studies on various GFRP bars from different manufacturers have shown bond coefficients ranging from 0.71 to 1.83, meaning that the bars can have bond characteristics superior or inferior to steel.[6] The average bond coefficient found for the GFRP-reinforced beams with FRC was 0.64. This is a considerable decrease in the bond coefficient, which is a result of a significant decrease in the maximum crack widths observed.

CONCLUSIONS

The cracking behavior of the reinforced beam specimens was measured using a digital imaging system. The following conclusions can be made from the results obtained:

1. The addition of polypropylene fibers to the concrete improves the cracking behavior (crack widths and spacing) of beams reinforced with FRP bars.
2. The addition of polypropylene fibers to the concrete does not improve the cracking behavior of beams reinforced with steel bars.
3. The modified Gergely-Lutz equation given by ACI Committee 440 to predict crack widths in FRP-reinforced beams cannot be used without additional modification to accurately predict the cracking response of FRP-reinforced beams with plain or fiber-reinforced concrete. A modification is proposed that provides a good prediction of the crack widths when using both FRP reinforcing bars and fiber-reinforced concrete.

REFERENCES

[1]B. Benmokrane, O. Chaallal and R. Masmoudi, R., "Flexural Response of Concrete Beams Reinforced with FRP Reinforcing Bars," *ACI Structural Journal*, **93** [1] 46-55 (1996)

[2]R. Masmoudi, M. Thériault, and B. Benmokrane, "Flexural Behavior of Concrete Beams Reinforced with Deformed Fiber Reinforced Plastic Reinforcing Rods," *ACI Structural Journal*, **95** [6] 665-676 (1998)

[3]H.A. Toutanji, and M. Saafi, "Flexural Behavior of Concrete Beams Reinforced with Glass Fiber-Reinforced Polymer (GFRP) Bars," *ACI Structural Journal*, **97** [5] 712-719 (2000)

[4]M. Thériault, and B. Benmokrane, "Effects of FRP Reinforcement Ratio and Concrete Strength on Flexural Behavior of Concrete Beams," *Journal of Composites for Construction*, Feb. 1998, pp. 7-16

[5]A.G. Razaqpur, D. Svecova, and M.S. Cheung, "A Rational Method of Deflection Calculation for FRP Reinforced Concrete Beams," *ACI Structural Journal*, **97** [1] 175-184(2000)

[6]W.K. Lee, "Flexural Behavior of Fiber-Reinforced Concrete Beams Reinforced with FRP Bars," *Masters Thesis*, Tufts University, Medford, MA (2003)

[7]American Concrete Institute Committee 440, "Guide for the Design and Construction of Concrete Reinforced with FRP Bars," American Concrete Institute, Farmington Hills, MI (2001)

[8]P. Gergely and L. Lutz, "Maximum Crack Width in Reinforced Concrete Flexural Members," *Causes, Mechanisms, and Control of Cracking in Concrete*, SP-20, American Concrete Institute, Detroit, MI, 87-117 (1968)

[9]S.S. Faza, and H.V.S. GangaRao, "Theoretical and Experimental Correlation of Behavior of Concrete Beams Reinforced with Fiber Reinforced Plastic Rebars," *Fiber-Reinforced-Plastic Reinforcement for Concrete Structures*, SP-138, A. Nanni and C. W. Dolan, eds., American Concrete Institute, Farmington Hills, MI, 599-614 (1993)

[10]R. Masmoudi, B. Benmokrane and O. Chaalal, "Cracking Behavior of Beams Reinforced with FRP Rebars," *First Int. Conf. on Composites in Infrastructures*, Tucson, AZ, Ed. by H. Saadatmanesh and M. Ehsani, 374 – 388 (1996)

[11]B. Tighiouart, D. Benmokrane, and D. Gao, "Investigation on the Bond of Fiber Reinforced Polymer (FRP) Rebars in Concrete," *Second Int. Conf. on Composites in Infrastructures*, Tucson, AZ, Vol II, Ed. by H. Saadatmanesh and M. Ehsani, 102 – 112 (1998)

INVESTIGATION OF WOOD PULP FIBER REINFORCEMENT FOR READY MIXED CONCRETE APPLICATIONS

H.J. Brown
Concrete Industry Management Program
Middle Tennessee State University
P.O. Box 19
Murfreesboro, TN 37132

J.H. Morton
Product Development
Buckeye Technologies
1001 Tillman Street
Memphis, TN 38108-0407

ABSTRACT

Wood pulp fibers have been investigated for use in cement based composites for applications such as cement pipe, siding and other building components, and fiber-reinforced concrete. Their higher elastic modulus and tensile strength than other natural fibers has propelled cellulose to the forefront of once common asbestos fiber applications. There is an understanding of the chemical, mechanical, density and shrinking/swelling characteristics of many wood pulp fiber species. The variability of these fibers can stem from the treatment process applied at the manufacturing plant. This ongoing research study investigates a treated wood pulp fiber for ready-mixed applications. The ready-mixed industry makes up nearly 75 percent of the U.S. consumption of cement so the market runs deep for possible wood pulp fiber applications.[1] Properties of the fiber composite that are discussed include bond between fiber and cement matrix, alkaline stability, physical fiber characteristics, durability during freezing/thawing cycles, plastic shrinkage cracking resistance, compressive strength and impact resistance. Full scale trials were also monitored for constructability, finishability and curing procedures as well as physical and mechanical properties.

INTRODUCTION

In 2002, U.S. portland cement consumption was 103.8 million metric tons which is now 81% owned by foreign companies. Approximately 75 percent of the U.S. cement capacity is sent to ready mix producers with only 13 percent being used for the manufacture of concrete products.[1] Alternatively, most of the research and development surrounding cellulose fiber-cement composites have focused on manufactured concrete products such as flat and corrugated sheets for cladding and roofing, non-pressure pipes and cable pits. Integration of cellulose fiber into the mainstream ready mix operation for value added benefits to the concrete mix became the focus of this project. An investigation of fiber reinforced concrete properties was performed on a specialty cellulose fiber versus control mixtures with no fibers and mixtures with standard cellulose and polypropylene fibers used in similar applications. Testing regimes were conducted at private laboratories, university research labs and commercial laboratories to obtain a wide neutral snapshot of the specialty cellulose fiber.

The properties that were the focus of this research concentrated in three areas. The first area was fiber properties looking at fiber count, spacing, surface area, alkaline stability and fiber surface bonding. The second area was plastic properties such as slump, temperature, air content and density. The third area was hardened properties including set time, compressive strength, flexural strength, impact resistance, average residual strength, plastic shrinkage resistance and freeze/thaw durability. Each of these properties is important to the success of a proprietary mix specified by the architect/engineering community and sold to concrete contractors.

LITERATURE SEARCH

Kraft pulps are the dominant wood fiber types used in cement-based materials. These fibers have minimum lignin contents and have exhibited resistance to alkaline environments in past applications.[2] ACI 544.1R-96 lists kraft pulps as having an average fiber length of 2.54 -5.08 mm, average diameter as 1-3 mil, average tensile strength of 101 ksi and water absorption of 50-75 percent.

Alkaline Stability: There have been concerns regarding the long-term performance of cellulose fiber reinforced cement composites; some natural fibers tend to disintegrate in the alkaline environment of cement.[3] The effect of exposure to alkaline environments on the strength of cellulose fibers can be determined by ASTM D6942. An alkaline environment is defined to be any matrix in which the pH is greater than 8 for a period of 2 or more hours. Concern over degradation of the fiber in this environment has been alleviated with special proprietary processing of the fiber. Figure I shows the standard cellulose fiber versus a specialty cellulose fiber with the coating on the surface.

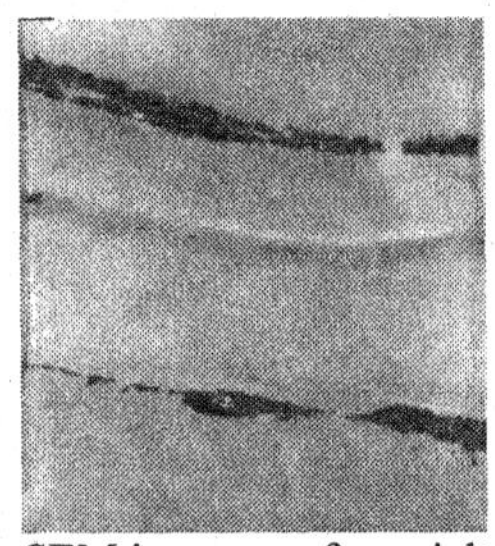

Figure I. SEM imagery of specialty (left) vs. standard (right) cellulose fibers.

Fiber Surface Bonding: Bonding of the cement matrix to the fiber surface is important in achieving many of the hardened properties necessary for quality concrete. Cellulose fibers have small effective diameters that are comparable to the cement particle size which promotes close packing and development of a dense bulk and interface microstructure in the matrix. Studies show that upon natural weathering, bond between fibers and cement matrix improved.[4] However, the growth of cement hydration products within the cellulose fibers (petrification) may lead to excessive fiber-matrix bonding and brittle failure after exposure to natural weathering.[3] Improvements in the bond between cellulose fibers and the cement matrix can be realized by coating the fibers with coupling agents such as alkoxides of titanium, titanates and silanes.[5]

Slump, Temperature, Air Content and Density: It has been reported that the addition of cellulose fiber reduces workability and the slump of a concrete mix. To achieve the same desired slump, an increase in water content was used.[5] This single property can be the bridge between the ready mix producer and concrete contractor to ensure that a good product has been delivered. No reported fluctuations in temperature have been determined in previous cellulose research. Heat of hydration spikes would be detrimental to placing and finishing operations due to an increase or decrease in set time. Air content can be a source of controversy for relations between producer and contractor. Air entrained concrete is specified for projects due to durability or workability issues. Therefore predictable air content is essential to the producer. An increase in air content has been observed in past research when cellulose was entered into the concrete mix. This might indicate that fibrous mixtures are not as compactable as the plain matrix.[2] The density, commonly referred to as unit weight, of the mix is directly related to the air content of the mix. The producer relies on density to determine proper yield of the concrete mixture for the manufacturing and distributing of the product.

Set Time: Set times have been determined to be slightly increased by the addition of cellulose fiber. Some impurities in the pulp could cause this tendency by playing the role of a set retarder.[2] If a mixture requires higher water content for a desired slump, then delayed set times can be expected. However, the decrease in workability can be offset by the addition of a high-range water reducer instead of increasing the water-to-cement ratio thus minimizing changes in set time.

Compressive Strength: Cellulose fibers are observed to increase the toughness of cementitious materials. The compressive properties of fiber-reinforced concrete (FRC) are relatively less affected by the presence of fibers as compared to the properties under tension and bending. The influence of fibers in improving the compressive strength of the composite matrix depends on whether mortar or concrete is used and on the magnitude of compressive strength. Studies have shown that with the addition of fibers there is an almost negligible increase in strength for mortar mixes; however for concrete mixes, strength increases by as much as 23 percent.[5]

Flexural Strength: The researched benefits of cellulose composites have been an increase in flexural strength and flexural toughness.[5] Cellulose fiber-cement composites subjected to repeated cycles of wetting and drying have shown increases in flexural toughness. The mode of failure of cellulose fiber-cement composites in flexure consists of a complex combination of multiple cracking, fiber debonding, stress redistribution to secondary cracks, fiber pullout, fiber fracture, and a shift of the tensile zone towards the compressive zone through the specimen thickness.[4] Improvement in flexural strength and toughness over plain mortar has been reported for cellulose fiber-cement composites with 2 percent fibers by mass. This improvement was more pronounced with the use of kraft fibers than mechanical fibers.[6] Up to a 500 percent increase in flexural strength and even greater increases in fracture toughness of fiber-cement composites has been reported with the use of kraft pulp at a fiber mass fraction of 2 percent.[2] Increased fiber content leads to slight increases in flexural strength and substantial increases in toughness dependent upon the water-to-cement ratio.[7, 8] It has also been shown that specialty cellulose fibers in fiber reinforced concrete have been highly effective in enhancing flexural strength at dosages of 0.24 percent volume fraction.[9]

Average Residual Strength: Reinforcing fibers will stretch more than concrete under loading. Therefore, the composite system of fiber reinforced concrete is assumed to work as if it were unreinforced until it reaches its first crack strength. At this point the fiber takes over and holds the concrete together. For fiber reinforcement, the maximum load carrying capacity is controlled by fibers pulling out of the composite because fiber reinforcement does not have a deformed surface like larger steel reinforcement. This condition limits performance to a point far less than yield strength of the fiber itself. This action will affect the overall toughness of the concrete product which is based on the total energy absorbed prior to complete failure. Toughness is dependent upon 1) type of fibers used; 2) mass/volume percent of fiber; 3) aspect ratio and; 4) orientation of the fibers in the matrix. To obtain relative data about toughness, average residual strength (ARS) was performed. ARS is the average stress-carrying ability of the cracked beams that is obtained by calculation, using the residual strength at the four deflections specified in ASTM C 1399.

Plastic Shrinkage Resistance: The relatively high surface area and the close spacing of cellulose fibers when combined with their desirable mechanical characteristics make them quite effective in the suppression and stabilization of microcracks in the concrete matrix.[9] Specialty cellulose fibers processed for the reinforcement of concrete offer relatively high levels of elastic modulus and bond strength while enhancing durability.[10] Having a finer denier than most normal synthetic fibers means there are many more cellulose fibers in a given weight. Cellulose fibers are more closely spaced to each other and the hydrophilic surface causes the fibers to be more uniformly dispersed in concrete. Cellulose fibers are microfibers that are very short and fine, therefore cellulose fibers are not readily noticeable on the surface of concrete. In low volume fractions (0.06 percent), cellulose fibers have statistically significant effects of reducing plastic shrinkage cracking of conventional and high performance concrete.[9]

Impact Resistance: Preliminary observations of cellulose systems have shown considerable increase in impact resistance over plain mortars and concretes with a fiber content of 2 percent by mass or higher. The increase in fracture toughness of cellulose fiber reinforced cement composite suggests that this system could be valuable in areas of application in which resistance to impact is a noted advantage. Examples of such areas might be walls in locations of high usage (e.g. schools, shops, factories)[2]

Freeze/Thaw Durability: When subjected to cycles of freezing and thawing (F&T), the relative dynamic modulus of elasticity of cellulose fiber-cement composites slightly increases initially, indicating slight improvement in F&T resistance, but "a gradual decrease in the relative dynamic modulus of elasticity" was observed at later cycles.[6] Moisture cycling results indicated that treated fiber-cement composites are more resistant to deterioration than neat cement specimens.[11] Both natural weathering and accelerated aging testing has shown a reduction in the porosity, water absorption, and nitrogen permeability in the cement matrix, and enhanced the durability of the cellulose fiber-cement composites.[12]

RESEARCH METHODOLOGY

Test methods were based on ASTM standards, TAPPI standards, and AC217[13] ("Acceptance Criteria for Concrete with Virgin Cellulose Fibers"). The following list summarizes the specifications of the research.

- Natural fibers from kraft pulping of slash pine (Pinus elliotti), Average properties: length 2.73 mm, coarseness 31 decigrex, wall thickness 9.7 um, 2.8 denier, and 1.18 million fibers/gram.
- Polypropylene synthetic monofilament fibers, 19 mm length, 15 denier
- #67 Stone, Ohio River Sand (FM = 2.67)
- Type I Portland cement
- 0.50 w/c ratio for non air entrained mixes, 0.42 w/c ratio for air entrained mixes
- 3 to 1 coarse aggregate-cement ratio, 2 to 1 fine aggregate-cement ratio

The specialty cellulose fiber was entered in an individualized fiber form either in a dry state or slurry state. The volume fraction of cellulose fiber was 0.06 percent and synthetic fiber was 0.08 percent and they were both tested at 0.9 kg/m^3. Slurry fiber mass was put into a water quantity equaling 3.0 percent fiber consistency. The fibers were allowed to rest in that state for a period of 12 hours. The fiber slurry was added into the mixer after the pure water was added, during the mix. The actual amount of water added to the mix was determined by subtracting the water used in preparing the slurry from the total water quantity needed to equal the target w/c ratio for the batch.

Alkaline Stability: This test was conducted in accordance with ASTM D6942. The test was run at time intervals of 1, 3, 7, 14, 21, 28 and 35 days, treating the fibers with saturated calcium hydroxide and 1N sodium hydroxide in open containers. Following the alkaline exposure, the fiber samples were washed to neutral, handsheeted (TAPPI T205), and tested for TAPPI T231, Zero-Span Tensile. Handsheeting is a method of dispersing pulp fibers in a large volume of water and depositing them on a wire mesh to form a piece of paper. The fibers are randomly oriented in order to test physical properties of the pulp fibers and derived paper. From the zero-span tensile readings, the "zero-span stability ratio" is determined. For each sample, the zero-span tensile value shall be the average from testing eight strips cut from four Tappi handsheets. Zero-span tensile testing was carried out with no gap between jaws clamping strips of paper, forcing most of the fibers between the jaws to break. This force directly measures the average individual fiber strength of thousand of fibers.

Fiber Surface Bonding: Samples of the fracture surfaces of the fiber were cut to a convenient size as to fit onto a 12.5 mm diameter aluminum stub. The samples were then mounted on the stubs using conductive tabs. The samples were then rimmed with carbon paint to ensure proper ground. After drying, the samples were Au coated for 2.0 min at 25 milliamps after a 10 min. evacuation period using an Emitech 550 sputter coater. The samples were then ready to be examined for cementitious bond to the fiber surface using a scanning electron microscope.

Slump, Temperature, Air and Density: Each laboratory mixture was tested for slump, temperature, air content and density following ASTM procedures C143, C1064, C231, and C138, respectively.

Hardened Concrete Properties

Set Time: ASTM C403 was utilized in determining initial time of set for each mixture. Samples were prepared and compacted into 150 x 150 mm cylindrical mold. The start of set time began when water and cement were in contact in the mixer. Once each specimen reached 35 kg/cm^2 compressive strength, initial set time was recorded in minutes.

Compressive Strength: The purpose of the test was to evaluate whether the addition of cellulose fibers to a concrete mix adversely affects the compressive strength. Comparative tests were conducted in accordance with ASTM C39. Tests were continually conducted on three concrete specimens with proprietary fibers and three specimens without fibers (control) at 7 and 28 days as well as comparative mixes with synthetic fibers. 150 x 300 mm and 100 x 200 mm cylinder specimens were utilized throughout the study. Specimens were cast and covered during the first day with plastic lids, then demolded and placed into a saturated lime bath for the remainder of the 28 day cycle at 23±2°C.

Flexural Strength: 150 x 150 x 500 mm specimens were prepared from each concrete mixture to evaluate flexural strength of the fiber reinforced concrete. ASTM C31 was followed for specimen preparation. These specimens were allowed to cure for one day in the laboratory under plastic and then demolded and entered into a saturated lime bath for the remainder of the 28 day cycle at 23±2°C. The test specimens were tested in accordance with ASTM C78.

Average Residual Strength: ASTM C1399 was followed for determining ARS. Five beam specimens were tested for each variety of fiber. The average width and depth of the beams, the loads obtained upon reloading at deflections of 0.5, 0.75, 1.00 and 1.25 mm, and the average residual strengths (ARS) were evaluated. The test provides data needed to obtain that portion of the load-deflection curve beyond which significant cracking damage has occurred and provides a measure of post-cracking strength, such as how strength is affected by the use of fiber-reinforcement.

Resistance to Shrinkage Cracking: Tests were conducted using both the Paul Kraai method and AC217 method.[13,14] Paul Kraai used 51 mm (2 in.) thick slabs that were 1 m long and 0.6 m wide. The slabs were restrained around the perimeter using wire mesh. Appendix A of AC217 uses a 102 mm thick mold with a minimum surface area of 0.16 m^2 and rectangular dimensions of 356 x 559 mm with internal restraints and stress risers. Both were subjected to prescribed wind velocities. Crack openings were measured at the surface of the panels.

Impact Resistance: Appendix B of AC217 was used to test impact resistance of the specialty cellulose fiber. This test method compares the impact resistance of concrete with and without cellulose fibers. Several parameters were chosen to evaluate impact resistance. Impact resistance was characterized by the measure of: (1) the energy consumed to fracture a specimen; (2) the number of blows in a "repeated impact" test to achieve a prescribed level of distress; and 3) the extent of damage. A 4.54-kg hammer with a 46 cm drop is used along with a 50 mm diameter hardened steel ball. A flat base plate with four lugs welded to the frame was used to secure the test specimen. Cylinder molds complying with ASTM C 31 or ASTM C 470 were used to obtain a 150 mm diameter by 50 mm concrete specimen. Specimens were tested at 7 and 28 days with five specimens for each test age and test condition.

Freeze/Thaw Durability: Comparative tests were conducted in accordance with Procedure A of ASTM C666. The number of cycles should be 300 or until the average relative modulus of elasticity of the fiber specimens or the control specimens reaches 60 percent of the respective initial modulus, whichever occurs first. Three beam specimens were made per each mixture and air content was recorded.

RESULTS

Alkalinity Stability: AC217 conditions of acceptance include the average Zero-Span Stability Ratio (ZSSR) determined by ASTM D6942 shall not fall below 90 percent for both alkali environments tested. The comparison of standard cellulose versus specialty cellulose in an alkaline environment proves to be superior for the experimental fiber as shown in Figure II.

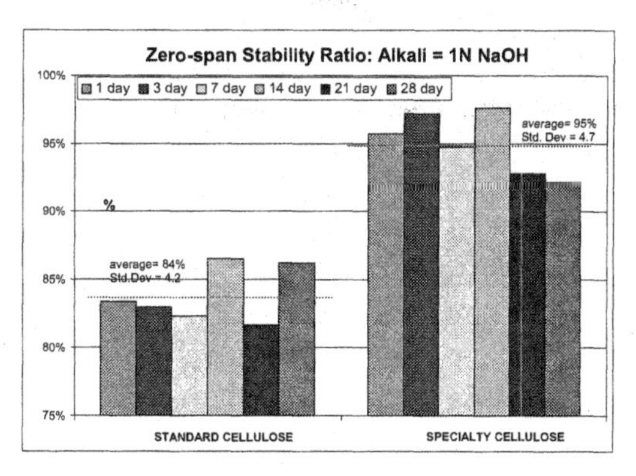

Figure II. ZSSR for Standard Cellulose vs. Specialty Cellulose

Fiber Surface Bonding: Figure III shows a microscopic view of cementitious material bonding to the specialty fiber surface. The left picture was taken from a laboratory sample cylinder while the right picture was taken from a core of cast in place slab concrete. The paste bonding is evident on the surface of both samples.

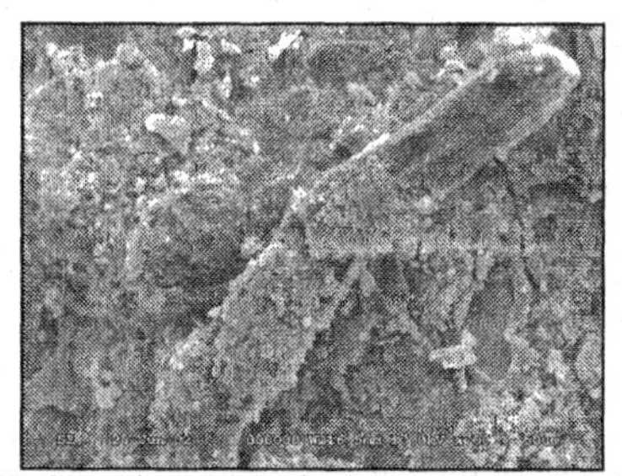

Figure III. SEM imagery of Fiber Surface Bonding of Laboratory (left) and Field (right) Samples

Plastic Properties: Table I shows the % difference between the Control mixtures as a baseline to the other variable mixtures. Air is noted in the table as a reference to air entrained mixes.

Table I. Plastic Properties % Differences for Various Mix Designs

Mix Design	Slump	Temperature	Air Content	Unit Weight
Control	0%	0%	0%	0%
Synthetic	34.7%	0.4%	-15.9%	-0.66
Standard Cellulose (Dry)	-26.6%	4.25%	7.1%	-2.6
Specialty Cellulose (Dry)	-16.7%	3.3%	-19.6%	0.37
Specialty Cellulose (Slurry)	4.2%	4.5%	-42.5%	-1.8
Specialty Cellulose (Dry, Air)	100.0%	-1.5%	185%	N/A
Standard Deviation	19 mm	0.87 ° C	1.75%	68 kg/m^3

Set Time, Compressive Strength and Flexural Strength: Compressive set time was analyzed based on time it took to reach 35 kg/cm^2 for each specimen. An increase in set time was presented as a positive percent difference. Average compressive and flexural strengths should be greater than or equal to control specimen average compressive strength at 28 days to be considered for application. Table II presents the percent differences for the three hardened properties discussed.

Table II. Hardened Properties % Differences for Various Mix Designs

Mix Design	Set Time	Compressive Strength	Flexural Strength
Control	0%	0%	0%
Synthetic	28.6	-5.6	22.7
Standard Cellulose (Dry)	N/A	-4.0	11.7
Specialty Cellulose (Dry)	9.5	2.9	8.6
Specialty Cellulose (Slurry)	14.3	-1.3	---
Specialty Cellulose (Dry, Air)	26.2	-4.2	---
Standard Deviation	25 min.	0.92 MPa	1.72 MPa

Impact Resistance: The minimum AC217 acceptance for the final results is that the cellulose fibers increase the impact of concrete by 40 percent at 7 days and 40 percent at 28 days. Table III lists the parameters and results for the impact resistance testing with standard deviations of 44 blows for 7 day testing and 19 blows for 28 day testing.

Table III. Impact Resistance % Increase for Three Test Cycles at 7 and 28 day

Test 1 – Dry vs. Slurry Addition	7 d %	28 d %	Test 2 – Fiber Dosage	7 d %	28 d %	Test 3 – 18 MPa	7 d %	28 d %
Control	---	---	Control	---	---	Control	---	---
Cellulose(Dry,Air)	62	39	1.2 kg/m^3	75	44	Cellulose	89	44
Cellulose (Dry)	62	43	1.5 kg/m^3	85	48			
Cellulose (Slurry)	67	47	1.8 kg/m^3	120	59			

Average Residual Strength: The average residual strength was computed for specified beam deflections beginning after the beam have been cracked in a standard manner. The test provides data needed to obtain that portion of the load-deflection curve beyond which significant cracking damage has occurred and provides a measure of post-cracking strength, such as how strength is affected by the use of fiber-reinforcement. It was found that the average residual strength of specialty cellulose fiber was lower than the synthetic fiber; however it maintained a 0.19 MPa average residual strength as seen is Table IV.

Plastic Shrinkage Resistance: Shrinkage cracking was quantified by computing the area of surface cracks in each panel specimen to obtain a value in square meters, which represents the assumed total area of crack opening on the surface of the concrete panels. The effect of fibers on the development of shrinkage cracking was determined by expressing the cracking value for the concrete with cellulose fibers as a percentage of the value for the control concrete (without fibers) when subjected to the identical and simultaneous drying conditions. Table V lists the results from both plastic shrinkage methods. AC217 acceptance conditions state that cracking must be reduced by a minimum of 50 percent.

Table IV. Average Residual Strength for Specialty Cellulose Fiber

Specimen	Load (kg) at Deflection (mm)				Breadth	Depth	ARS
	0.5	0.75	1.00	1.25	mm	mm	MPa
C1	69	64	64	58	102	105	0.17
C2	89	79	74	64	101	103	0.22
C3	76	71	62	53	103	105	0.17
C4	87	66	51	44	100	104	0.17
C5	90	84	71	59	101	102	0.22
						Average	0.19
						Std. Dev	0.02

Table V. Plastic Shrinkage Reduction Due to Addition of Fibers

Method	Fiber	Dosage (kg/m^3)	Crack Area Reduction (%)
Paul Kraai	Synthetic	0.6	54.1
Paul Kraai	Specialty	0.6	60.3
AC217	Specialty	0.6	76.8
AC217	Specialty	0.9	80.0
Std. Dev.			12.6

Freeze Thaw Durability: AC217 conditions of acceptance are that the average durability factor of the three specimens containing the fibers shall be at least equal to the average durability factor of the control specimens. Table VI lists the durability factor difference for the mixtures tested which resulted in an 8.85% standard deviation. Air entrained concrete should reside above 80% durability factor however that is dependent upon state specifications.

Table VI. Durability Factor (DF) in % for Various Tested Mixtures

	Control	Slurry 1	Slurry 2	Dry 1	Dry 2	Synthetic 1	Synthetic 2
DF	68.45	80.34	91.05	96.44	85.00	80.97	85.06

CONCLUSIONS AND RECOMMENDATIONS

- ZSSR was determined to be greater than 90% (average = 95%) for the specialty cellulose product.
- It is possible to incorporate the specialty cellulose fiber in the specified dosages in concrete without causing any balling, clogging, or segregation.
- Satisfactory workability was maintained with the addition of the specialty cellulose fibers at the given dosages.
- In all fiber reinforced concretes the mode of failure was changed from a brittle to a ductile failure when subjected to compression or bending.
- Hardened property results were found to have a small increase in set time, an increase in compressive and flexural strength for dry specialty cellulose fiber.
- The impact resistance met AC217 acceptance conditions in all scenarios except when air entrainment was used.

- The Average Residual Strengths were found to be 0.53 MPa for synthetic fiber, 0.21 MPa for standard cellulose fiber and 0.19 MPa for specialty cellulose fiber.
- Plastic shrinkage crack reduction was determined as 60.3 percent (0.6 kg/m^3) with the Paul Kraai method and 76.8 percent and 80.0 percent for the AC217 method (0.6 kg/m^3 and 0.9 kg/m^3, respectively) which met AC217 conditions of acceptance.
- The freeze/thaw resistance of concrete was improved using dry and slurry specialty cellulose products over synthetic fiber and control specimens and met AC217 acceptance conditions.

REFERENCES

[1]Czechowski, David E., Economics of the U.S. Cement Industry, July 15, 2004, www.cement.org

[2]Soroushian, P. and Marikunte, S., "Reinforcement of Cement-Based Materials with Cellulose Fibers," Thin-Section Fiber Reinforced Concrete and Ferrocement, ACI Special Publication SP-124, ACI, Detroit, MI 1990, pp. 99-123

[3]Soroushian, P. and Marrikunte, S., "Statistical Evaluation of Long-Term Durability Characteristics of Cellulose Fiber Reinforced Cement Composites," ACI Materials journal, Vol. 91, No. 6, Nov./Dec. 1994. pp. 1-10

[4]Akers, S. A. S.; Crawford, D.; Schultes, K.; and Gerneka, D. A., "Micromechanical Studies of Fresh and Weathered Fiber Cement Composites. Part 1: Dry Testing," The International Journal of Cement Composites and Lightweight Concrete, Vol. 11, No. 2, Construction Press, Harlow, England, UK, May 1989, pp. 117-124

[5]Naik, T., Chun, Y., and Kraus, R., "Use of Residual Solids from Pulp and Paper Mills for Enhancing Strength and Durability of Ready-Mixed Concrete," Department of Energy Final Report, UWM Center for Byproducts Utilization, Milwaukee, Wisconsin 2002, 300 pages

[6]Soroushian, P.; Marikunte, S.; and Won, J., "Wood Fiber Reinforced Cement Composites under Wetting-Drying and Freezing-Thawing Cycles," Journal of Materials in Civil Engineering, Vol. 6, No. 4, ASCE, New York, NY, November 1994, pp. 595-611

[7]Vinson, K. D. and Daniel, J. I., "Specialty Cellulose Fibers for Cement Reinforcement," Thin Section Fiber Reinforced Concrete and Ferrocement, ACI Special Publication SP-124, ACI, Detroit, MI, 1990, pp. 1-18

[8]Coutts, R.S.P., Warden, P.D. "Effect of Compaction on the Properties of Air-Cured Wood Fibre Reinforced Cement." Cement and Concrete Composites 12 (1990): 151-156

[9]Soroushian, P., Control of Plastic Shrinkage Cracking with Specialty Cellulose Fibers, ACI Materials Journal 95-M40, 1998, pp 429-435

[10]Soroushian, P., "Secondary Reinforcement—Adding Cellulose Fibers," Concrete International: Design & Construction, Vol. 19, No. 6, American Concrete Institute (ACI), Farmington Hills, MI, June 1997, pp. 28-34

[11]Blankenhorn, P.R., Silsbee, M.R., Blankenhorn, B.D., DiCola, M., Kessler, K., "Temperature and moisture effects on selected properties of wood fiber-cement composites" *Cement and Concrete Research, Volume 29, Issue 5, May 1999, Pages 737-74*

[12]MacVicar R, Matuana LM, Balatinecz JJ , "Aging mechanisms in cellulose fiber reinforced cement composites" *Cement and Concrete Composites* 21 (3): 189-196:1999

[13]AC217 Acceptance Criteria for Concrete with Virgin Cellulose Fibers, ICC Evaluation Services, Whittier, California, Copyright 2003

[14] Kraai, Paul, "A Proposed Test to Determine the Cracking Potential Due to Drying Shrinkage of Concrete", Concrete Construction, Vol. 30 No. 9, September 1985, pp 775-78

CONFINEMENT ANALOGY -- TOWARDS TOUGHNESS-BASED DESIGN FOR REINFORCED FIBER REINFORCED CONCRETE BEAMS

V.S. Gopalaratnam
Dept. of Civil Engineering
E2509 Engr. Bldg. East
University of Missouri-Columbia
Columbia, Missouri 65211-2200

Z. El-Shakra
Program Manager for Palastine
United Nations - Habitat
Gaza, Palastine

H. Mihashi
Division of Building Materials
Tohoku University
Sendai, Japan

ABSTRACT

The single-most important attribute of incorporating fibers in concrete is the resultant enhancement in its toughness. This is also perhaps the only mechanical property that significantly differentiates its behavior from that of unreinforced concrete. It is generally accepted that improved resistance to crack-growth, improved ductility, improved resistance to fatigue and impact loads are directly or indirectly associated with this enhancement in it's toughness. Presently no design procedures are available to make use of the composite's superior energy absorption capacity. Conventional strength-based design procedures are inadequate to demonstrate the benefits of fiber incorporation. The benefits of incorporating fibers can be demonstrated convincingly if in addition to the strength parameter, design procedures explicitly or implicitly incorporate toughness or other related design parameter(s). The article presents results on how toughening due to fiber incorporation influences the flexural performance of reinforced concrete. Model predictions are compared with results reported in the published literature as well as from experiments on practical size beams. Some practically-viable and promising toughness-based design approaches are discussed.

INTRODUCTION

The benefits of incorporating fibers in a concrete matrix have been recognized since a very long time. While the incorporation of fibers in concrete marginally improves its strength, its more significant contribution is in the enhancement of its energy absorption capacity. A non-dimensional index intended to reflect the toughness of the composite was used by Henager (1978) to capture the enhancement in energy absorption capacity. The test procedure proposed by him was accepted by the American Concrete Institute's (ACI) Committee 544 on Fiber Reinforced Concrete as early as 1978 (ACI, 1978). Since then, there have been a significant research and understanding on how toughness should be measured and interpreted (Barr and Hasso 1985, JCI, 1984, Johnston, 1985, Gopalaratnam et. al, 1991, ASTM, 1992, El-Shakra and Gopalaratnam, 1993, Mindess, Chen and Morgan, 1994, Trottier and Banthia, 1994, Gopalaratnam and Gettu, 1995). However, little progress has been made on use of toughness in design. It should be recognized that just like for the purposes of material characterization where a special test is used for FRC to quantify parameters other than strength, design procedures should also include energy-related/toughness parameters, if the benefits of fiber incorporation are to be fully realized. The many benefits from fiber-related toughening in the flexural response of beams that are also conventionally reinforced, is clearly demonstrated in this article. Observations from this analytical and experimental study are used to report on the viability of two potential pseudo-toughness-based design procedures.

A fairly exhaustive review of toughness test standards for FRC from around the world has been prepared by Gopalaratnam and Gettu (1995). The conventional wisdom has been that fibers enhance the tension response of plain concrete and hence toughness testing should be undertaken in a tensile or flexural loading configuration. Since the flexural configuration duplicates the state-of-stress in many practical situations and also since it is more convenient from a testing stand point, it is widely used in most of the test standards. It will be demonstrated in this article that, for flexural applications where both continuous and fiber reinforcement are present, a toughness test using a compressive configuration provides more practically relevant information for design purposes.

Until recently, only strength-based approaches were available for the analysis and design of reinforced concrete members. Typically, analysis and design procedures used for conventional reinforced concrete have been modified for use with FRC (Hannant, 1978, Bentur and Mindess, 1990, Balaguru and Shah, 1992). There also has been increased interest in and progress with energy-based fracture mechanics approaches to analyze performance of reinforced concrete members (ACI 446, 1992). Issues related to ductility and energy absorption capacity of structural members, increasingly important to high performance cement composites, have contributed to the interest in alternate approaches to conventional strength-based analysis and design. Naaman and Reinhardt (1993) have proposed a method to predict the load-deflection response of SIFCON (Slurry InFiltrated CONcrete) beams that are also conventionally reinforced. An alternate approach developed here (El-Shakra, 1995) using confinement-analogy and deformation-controlled loading procedure allows predictions of load-deflection response for FRC also reinforced with conventional reinforcing bars. Results from the analysis have been used to develop the rationale for toughness-based design of reinforced fiber reinforced concrete (RFRC) beams.

CONFINEMENT-ANALOGY FOR FRC IN COMPRESSION

It is proposed here that the macroscopic response of FRC in uniaxial compression can be simulated using confinement-analogy (Gopalaratnam and Shah, 1985). Uniaxial compressive loading produces tensile strains in the transverse directions. Steel fibers present in a concrete matrix subjected to uniaxial compression provide restraint to transverse deformations. Both in the uncracked stage and more so in the cracked stage, the restraint to transverse tensile deformations provided by the fibers has an effect comparable to lateral confinement in plain concrete. The level of confinement mobilized depends upon the fiber volume fraction, V_f, fiber aspect ratio, ℓ/φ (where ℓ and φ are the length and the diameter of the fiber, respectively), interface properties, fiber type and distribution parameters. The relationship between the level of confinement and these parameters has been derived using a semi-empirical approach (El-Shakra, 1995), a simple fiber pull-out model and assuming a 3-D random Normal distribution of fibers, and appropriate efficiency factors to account for angular and length efficiency across a matrix crack.

The level of confinement mobilized is expected to be different during different stages in the pull-out of individual fibers. Peak levels of confinement have normally been observed at very small average pull-out displacements. The level of confinement mobilized has been observed to be relatively constant except at very large average pull-out displacements. Consequently, the equivalent lateral confinement provided by the fibers in the matrix, f_r, has been derived as

$$f_r = c_1 \sqrt{f_c^{'}} V_f \frac{\lambda}{\phi} \tag{1}$$

where c_1 is an empirical constant(El-Shakra, 1995). An assumption similar to the strength enhancement in reinforced concrete columns confined by spiral reinforcement is made here (Shah, Fafitis and Arnold, 1983).

$$f_c^{*} = f_c^{'} + 6 f_r \tag{2}$$

where f_c^{*} is the compressive strength of FRC (analogous to strength of confined concrete), $f_c^{'}$ is the compressive strength of plain concrete (unconfined uniaxial strength), and f_r is the equivalent lateral confinement. The strain corresponding to f_c^{*}, ε_c^{*} is obtained in the form

$$\varepsilon_c^{*} = \varepsilon_c^{'} + c_2 f_r \frac{\varepsilon_c^{'}}{f_c^{'}} \tag{3}$$

where $\varepsilon_c^{'}$ is the strain corresponding to $f_c^{'}$ in plain concrete (uniaxial unconfined), and c_2 is a constant obtained from experimental results (El-Shakra, 1995).

The model for confined concrete developed by Shah, Fafitis and Arnold (1983) has been used in the present investigation to establish the compressive response of FRC. f_c^{*} and ε_c^{*} values reported here correspond to the strength and associated strain capacity of confined concrete in their model. f_r for this study is given by Eqn 1.

Figure 1 shows a comparison of the compressive stress-strain response predictions for FRC using the confinement-analogy model with experimental results reported by Gopalaratnam and Shah (1985). It should be noted that confinement-analogy model with the above two calibration constants is valid for unreinforced concrete (f_r = 0) as well as for conventional steel fiber concrete (V_f in the range 0.0 to 2.0% and ℓ/φ values up to 150). In principle, similar analogy with different constants may hold for materials like SIFCON or FRC made with other types of fibers.

FRC IN TENSION

In recent years, a number of micromechanics-based models have become available for characterizing the tensile response of FRC. For the purposes of discussions in this paper, a very simple yet effective model is adequate to characterize FRC behavior in tension. An elastic-frictional model has been used here. The tensile strength of conventional FRC, f_t, is assumed as $7.5\sqrt{f_c^{'}}$ (both f_t and $f_c^{'}$ in psi). The elastic modulus in tension is assumed to be identical to the elastic modulus in compression. The post-cracking frictional strength in tension is assumed as αf_t, where α is a constant in the range $0 \le \alpha \le 1$. The sensitivity of results from the analysis to α values is discussed later. An α value of 0 represents an elastic-brittle material while an α value of 1 represents an elastic-ideally plastic material.

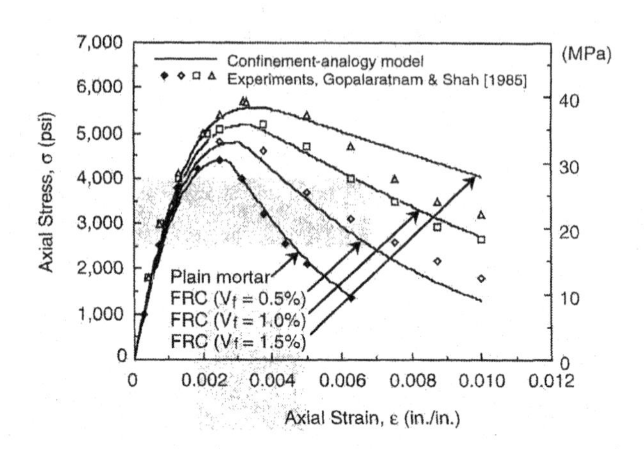

Fig.1 Confinement-analogy model predictions of stress-strain response in uniaxial compression compared with experimental results from different steel fiber concrete mixes.

MOMENT CURVATURE RESPONSE OF RFRC BEAMS

In many practical applications it is expected that fiber reinforcement would be used in conjunction with conventional continuous steel reinforcing bars. In these situations, it is useful and necessary to understand the roles, at the various loading stages, of the primary continuous reinforcement and the secondary discontinuous fiber reinforcement. In the present study, a computer program has been developed for establishing the complete inelastic moment-curvature (M-κ) response of RFRC cross-sections using a layered-beam approach. The FRC matrix in compression is modeled using confinement-analogy while an elastic-frictional model is used for its tension response. Steel rebars have been assumed to obey an identical elastic-plastic constitutive law in both tension and compression. The bond between the matrix and reinforcement is assumed to be perfect. For any top fiber strain value, ε_c, the computer program is designed to iteratively establish the depth of the neutral axis, c, by enforcing compatibility, constitutive laws and equilibrium. The program is designed so as to allow arbitrary number of layers of steel reinforcement, prescribed nonlinear constitutive laws for the constituent materials and arbitrary cross-sectional shapes. It can be used to establish nonlinear moment-curvature relationships with different prescribed levels of axial force. The prescribed axial force is set to zero while dealing with pure bending problems such as the ones discussed here.

Figures 2a and 2b show the predicted moment curvature response for practically sized RFRC beams that are "under-reinforced" in the conventional reinforced concrete sense (longitudinal tension steel ratio, $\rho <$ balanced steel ratio, ρ_b). All further references to "under-reinforced" and "over-reinforced" in this article are also based on the conventionally-defined balanced reinforcement ratio, ρ_b. As has been discussed later, the definition of the balanced reinforcement ratio, ρ_b is influenced by fiber incorporation. The beam cross-section (15 in. (381 mm) width, 30 in. (762 mm) height and 28 in. (711 mm) effective depth) and material properties (unconfined strength of plain concrete of 5,000 psi (34 MPa) and Grade 60 (414 MPa) steel) with the exception of fiber volume content are identical in Figs. 2a and 2b. Fig. 2a shows the responses for V_f = 0.5% ($\lambda/\phi = 100$), while Fig. 2b shows the response for V_f= 2.0% ($\lambda/\phi = 100$), two cases which represent the limits of the practical range of interest as far as current use of steel fiber concrete is concerned. The constant α, defining the post-cracking frictional capacity of stress transfer across a cracked section is varied from 0 (elastic-brittle response) to 1 (elastic-perfectly plastic response) to investigate the influence of this constant on the overall response. For the low fiber volume case, there is only a small increase in the moment capacity (< 11%) and no significant change in the curvature at ultimate, even when the α value is changed so as to produce dramatic difference in FRC's tensile response. For the high fiber volume case, the moment capacity increase and the ductility improvement with increases in α value is slightly more, but in any case, not of practical significance. What is of interest however, is that the level of ductility in the high fiber volume case is significantly improved (Fig. 2a vs. Fig. 2b). This improvement is due mainly to the improved compressive response of FRC. A detailed look at the results from the analysis show that this improvement is only partly due to the small improvement in compressive strength. It is largely attributable to the larger top fiber strain capacity at ultimate loads, resulting from the better post-peak response of FRC. Figures 3a and 3b highlight this observation even better where an "over-reinforced" beam has been analyzed with the same two fiber contents. Note the beam cross-section and material properties are identical to the beam shown in Fig. 2. The longitudinal steel content has been doubled from the previous case. While changes in the tensile response marginally alters the moment capacity and resultant ductility for each of the two cases, a comparison of Fig. 3b with Fig. 3a demonstrates the difference in ductility with fiber addition that comes about because of the improved compression response of FRC. These plots also highlight the fact that the incorporation of fibers in conventional reinforced concrete has an effect similar to that of the compression reinforcement. While there is some enhancement of the moment capacity, fiber incorporation plays a major role in enhancing the ductility of the beam. Indirectly, it also has the ability, similar to the conventional compression reinforcement, to redefine balanced reinforcement ratio. This is clearly evident from Fig. 3, where a conventionally "over-reinforced" beam responds in a ductile manner with sufficient fiber addition.

LOAD-DEFLECTION RESPONSE

An analytical approach that simulates a deflection or strain-controlled stable test of a post-peak softening structure is used here. The parameters needed for this analysis include: material properties, reinforcement details, geometric properties and loading configuration along the length of the beam. In principle, it is not necessary that beams analyzed using this method be prismatic. Each section along the length can have different moment-curvature response. Additionally, the loading could also vary along the beam length in a more general formulation. For the purposes of illustration of the analysis procedure, it has been assumed that the beam is prismatic with identical moment-curvature response for all the beam cross-sections. Like in a deflection-controlled test, there is no loss of stability soon after the critical section enters the post-peak regime. At deflections (curvatures, strains) larger than that at peak load, the inelastic zone spreads away from the critical section. This smeared treatment of inelastic behavior is analogous to multiple cracking of a ductile reinforced concrete beam (distributed cracking modeled as a spreading inelastic hinge). The length of the inelastic zone is established implicitly by ensuring that equilibrium, constitutive relationships and compatibility of displacements along the axis of the beam are exactly satisfied.

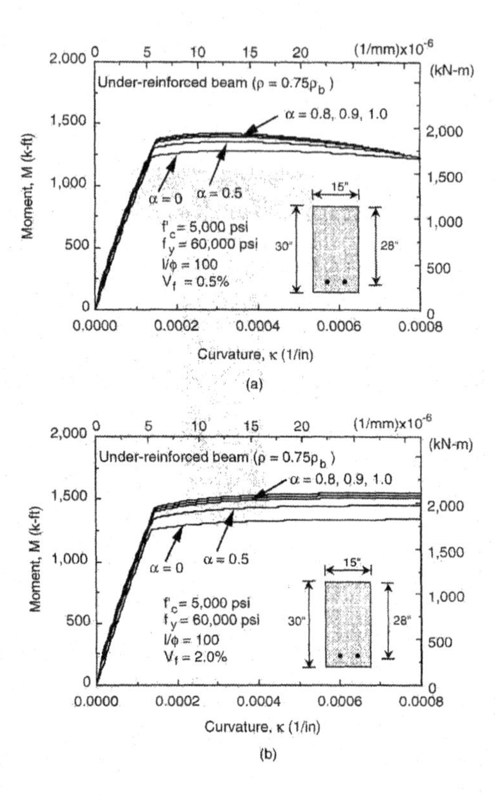

Fig. 2 Predicted moment-curvature responses of under-reinforced beams for different α values for matrices containing (a) 0.5%, and (b) 2.0% steel fibers

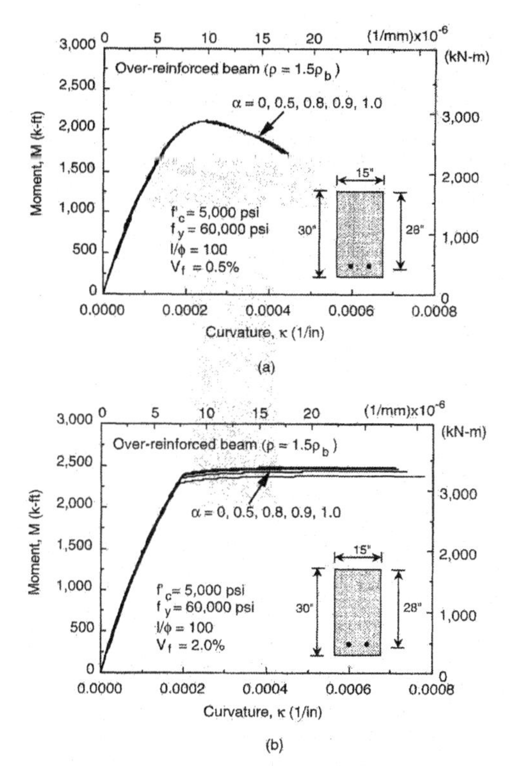

Fig. 3 Predicted moment-curvature responses of over-reinforced beams for different α values for matrices containing (a) 0.5%, and (b) 2.0% steel fibers

Other investigators Naaman and Reinhardt (1993) have used empirical estimates of the length of inelastic hinge for predicting the load/moment-deflection response from the moment-curvature behavior of a typical cross-section. Note that the approach presented here treats displacements discontinuities along the length of the beam in a smeared or averaged sense. This simplification has been observed to provide realistic predictions of the macroscopic load-deflection response for composite beams that exhibit multiple cracking and a reasonable level of ductility. The simplification however is inappropriate for composite beams that fail in a brittle manner due to the catastrophic propagation of a single crack (i.e., low-volume FRC that have very little or no conventional reinforcement).

COMPARISON OF RESULTS FROM THE MODEL

Results of the experimental moment-deflection response reported by Craig (1987) for a conventionally reinforced concrete beam and a RFRC beam of identical geometrical and cross-sectional detail are compared in Fig. 4a with model predictions. It should be noted that the reinforced concrete beam used in this comparison is an under-reinforced beam.

The following model input parameters were used for the conventionally reinforced concrete beam: $f_c^{'}$ = 7,900 psi (54 MPa), f_y = 65,000 psi (448 MPa), b = 8 in. (203 mm), h = 12 in. (305 mm), d = 10 in. (254 mm), A_s = 1.57 in^2 (2 Number 8 bars, 1,013 mm^2), outer span = 108 in. (2.74 m) and inner span = 38 in. (0.97 m).. The same model input parameters were used for the RFRC beam with the exception of the matrix strength, which was reported as 10,000 psi (69 MPa). All of these values are identical to that reported by Craig (1987). It is interesting to note that using the confinement-analogy model, a FRC compressive strength of 10,000 psi (69 MPa) correlates to a plain concrete compressive strength of 7,235 psi (50 MPa) for the fiber reinforcement parameters used by Craig. The FRC matrix in the RFRC beam comprised hooked-end steel fibers with an aspect ratio of 100 and a fiber volume fraction of 1.75%. A reduction in the plain matrix strength due to fiber incorporation (ACI, 1988) has been noted previously and has been attributed to poor compaction, a consequence of loss of workability due to steel fiber incorporation.

Except at very large deflections, the model predictions are acceptably accurate both for the conventional reinforced concrete (RC) and the RFRC beams. As noted earlier, at large deflections the effectiveness of the confinement provided by fibers diminishes. This reduction is not incorporated in the confinement-analogy model, as it often occurs at displacement levels that are well beyond the range of normal practical interest.

Figure 4b and Fig. 4c show comparisons of the experimental and analytical moment-midspan deflection responses of two different sizes of RFRC beams tested at the University of Missouri-Columbia. The experimental results were obtained in related studies on the effect of structural size on the failure characteristics of reinforced concrete beams (Ghazavy, 1994, El-Shakra, 1995). Figure 4b shows results from four-point flexural test on an RFRC beam 36 in. (0.91 m) deep, 6 in. (152 mm) wide and 18 ft (5.49 m)long, tested over an outer span of 16 ft (4.88 m) and an inner span of 8 ft (2.44 m). The beam was reinforced with two Grade 60 (414 MPa) Number 8 bars (25 mm diameter) at an effective depth of 32 in (0.81 m). The response shown in Fig. 4c was obtained for a one-fourth scale beam compared to that described in Fig. 4b. The longitudinal reinforcement for this beam comprises two Number 4 (13 mm diameter) bars.

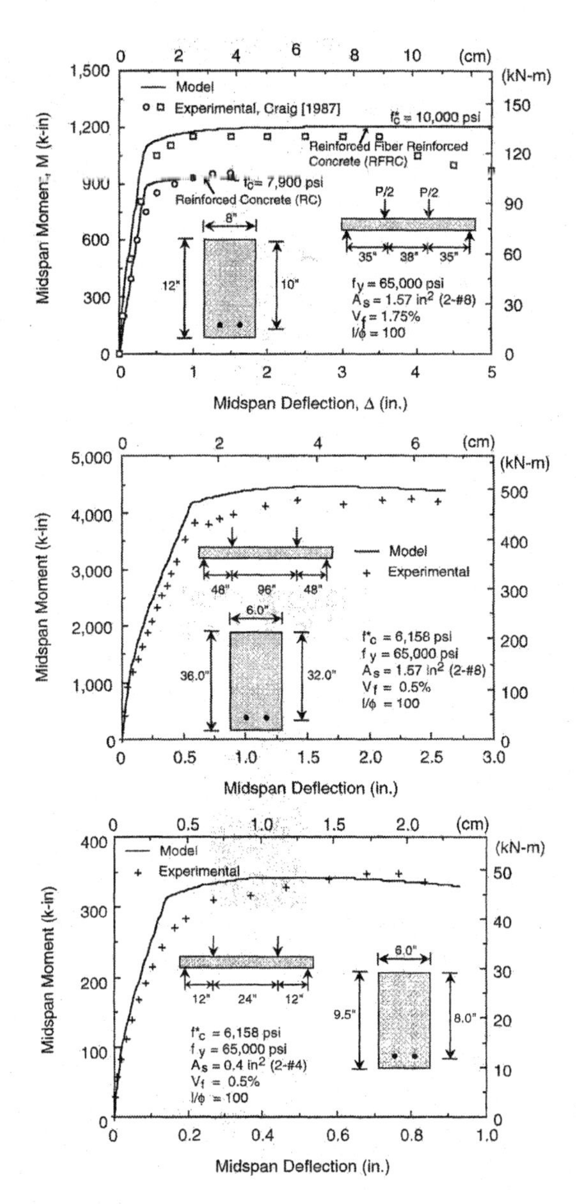

Fig. 4 Comparison of model predictions with experimental results for RC versus RFRC beams (top), large size RFRC beam (middle) and small size RFRC beam (bottom).

Constituent materials and their properties were identical for the two beams. The FRC matrix was reinforced with 2 in. (51 mm) long hooked-end steel fibers (Bekaert fibers, V_f= 0.5% and λ/ϕ = 100). Compressive strength of the FRC matrix of 6,158 psi (42 MPa) and the tensile yield strength of steel rebars of 65,000 psi (448 MPa) reported in the figure are based on constituent material tests conducted as a part of the study (Ghazavy, 1994, El-Shakra, 1995). The flexural tests were conducted using closed-loop electro-hydraulic testing machines operated under displacement-controlled conditions. Both beams, results for which are illustrated in Figs. 4b and 4c failed in a flexural mode initiated by yielding of the tension steel. The overall analytical prediction of the moment-deflection response for the under-reinforced beams in both cases is good and compares well with the experimental observations at all stages of loading. Model predictions marginally overestimate the post-cracking stiffness of the composite beams. Ultimate moment capacity predictions are typically within 3.3% for a number of replicate beams including other beams similar to the ones shown in Figs. 4b and 4c (El-Shakra, 1995).

ROLE OF FIBER REINFORCEMENT IN RFRC

It should be noted that the role of fibers when fibers are accompanied by continuous conventional reinforcement is significantly different from when fibers are the only form of reinforcement used. As demonstrated earlier in Figs. 2 and 3, fibers in RFRC beams enhance the overall response more through improved behavior in compression than through better tensile response. The confining effect in compression attributable to fiber incorporation provides benefits through three distinct but interrelated mechanisms; (1) at comparable levels of strain, the presence of fibers increases the compressive strength; (2) as far as the structural response is concerned, the more significant influence can be attributed to the enhanced top-fiber strain at ultimate resulting from fiber incorporation; and (3) additionally, depending upon the top-fiber strain at ultimate, the centroid of the compressive force itself moves farther away from the neutral axis.

Improvements in the tensile response of plain concrete through incorporation of fiber are quite significant. However, when continuous reinforcement is simultaneously present, the supplemental contribution by fibers to the tension capacity or to member ductility is negligible. Fibers in the tension zone of an RFRC beam serve very effectively in a secondary role by limiting crack-widths, distributing cracks so that there are more numbers of finer cracks, and improving bond characteristics by locally arresting splitting and secondary radial cracks in the vicinity of the reinforcing bar.

As far as the overall influence of fibers in a RFRC beam is concerned, there are striking similarities between fiber incorporation and the use of conventional compression reinforcement. In an under-reinforced beam, the presence of fibers does not enhance the moment capacity significantly, even while measurable improvement in the member ductility can be realized, Fig. 2a. However, in an over-reinforced beam, the presence of fibers tremendously improves the member ductility while simultaneously providing for moderate increases in member capacity, Fig. 2b. In the over-reinforced case, it is also possible to effect a change in the failure mode itself from a brittle compression-initiated failure to a ductile tension-governed failure, depending upon conventional and fiber reinforcement contents. Again, analogous to the influence of conventional compression reinforcement, the presence of fibers influences the balanced reinforcement ratio, ρ_b. The presence of fibers allows placement of larger amounts of tension steel more efficiently.

In an RFRC beam the fibers implicitly also contribute to the shear capacity of the beam even if their intended role was to enhance the flexural performance of the beam. Enhanced shear performance resulting from fiber incorporation has been recognized in numerous previous investigations and has also been studied by El-Shakra (1995). Improved fatigue, impact and bond performance due to fiber incorporation have been adequately documented (ACI, 1988). These additional benefits from fiber incorporation make it more attractive than conventional compression reinforcement.

TOWARDS TOUGHNESS BASED DESIGN

It is logical and important that measures of engineering toughness values be correlated and defined in terms of fundamentally more relevant parameters such as fracture energy. Similarly, it is appropriate given the results presented here, that design approaches developed for specific applications "explicitly" use energy absorption or engineering toughness of FRC as a design parameter in addition to conventional parameters. Work toward this long-term objective is presently being carried out under the auspices of the ACI Committee 544. In the interim, two pseudo-toughness approaches briefly described below have been proposed. These preliminary proposals that appear to hold promise require further study before they can be practically implemented.

One approach is based on the equivalent post-cracking flexural strength. The equivalent post-cracking flexural strength can be computed up to prescribed levels of post-cracking displacement or between prescribed limits of post-cracking displacement using the flexural load-deflection response (Gopalaratnam and Gettu, 1995). This design approach has already been used for slab-on-grade applications (Moens and Nemegeer, 1991) and tunnel linings (CDCECL, 1993). It is attractive from a practical stand-point where limits on crack-widths and deflection require a serviceability-dominated design procedure.

The other approach is suitable for flexural design of beams where FRC is used in conjunction with conventional steel rebars. This procedure is the "equivalent compression steel approach" to the design of RFRC beams is currently being studied by the first author. Fiber volume fraction, type and aspect ratios are empirically related to equivalent amounts of compression steel. Conventional procedures used for the design of doubly reinforced concrete beams can then be used for the design of RFRC beams. One major advantage of this approach is that a practicing engineer who is well-versed in conventional reinforced concrete practice can readily incorporate concepts of energy absorption and toughening in design without the need for a complete grasp of nonlinear fracture mechanics and related concepts.

CONCLUSIONS

1. The constitutive behavior of steel fiber reinforced concrete in uniaxial compression can be modeled using the analogy of laterally confined unreinforced concrete. This observation allows for an elegant and phenomenologically consistent incorporation of the compression behavior of FRC in more complex analytical and numerical models for the structural response.
2. When steel fibers are used in beams in conjunction with conventional tension reinforcement, fiber-related toughening in the compression zone has a more significant influence on the overall flexural response of the RFRC beam than its contribution in the tension zone. Contributions in the compression zone are realized through the facts that (a) the top-fiber strain at ultimate for the FRC matrix is significantly greater than that for unreinforced concrete, (b) the compressive strength of the FRC matrix is greater than that of the unreinforced concrete, and (c) the centroid of the compressive force in a FRC matrix moves

away from the neutral axis when the section reaches its ultimate strength capacity. Fibers in the tension zone effectively serve in a secondary capacity to limit crack-widths, distribute cracks, and improve bond characteristics of the primary continuous reinforcement.

3. Fiber incorporation in beams that have conventional tension reinforcement alters overall flexural response of the resultant composite beam in a manner similar to when compression reinforcement is incorporated. The enhancement in ductility is more pronounced than the enhancement in strength for conventionally under-reinforced beams. For conventionally over-reinforced beams, it is possible to obtain large increases in strength capacity as well. Fibers, like conventional compression reinforcement, alter the balanced reinforcement ratio. This allows use of larger amounts of tension steel more efficiently.

4. Two potential pseudo-toughness approaches to design using FRC appear to hold promise. One uses the concept of equivalent post-cracking flexural strength to incorporate toughness in the design of on-grade slabs and tunnel linings. The other based on equivalent compression reinforcement indirectly incorporates fiber-related toughening in the flexural design of beams that use both fiber and conventional steel reinforcement. Both methods require further extensive verifications before they can be practically implemented in design guidelines.

ACKNOWLEDGMENT
This work was supported in parts by the following grants: Grant No. MSM 88190803 from the Structures and Building Systems Programs (Director Dr. Ken P. Chong), Grant No. INT 9311821 from the U.S. - Japan Program (Director M.S. Patricia Tsuchitani) from the National Science Foundation and Grant No. UM-RB-13 from the University of Missouri Research Board. Support received from Tohoku University, Sendai, Japan and the Architectural Institute of Japan while the first author was on research leave is gratefully acknowledged.

REFERENCES

ACI Committee 446 (1992), "Fracture Mechanics of Concrete: Concepts, Models and Determination of Material Properties," *ACI* 446-1R, 140 pp

ACI Committee 544 (1978), "Measurements of Properties of Fiber Reinforced Concrete," *ACI J.*, 75(7), pp. 283-289 (see also 1988, *ACI Mat. J.*, 85(6), pp. 583-593)

ASTM (1992), "Standard Test Method for Flexural Toughness and First-Crack Strength of Fiber-Reinforced Concrete (Using Beams With Third-point Loading)," ASTM C 1018-92, ASTM Annual Book of Standards, 04.02, ASTM, pp. 510-516

Balaguru, P.N., and Shah, S.P. (1992), "Fiber Reinforced Cement Composites," McGraw-Hill, Inc., 530 pp

Barr, B.I.G., and Hasso, E.B.D. (1985), "A Study of Toughness Indices," *Mag. Conc. Res.*, 37(132), pp. 162-173

Bentur, A. and Mindess, S. (1990), "Fibre Reinforced Cementitious Composites," Elsevier Applied Science, London and New York

CDCECL (1993), Recommendation for Design and Construction of Extruded Concrete Lining Method, Committee on Design and Construction of Extruded Concrete Lining (CDCECL), City Tunnel Series, Yoshii Shoten Publishers, Tokyo Japan (in Japanese)

Craig, R. (1987), "Flexural Behavior and Design of Reinforced Fiber Concrete Members," Fiber Reinforced Concrete-International Symposium, ACI Publication SP-105, pp. 517-563

El-Shakra, Z.M. (1995), "Toughness of Fiber Reinforced Concrete - Potential Applications in Reinforced Concrete Beams," Ph.D. Dissertation, University of Missouri-Columbia, 215 pp

El-Shakra, Z.M., and Gopalaratnam, V.S. (1993), "Deflection Measurements and Toughness Evaluations for FRC," *Cem. Conc. Res.*, 23(6), pp. 1455-1466

Ghazavy-Khorasgany, M. (1994), "Size Effect in the Shear Failure of Normal and High Strength Reinforced Concrete," Ph.D. Dissertation, University of Missouri-Columbia, 222 pp.

Gopalaratnam, V.S., and Gettu, R. (1995), "On the Characterization of Flexural Toughness in Fiber Reinforced Concretes," *Int'l J. of Cem. Conc. Comp.*, 17, pp. 239-254

Gopalaratnam, V.S., Shah, S.P., Batson, G., Criswell, M., Ramakrishnan, V., and Wecharatana, M. (1991), "Fracture Toughness of Fiber Reinforced Concrete," *ACI Mat. J.*, 88(4), pp. 339-353

Hannant, D.J. (1978), "Fibre Cement and Fibre Concretes," John Wiley and Sons, Chichester, 219 pp

Henager, C.H. (1978), "A Toughness Index of Fibre Concrete," Testing and Test Methods of Fiber Cement Composites (RILEM Symposium), Construction Press Ltd., Lancaster, pp. 79-86

Johnston, C.D. (1985), "Toughness of Steel Fiber Reinforced Concrete," Steel Fiber Concrete, US-Sweden Joint Seminar (NSF-STU), Stockholm, Elsevier, pp. 333-36.

Mindess, S., Chen, L. and Morgan, D.R. (1994), "First Crack Strength and Flexural Toughness of Steel Fibre Reinforced Concrete," *J. Adv. Cem. Based Mat.*, 1(5), pp. 201-208

Moens, J., and Nemegeer, D. (1991), "Designing Fiber Reinforced Concrete based on Toughness Characteristics," *Conc. Int'l*, pp. 38-43

Naaman, A.E., Reinhardt, H.W., Fritz, C., and Alwan, J. (1993), "Non-linear Analysis of RC Beams Using a SIFCON Matrix," *Mat. and Struct.*, 26, pp. 522-531

Shah, S.P., Fafitis, A. and Arnold, R. (1983), "Cyclic Loading of Spirally Reinforced Concrete," *J. Struct. Div.*, ASCE, 109, pp. 1695-1710

ENHANCED PERFORMANCE OF FIBER REINFORCED CONCRETE WITH LOW FIBER VOLUME FRACTIONS.

M. Lopez de Murphy, T. Hockenberry, and A. Achenbach
Penn State University
212 Sackett Building
University Park, PA 16802

ABSTRACT

This paper presents the results of an experimental program wherein fiber reinforced concrete specimens (FRC) were tested under flexure, compression and indirect tension. In particular, the performance of a new type of high strength twisted polygonal steel fiber was evaluated at a low volume fraction dosage. Comparisons were made with the behavior of FRC using commercially available fibers (two hooked steel fibers, and two PVA fibers) at low volume fraction dosages. Parameters investigated were: the matrix strength (f'_c), the type of fiber, the volume fraction of fibers (V_f), the fiber mechanical and geometric properties, including the fiber intrinsic efficiency ratio (FIER). Results from this experimental program showed that fibers with improved geometric features (hook-ends or twists, and large FIER) and high Young's modulus can exhibit an enhanced performance in the post cracking stage even at low volume fractions (0.75%).

INTRODUCTION

The use of fiber reinforced concrete (FRC) both in structural and non-structural applications has gained momentum in recent years. The improved mechanical properties of FRC, with respect to conventional concrete, are shown on its tensile behavior, crack toughness, ductility and energy absorption. Enhanced performance of these cementitious composites is achieved by careful design of the different components of the composite, which will exhibit strain hardening behavior in the post-cracking stage (e.g. increasing load and multiple cracking), as well as increased ductility and toughness with respect to conventional FRC. Unfortunately, this enhanced behavior or "high performance" is usually associated with a large fiber volume fraction in the composite which limits the cost effectiveness and practical applications of FRC[1]. Researchers[2], however, have identified fiber mechanical and geometric properties that can effectively enhance the performance of FRC. Mechanical properties are defined by the fiber tensile strength, young's modulus, stiffness, ductility and bond; whereas geometric properties have been related to the fiber cross sectional shape, length, equivalent diameter, and surface deformation. The fiber aspect ratio (L/d) has been associated as one of the geometric parameters controlling the post-cracking behavior of FRC. Recently, the use of fibers with improved geometries have lead to the use of a new parameter to better qualify the effect of the geometry, the fiber intrinsic efficiency ratio (FIER)[2]. The FIER is defined as the ratio of lateral surface area to cross section.

The objective of this study was to investigate the possibility of obtaining enhanced performance in fiber reinforced concretes with a low volume fraction of fibers (less than 1%) using conventional concrete mix procedures and a mid-range concrete strength. The effect of the fiber mechanical and geometric properties was evaluated for five different types of fibers.

TEST PROGRAM

Parameters investigated in this study were the matrix strength (f'c), the fiber material, the volume fraction of fibers (V_f), the fiber aspect ratio (L/d), and the fiber intrinsic efficiency ratio (FIER).

Material Selection

Materials used for the concrete matrix were: Type I Portland cement, No. 8 coarse (9.5 mm nominal size) and fine aggregates complying with ASTM standards[3], water reducing admixture, and five different types of fibers. Fiber materials used were: steel with two ranges of tensile strength (100 MPa, 2500 MPa), and PVA (polyvinyl alcohol). Fiber properties are shown in Table I. Each FRC mixture will be associated to the fiber type (acronyms indicated in parenthesis in this table).

Table I. Fiber Properties

Fiber Type ()	Aspect Ratio	Unit Weight	Tensile Strength	Youngs Modulus	FIER
	(l/d)	(kg/m3)	(MPa)	(MPa)	
30mm Hooked-End Steel Fibers (SF30)	40	7880.856	1100	195000	75
60mm Hooked-End Steel Fibers (SF60)	80	7880.856	1035	195000	151.90
30mm PVA Fibers (PVA30)	45.45	20.8234	900	23000	200.00
18mm PVA Fibers (PVA18)	90	20.8234	1000	30000	451.41
30mm Twisted Steel Fibers (Twisted)	100	7848.82	2570	195000	390-540

The PVA and hooked-end steel fibers selected for this study had similar length or aspect ratios to the high strength twisted steel fiber. One of the main objectives of this study was to evaluate the effect of the optimized fiber geometry[4] of the high strength twisted steel fiber in a matrix with low volume fraction. According to research conducted at the University of Michigan[5] the FIER value for this twisted steel fiber ranges between 390 and 540. For all the other types of fibers, values of FIER were calculated based on the average cross section and length of a representative sample.

Specimen Fabrication

A total of 14 concrete batches were made (13 FRC plus one control batch with no fibers). Two different water/cement ratios (w/c) and up to four different volume fractions (V_f) were tested. Table II shows relevant information of the mixture proportions. The batches were mixed using a small drum mixer and the following mixing procedure. As a first step, all of the coarse aggregate, two thirds of the fine aggregate, and half of the water was added to the mixer and blended for twenty seconds. Then the cement was added to the wet aggregates, followed by the rest of the fine aggregates. This mixture was then blended for twenty seconds before the rest of the water was added and the entire concrete mixture was mixed for three minutes. Fibers were then slowly added to the mixture while the drum was rotating. This was done to eliminate the formation of clumps of fibers throughout the concrete matrix. Once all the fibers were added to the mixture the drum continued to rotate for additional minutes, shown in Table II as extra time, to ensure proper dispersion of the fibers.

As shown in Table II, the steel fibers needed more time to separate and disperse in the mixture than the other fibers. This was due to the way the fibers were packaged by the manufacturer. The steel fibers were glued together in bundles that would brake apart when introduced into the mixing process. The mixture with twisted fibers was observed to have the best workability of all the FRC batches with no need for a water reducing agent.

Table II. Selected Mixture Properties

Mixture	w/c	Helix Fibers	30mm Hooked End Steel Fibers	60mm Hooked End Steel Fibers	30mm PVA Fibers	18mm PVA Fibers	Water Reducing Admixture	Extra Time in Mixer	Slump
		(% volume)	(% volume)	(% volume)	(% volume)	(% volume)	(mL/m3)	(min)	(mm)
CONTROL	0.41	—	—	—	—	—	—	—	114.3
SF30-0.25	0.52	—	0.25	—	—	—	—	10	190.5
SF30-0.50	0.53	—	0.50	—	—	—	—	10	165.1
SF30-0.75	0.48	—	0.75	—	—	—	—	20	203.2
SF30-1.0	0.55	—	1.00	—	—	—	—	20	203.2
SF30-0.75(2)	0.41	—	0.75	—	—	—	627.114	8	50.8
SF60-0.25	0.46	—	—	0.25	—	—	—	10	177.8
SF60-0.50	0.41	—	—	0.50	—	—	303.105	10	139.7
SF60-0.75	0.41	—	—	0.75	—	—	223.671	10	88.9
SF60-1.0	0.41	—	—	1.00	—	—	223.671	10	82.55
PVA30-0.50	0.41	—	—	—	0.50	—	—	5	82.55
PVA30-0.75	0.41	—	—	—	0.75	—	—	5	88.9
PVA18-0.75	0.41	—	—	—	—	0.75	627.114	5	38.1
TWISTED-0.75	0.41	0.75	—	—	—	—	—	5	38.1

For each batch, six beams (350 x 100 x 100 mm) were fabricated for flexural testing and five cylinders (100 x 200 mm) were made for compressive and split tensile testing. As per ASTM C 1018[6], all specimens were molded using a vibrating table (instead of using a rod) in order to ensure continued dispersion of the fibers throughout the matrix avoiding possible segregation problems. It was observed that fibers with a density lower than the concrete matrix (the PVA fibers) tended to float themselves to the surface of the specimen, during the vibration period. The overall workability of the mixture was also affected (to different degrees) by the addition of fibers to the concrete. Mixtures containing 60mm steel fibers were especially hard to transfer to the molds.

After vibration, the top surfaces were smoothed to remove air bubbles, and the specimens were placed in an environmental chamber, as per ASTM standards, to ensure adequate curing. After ten days in the curing chamber the cylinders and beams were removed for testing under compression, split tensile, and flexure loading.

Compression Tests

To determine the compressive strength of the each batch of concrete, cylindrical specimens were loaded axially in accordance to ASTM C 39[7]. The average of the compressive strength of the cylinders for each batch can be found in Table III. It was found that adding fibers did not affect the compressive strength of the composite. During testing of the FRC specimens it was noted that the fibers within the concrete bridged the cracks, avoiding the presence of full splitting failure. Most specimens exhibited crushing and spalling at the ends during the final failure stage, as shown in Figure I (a). The 30mm PVA fiber reinforced specimens differed from this trend by failing in shear, as shown in Figure I (b).

Split Tensile Tests

The tensile strength of the concrete was determined by loading cylinders along their length in accordance with ASTM C 496[8]. The average splitting tensile strength of the cylinders can also be found in Table III. The bridging effect of the fibers was also noted in this type of test. This effect can be observed in Figure I (c). The fibers prevented the cylinder from splitting like the control mixture shown in Figure I (d). It can be observed that the addition of fibers increases the splitting tensile strength of the composite. Similarly, higher fiber volume fractions lead to higher tensile strength. The largest splitting/compressive strength ratio was obtained in 60 mm steel fiber mixtures with volume fractions larger than 0.5% and in the twisted steel fiber mixture with 0.75% fiber volume fraction.

Table III. Mechanical Properties of Specimens Tested

Mixture	Compressive Strength, f'c	Splitting Tensile Strength, f_t	f_t/f'c	First Crack Strength	Maximum Post-Crack Strength	I_5	I_{10}	I_{20}
	(MPa)	*(MPa)*		*(MPa)*	*(MPa)*			
CONTROL	34.8	2.552	7%	5.20	5.20	1.0	1.0	1.0
SF30-0.25	28.0	3.051	11%	4.30	4.30	3.1	6.5	9.4
SF30-0.50	30.0	2.928	10%	4.30	4.40	3.6	6.9	13.3
SF30-0.75	25.2	3.042	12%	4.10	4.25	3.7	8.3	14.7
SF30-1.0	26.9	3.413	13%	4.35	5.20	3.8	10.6	20.0
SF30-0.75(2)	33.2	3.410	10%	*4.90*	*4.90*	4.2	8.9	17.0
SF60-0.25	29.5	2.884	10%	5.15	5.15	1.5	2.6	8.2
SF60-0.50	32.9	5.282	16%	5.05	7.50	5.2	11.5	24.7
SF60-0.75	36.1	5.829	16%	5.10	9.45	5.8	14.0	28.5
SF60-1.0	35.3	6.181	17%	5.65	10.00	7.8	18.2	37.9
PVA30-0.50	39.5	3.511	9%	5.20	5.20	3.5	7.1	11.4
PVA30-0.75	36.2	3.697	10%	5.48	5.60	4.1	8.8	16.3
PVA18-0.75	32.2	3.367	10%	5.05	5.05	3.7	6.6	10.2
TWISTED-0.75	37.8	5.629	15%	5.85	8.55	5.0	11.3	23.4

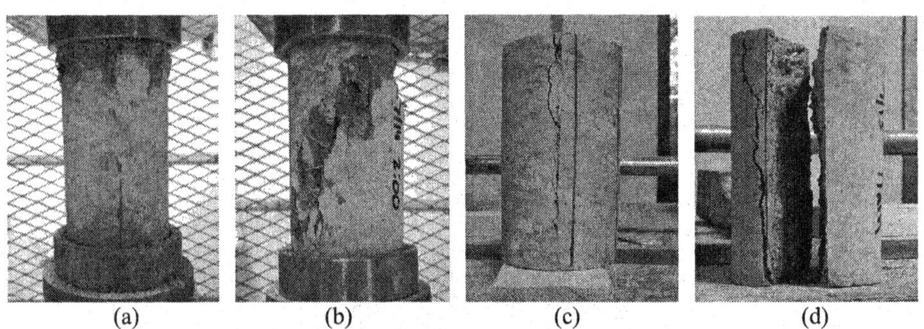

(a) (b) (c) (d)

Figure I. Compression and Split Tensile Failure Modes: a. Typical Compression Failure; b. 30mm PVA Shear Failure; c. Typical Splitting Tensile Failure for FRC; d. Splitting Tensile Failure for Control Mixture

Flexural Testing

The flexural toughness and first-crack strength (modulus of rupture) of the 350 x 100 x 100 mm beams were determined in accordance to ASTM C 1018[8]. Third point loading was applied to the beam over a 300 mm simply supported span. Figure II (a and b) shows the testing setup before load was applied to the beam. Four potentiometers were used to measure deflection at both the mid-span and end support. Vertical guides (not shown) were used to guaranty that only the vertical deflection of the beam was captured by the potentiometers. This set-up allows to determine the net deflection at any load level, as required by ASTM C 1018.

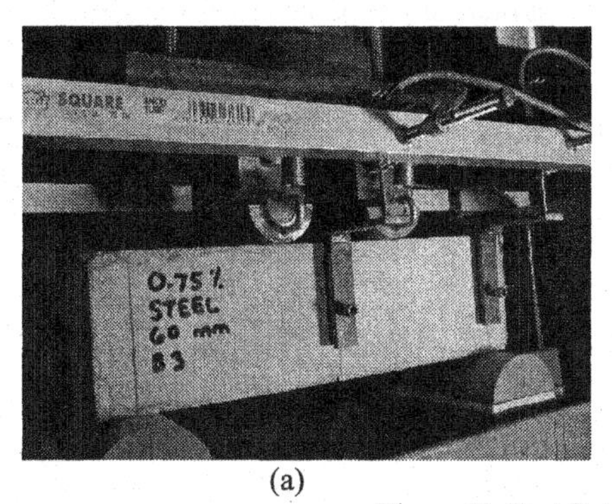

(a)

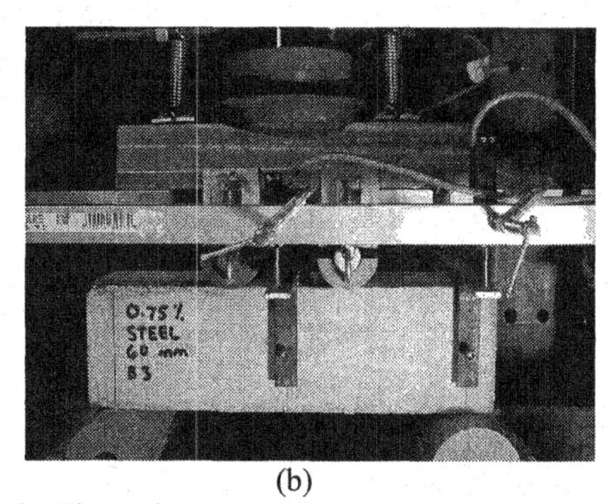

(b)

Figure II. Test Setup for Flexural Tests

Failure Mechanisms. All the specimens tested exhibited an elastic behavior up to the first crack. After cracking, the fibers bridging the crack were activated. Most specimens experienced a load drop after cracking, except for specimens SF60-0.75, SF60-1.0 and TWISTED-0.75. These three mixtures exhibited a quasi-strain hardening behavior, with increase in load after first crack as well as development of more than one flexure crack in the middle third of the span. Examination of the main flexure crack after failure showed that fiber pull-out occurred in all the mixtures tested. Figure III (a) shows a detail of the crack of a hooked-end steel fiber. For the twisted high strength steel fibers, untwisting and pull-out were observed, Figure II (b). For some of the specimens containing PVA fibers, a combination of pull-out and partial damage of the fiber ends was observed, Figure III (c).

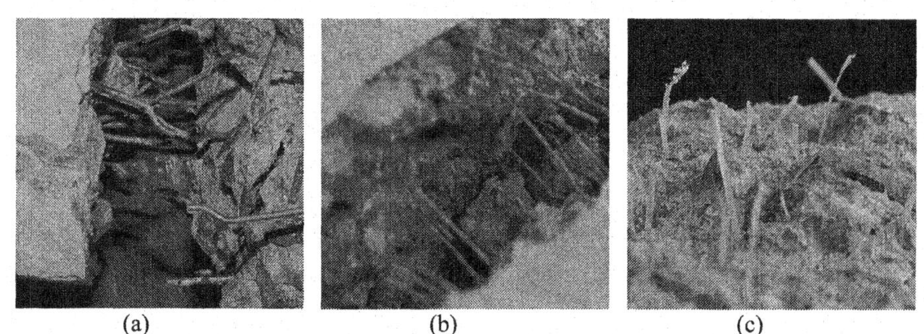

(a) (b) (c)

Figure III. Effect of Fibers Bridging Main Flexure Crack: a. Steel Fibers; b. Twisted Steel Fibers; c. PVA Fibers

Effect of fiber volume fraction. Figure IV shows the effect of the volume fraction on specimens with 30 and 60 mm hooked-end steel fibers. It can be observed that all the specimens with 30 mm fibers experienced a drop of load after the first crack. Specimens with a fiber volume fraction (V_f) less or equal than 0.75% had a post-cracking strength lower than the strength at first cracking (modulus of rupture). Only specimens SF30-1.0 exhibited a higher post-cracking strength (1.4 times larger than the strength at first cracking). In the case of the specimens with 60 mm, only the specimen with $V_f = 0.25$ % did not experience a higher post-cracking strength. This second series of specimens shows the improvement of performance that can be obtained by using a fiber with a larger aspect ratio (L/d) and FIER value. Specimen SF60-1.0 had a post-cracking strength 3.6 times higher than the value for the modulus of rupture reported in Table III.

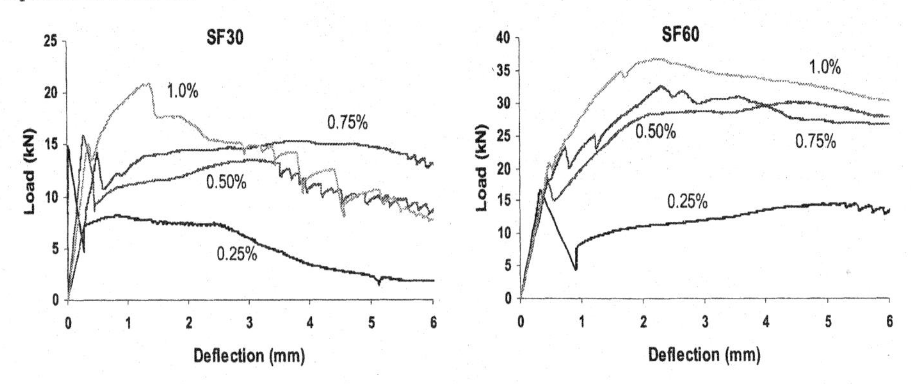

Figure IV. Effect of Volume Fraction on Hooked-end Steel Fibers: a. 30 mm length; b. 60 mm length

Toughness and Modulus of Rupture. Table III shows the results for values of modulus of rupture and toughness indexes calculated per ASTM C1018. It can be clearly seen that an addition of fibers increases the toughness of the composite material, with respect to the control mixture. An increase in the fiber volume fraction increases the ductility of the FRC (reflected in an increase in the toughness indexes).This increase in ductility is higher for longer fibers (60 mm steel) in particular for I_{10} and I_{20} indexes. It was interesting to note that although mixtures with enhanced performance (exhibiting quasi-strain hardening behavior) had high modulus of rupture and toughness indexes, similar values were also obtained by mixture that did not perform as well (SF30-1.0, SF30-0.75). It can be concluded that the effectiveness of a fiber in improving FRC performance cannot be based solely in the values of toughness indexes.

Effect of fiber geometry. Figure V shows typical load–deflection curves of specimens with same w/c ratio and a 0.75% fiber volume fraction. It can be observed that only the 60 mm hooked-end and the twisted high strength steel fibers exhibited an enhanced performance with a quasi-brittle strain hardening in the post-cracking state. The right plot shows how specimens with the twisted fibers had a smoother curve than the 60 mm steel fiber specimens.

It is also important to note that both PVA fibers (30 mm and 18 mm) did not exhibited strain hardening behavior, shown in Figure VI, even though their FIER was of the same order than the 60 mm and the twisted steel fibers. The observed behavior is attributed to the fact that PVA fibers have a smooth surface that limits the bond that can be developed with the matrix as well as a Young's modulus that is very close to that of the concrete matrix (E_c = 28,000 MPA for control specimen, per reference 9). Naaman[2] indicates that in order for fibers to be effective in concrete matrices, the elastic modulus of the fiber should be at least three times higher than that of the matrix.

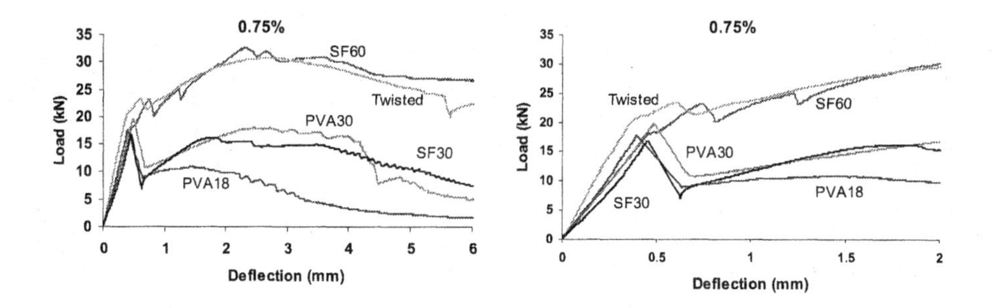

Figure V. Comparison of load-deflection curves for FRC with 0.75% volume fraction.

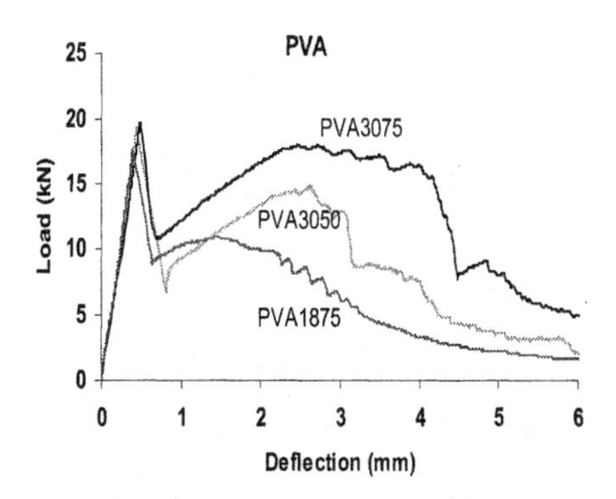

Figure VI. Typical Load-Deflection Curves for PVA Specimens.

Enhanced performance. From the analysis of the results of the flexure tests conducted it can be concluded that specimens with 60 mm hooked-end steel fibers and 1% fiber volume fraction were the only ones to exhibit a well defined strain hardening behavior in the post cracking stage. When only fibers with 30 mm length are considered, the high strength twisted fibers showed enhanced performance over conventional steel and PVA fibers.

CONCLUSIONS

The FRC mixture with 0.75% twisted fibers was observed to have the best workability of all the batches tested. It was found that adding fibers did not affected the compressive strength of the composite. On the other hand addition of fibers increased the splitting tensile strength of the composite. The largest splitting/compressive strength ratio (>15%)was obtained in 60 mm steel fiber mixtures with V_f larger than 0.5% and in the twisted steel fiber mixture.

All the FRC specimens exhibited fiber pull-out in the flexure tests. In the case of the twisted high strength steel fibers, untwisting was observed.

Hooked-end steel FRC improved performance by using a fiber with a larger FIER value. Specimens with 60 mm fibers and 1% fiber volume fraction had a post cracking strength 3.6 times higher than their modulus of rupture.

Increasing the fiber volume fraction increases the toughness of the composite material. FRC with longer fibers (60 mm) showed the most significant increase in toughness in particular for I_{10} and I_{20} indexes.

FRC exhibiting enhanced performance had similar toughness indexes than other FRC mixtures that did not exhibited this improved behavior, indicating that the effectiveness of a fiber in improving FRC performance cannot be based solely in the values of toughness indexes.

Only the 60 mm hooked-end and the twisted high strength steel fibers exhibited an enhanced performance with quasi-brittle strain hardening in mixtures where $V_f > 0.75\%$. All three mixtures had fibers with improved geometric features (hook-ends, twists, and large FIER values) as well as a Young's modulus one order of magnitude higher than the concrete matrix.

When only fibers with 30 mm length are considered, the high strength twisted fibers showed enhanced performance over conventional steel and PVA fibers

ACKNOWLEDGEMENTS

The research activities described in this paper have been supported by a CAREER grant from the National Science foundation (CMS-0330592). A supplemental award was provided in April 2004 for a Research Experience for Undergraduates (REU). This support is gratefully acknowledged.

REFERENCES

[1]P. Balaguru and H. Najm, "High-Performance Fiber Reinforced Concrete Mixture Proportions with High Fiber Volume Fractions," ACI *Materials Journal*, **101** [4] 281-286 (2004)

[2]A. Naaman, "Engineering Steel Fibers with Optimal Properties for Reinforcement of Cement Composites," *Journal of Advanced Concrete Technology*, **1** [3] 241-252 (2003)

[3]ASTM C 33 – 02a. Standard Specification for Concrete Aggregates. American Society for Testing and Materials, West Conshohocken, PA, 2002

[4]A. Naaman, "Optimized Geometries of Fiber Reinforcements of Cement, Ceramic and Polymeric Based Composites," U.S. Pat. No. 5 989 713, Nov. 23, 1999

[5] C. Sujiravorakul, "Development of High Performance Fiber Reinforced Cement Composites Using Twisted Polygonal Steel Fibers," Ph.D. Dissertation, University of Michigan, Ann Arbor, 2001

[6]ASTM C 1018 – 97. Standard Test Method for Flexural Toughness and First-Crack Strength of Fiber-Reinforced Concrete (Using Beam with Third-Point Loading). American Society for Testing and Materials, West Conshohocken, PA, 1997

[7]ASTM C 39/C 39M – 01. Standard Test Method for Compressive Strength of Cylindrical Concrete Specimens. American Society for Testing and Materials, West Conshohocken, PA, 2001

[8]ASTM C 496 – 96. Standard Test Method for Splitting Tensile Strength of Cylindrical Concrete Specimens. American Society for Testing and Materials, West Conshohocken, PA, 1996

[9]ACI 318-02. Building Code Requirements for Structural Concrete (ACI 318-02) and Commentary (ACI 318R-02). Reported by ACI committee 318. 2002

Next-Generation Cement Blends for High Performance

HYDRATION KINETICS OF PORTLAND CEMENT CONTAINING SUPPLEMENTARY CEMENTITIOUS MATERIALS

Y. Peng, W. Hansen, C. Borgnakke
University of Michigan
2350 Hayward
Ann Arbor, MI 48105

J.J. Biernacki
Tennessee Technological University
Campus Box 5013
Cookeville, TN 38505

ABSTRACT

Hydration kinetics was investigated for Type I ordinary portland cement (OPC) paste containing supplementary cementitious materials (SCMs) based on isothermal calorimetry at three temperatures (15°C, 23°C, and 35°C) and thermal gravimetric analysis (TGA) at 23°C. The SCMs were ground granulated blast furnace slag and fly ash (class F). Cement replacement levels were 25% and 50% by weight. Hydration kinetics of the OPC with SCM varies with SCM type (slag versus fly ash). Class F fly ash initially retards hydration, with retardation more pronounced with decreasing temperature. This retardation is followed by an acceleration of the OPC hydration. Pozzolanic reaction is very slow and was found to develop only after about 1000 hours at room temperature. Slag, on the other hand, was found to accelerate OPC hydration with the acceleration increasing with slag-cement replacement level. Hydraulic reactivity of the slag became apparent after about 72 hours at room temperature. The cement-slag system has a similar overall hydration rate as the OPC system, whereas the cement-fly ash system has a lower rate than the OPC system at any temperature. The methodology used in this study is well suited as a set of tools for analyzing the hydration kinetics of OPC-SCM systems and the factors controlling compatibility.

INTRODUCTION

The use of supplementary cementitious materials (SCMs) in ordinary portland cement (OPC) concrete is gaining acceptance as it has several advantages including reduced waste disposal, reduction in overall CO_2 levels, lower cost, and most important a well established improvement in later age strength and durability. However, the compatibility of the SCMs with portland cement depends on several factors including gypsum and alkali contents, which in turn affects the optimum cement replacement level for a given cement-SCM system. The influence of temperature is significant with respect to early-age strength development if the SCM initially retards the cement hydration, and if the retardation effect is more pronounced at lower temperatures. The rate of SCM hydration is also critical, particularly if the SCM hydrates significantly slower than portland cement, since strength development depends upon the relative proportions of cementitious and latent hydraulic/pozzolanic components in the mix and the rate of hydration of the constituents.

Cement hydration is a strongly exothermal reaction with several well-defined stages, I-V [1]. Based on rate of heat of hydration, these stages are defined in order of occurrence as the rapid initial stage (I), the dormant stage (II), the acceleration stage (III), the retardation stage (IV) and the long-term reaction stage (V).

The objective of this study is to evaluate the influence of two different SCM types and replacement levels on hydration kinetics (i.e. rate of reaction as affected by time and temperature) for hydration stages II-V, with V only covered partially.

EXPERIMENTAL

Isothermal calorimetry tests were conducted at three temperatures, 15 °C, 23 °C, and 35°C, starting immediately after mixing and continuing for three weeks. The water/binder ratio (w/b) was 0.45. Slag and fly ash (ASTM class F low calcium fly ash) were used as SCMs at two replacement levels: 25% and 50% by weight. The experimental matrix is shown in Table 1 and the chemical compositions of materials are listed in Table 2.

Table 1. Experimental matrix

	Type I cement (%)	Slag (%)	Fly Ash (class F) (%)
OPC	100		
25S	75	25	
50S	50	50	
25F	75		25
50F	50		50

Table 2. Chemical compositions of materials

% by weight	Type I cement	Slag	Fly Ash (class F)
SiO_2	20.4	37.4	56.2
Al_2O_3	5.04	7.77	17.6
Fe_2O_3	2.51	0.43	6.0
CaO	62.39	37.99	8.9
MgO	3.43	10.69	
Na_2O	0.25	0.28	1.2
K_2O	0.67	0.46	0.45
Cl	0.03		
SO_3	2.75	3.21	0.8
Total as Oxides	97.47	98.23	91.15
C_3S	53.66	-	
C_2S	18.01	-	
C_3A	9.11	-	
C_4AF	7.64	-	
Total as Clinker Phases	88.42	NA	NA
Blaine (cm^2/g)	4290	6020	NA

Thermal gravimetric analysis (TGA) was used to monitor the chemically bound water and calcium hydroxide (CH) content in the OPC paste and OPC pastes containing SCMs. TGA results were obtained only at one temperature (23 °C).

The calorimeter used in this study was a Thermometric model TAM Air. Two samples were tested and the average results were reported. The standard deviation is always less than 5% during the first 200 hours. After 200 hours, the hydration rate is so low that the standard deviation may increase to 20% for some batches. Since during the first 200 hours 70 to 80% hydration has completed, the results are considered having adequate accuracy. Thermal analysis was conducted using a Mettler TGA/SDTA model 851 LF. Also, two samples were run and the standard deviation was less than 3%.

RESULTS

The rate of heat generation dQ/dt and the cumulative heat Q from the isothermal calorimetry are shown in Figure 1 for the OPC system illustrating expected trends with temperature and time of hydration. A major feature of this work is that heat of hydration was monitored for three weeks. Typically, in the literature, heat of hydration is only monitored for the first 48-100 hours. The rate dQ/dt is shown in linear scale for the first 100 hours, then illustrated in logarithmic scale since the temperature-dependent difference is not discernable in linear plot for the total test duration of three weeks. The same plots are done for the cumulative heat curves for the same reason. The results shown in logarithmic scale demonstrate that the rate of reaction varies by three orders of magnitude (from ~10 to ~ 0.05 kJ/kg/hr) during a three week period, and is constantly decreasing beyond the maximum peak rate. The cumulative heat curves demonstrate a late stage *cross-over effect,* similar to that known from strength-time curves between low and high temperature hydration, and agree with the findings from hydration and strength testing by other researchers [2, 3].

The measured chemically bound water w_b is shown in Figure 2. The same trend observed from the isothermal heat data is seen for the w_b, with higher early age hydration at high temperature, and a cross-over effect at later age.

With increasing temperature the later-age rate of hydration is reduced; an observation that is most likely the result of the terminating processes resulting in an interplay between the temperature-dependent extent of hydration, microstructure and diffusion/reaction effects.

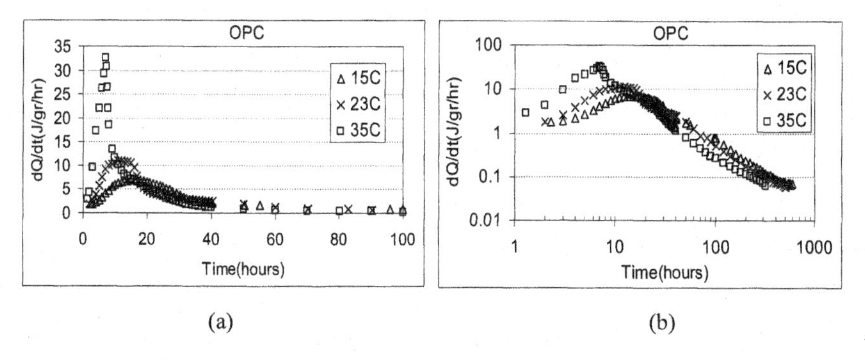

Figure 1. Typical heat of hydration and cumulative heat development curves at three temperatures, (a) dQ/dt vs. time, linear plot for the first 100 hours (b) dQ/dt vs. time, logarithmic plot for the whole test (c) Q vs. time, linear plot (d) Q vs. time, logarithmic plot. ((c) and (d) are on the following page)

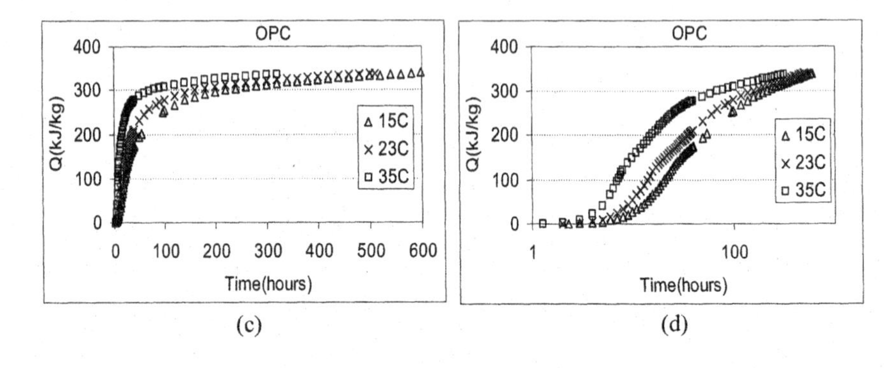

(c) (d)

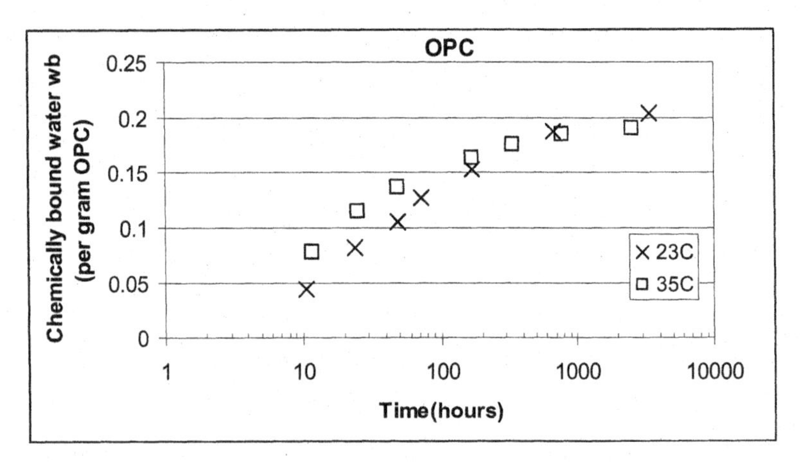

Figure 2. Chemically bound water w_b vs time

DATA ANALYSIS

Degree of cement hydration determined from chemically bound water w_b is a well-established hydration property. In order to use the heat of hydration to study kinetics, it must be first proven that the heat of hydration is also representative of the hydration process. Degree of cement hydration for OPC as obtained from TGA results is plotted against the relative isothermal heat of hydration for OPC as seen in Figure 3, where the long-term heat of hydration is predicted by Equation 1, and the ultimate value of chemically bound water w_{b-ulti} is 0.23.

$$Q = Q_\infty \exp[-(\frac{\tau}{t})^a] \qquad (1)$$

where τ , a =curvefitting constants.

In this study Q_∞ was assumed to be the 100-day heat of hydration value as this can be reliably determined from the 21-day test results. Clearly, this value is not the true value corresponding to infinite hydration time, however, it does provide a practical normalization basis.

The results in Figure 3 demonstrate that a unique relationship exists between the degree of hydration and the relative heat of hydration. This relation does not appear to be linear at the later stages of hydration suggesting that heat of hydration is lower. However, it should be mentioned that the uncertainty in the heat measurements increase as the signal decreases. For stage V, the rate of heat of hydration is nearly 3 orders of magnitude less than the peak acceleration value for stage III. At that stage, we have reached the resolution of the instrument.

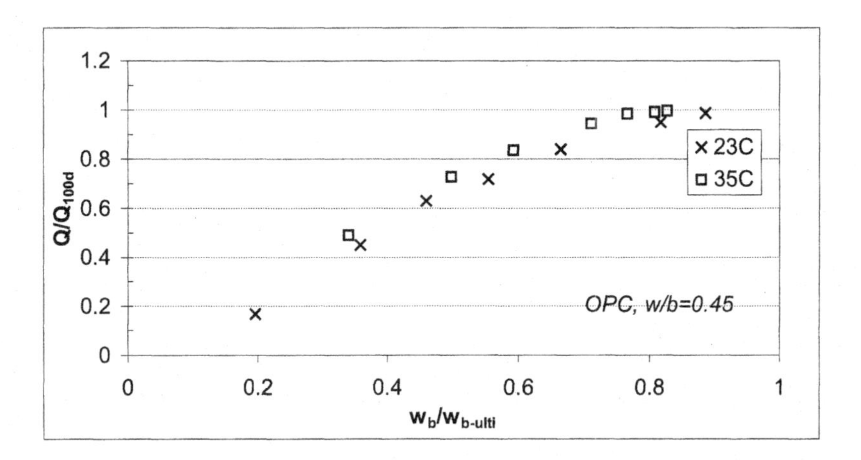

Figure 3. Relative heat development vs. degree of hydration for OPC at a w/b=0.45

In terms of SCM reactivity, slag and fly ash are very different due to the differences in their chemical compositions, morphology and particle size. Generally speaking, class F fly ash systems have lower reactivity than slag systems. Both materials consist mostly of a silicate glass, however, slag is considerably modified by large quantities of calcium and magnesium whereas low-calcium class F fly ash is modified with aluminum and iron and appears to be less reactive [4, 5]. When mixed with portland cement, a relatively small amount of portland cement is needed to activate (speed up the reaction of) the slag, whereas the alkalinity of cement hydration is apparently not enough to activate fly ash hydration at early ages [6]. The difference in reactivity of slag and fly ash may also be related to the size distribution of particles as suggested by Mehta [5] where typical surface areas measured by nitrogen adsorption for slag and fly ash are on the order of 4000 to 6000 cm^2/g, and 3000 to5000 cm^2/g, respectively. In general, the low-calcium fly ash (class F) tends to be coarser than high-calcium fly ashes (class C) [5]. Finer materials usually begin to contribute to hydration at an earlier time, although a comparison based on particle size alone has not been done and cannot be assumed since the calcium content likely plays a significant role as well.

The purpose of this paper is to evaluate the reactivity and effect of temperature on hydration of OPC pastes containing SCMs. This was facilitated by determining the rate function for cement hydration. Based on previous studies [7, 8, 9], the *absolute* rate of hydration can be written as a multiplicative function of temperature, degree of hydration, and water-to-binder ratio, as illustrated by Equation 2:

$$\frac{dQ}{dt} = k(T)g(\alpha)f(w/b) \tag{2}$$

where $\frac{dQ}{dt}$ = absolute rate of hydration heat, kJ/kg/hr,

$k(T)$ = *absolute* rate constant,

α = degree of hydration ranging between 0 and 1,

$g(\alpha)$ = function of relative degree of hydration,

and $f(w/b)$ = function of water binder ratio.

When heat of hydration is used as a measure of degree of hydration, α can be substituted with $\hat{\alpha}$, the relative heat development Q/Q_{100d}.

The rates of heat release were plotted against Q/Q_{100d} using a semi-log scale in Figure 4, where it is shown for the OPC system and the systems containing 25% slag and 25% fly ash (class F), respectively. The difference in hydration rates between any two temperatures was found to be nearly constant, thereby suggesting that the relative temperature sensitivity is also constant throughout the hydration process. In another word, it is not a function of the extent of hydration. This makes it possible to normalize the rates by their peak values, and to obtain a unique curve (which is a function of degree of hydration) for each system, as in [8, 9,]. This normalization procedure is performed on the same systems and shown in Figure 5.

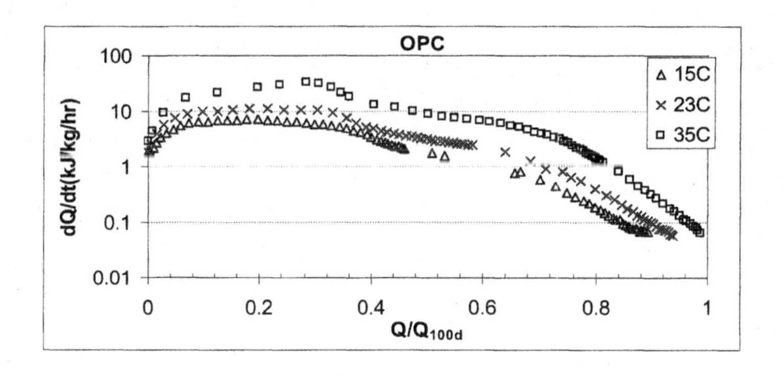

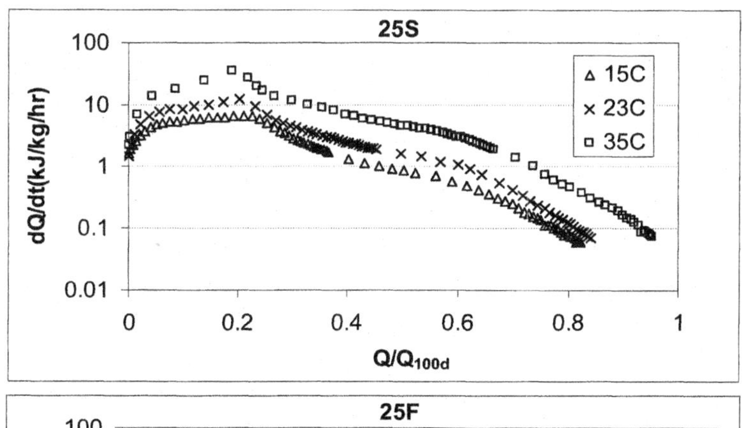

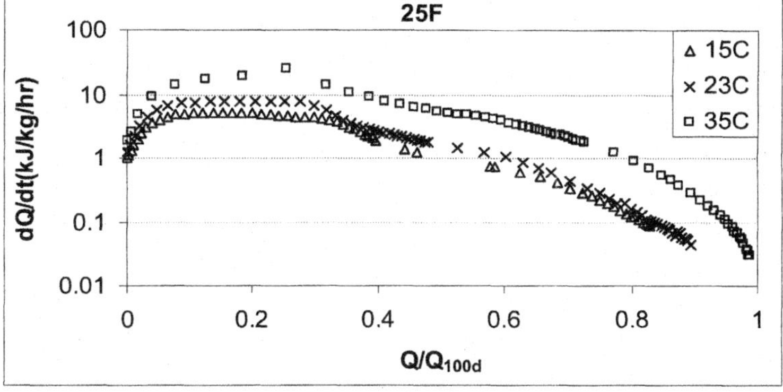

Figure 4. Hydration rate vs. Q/Q_{100d} at different temperatures (a) 100%OPC (b)75%OPC+25%slag (c) 75%OPC+25%fly fly ash (F)

(a)

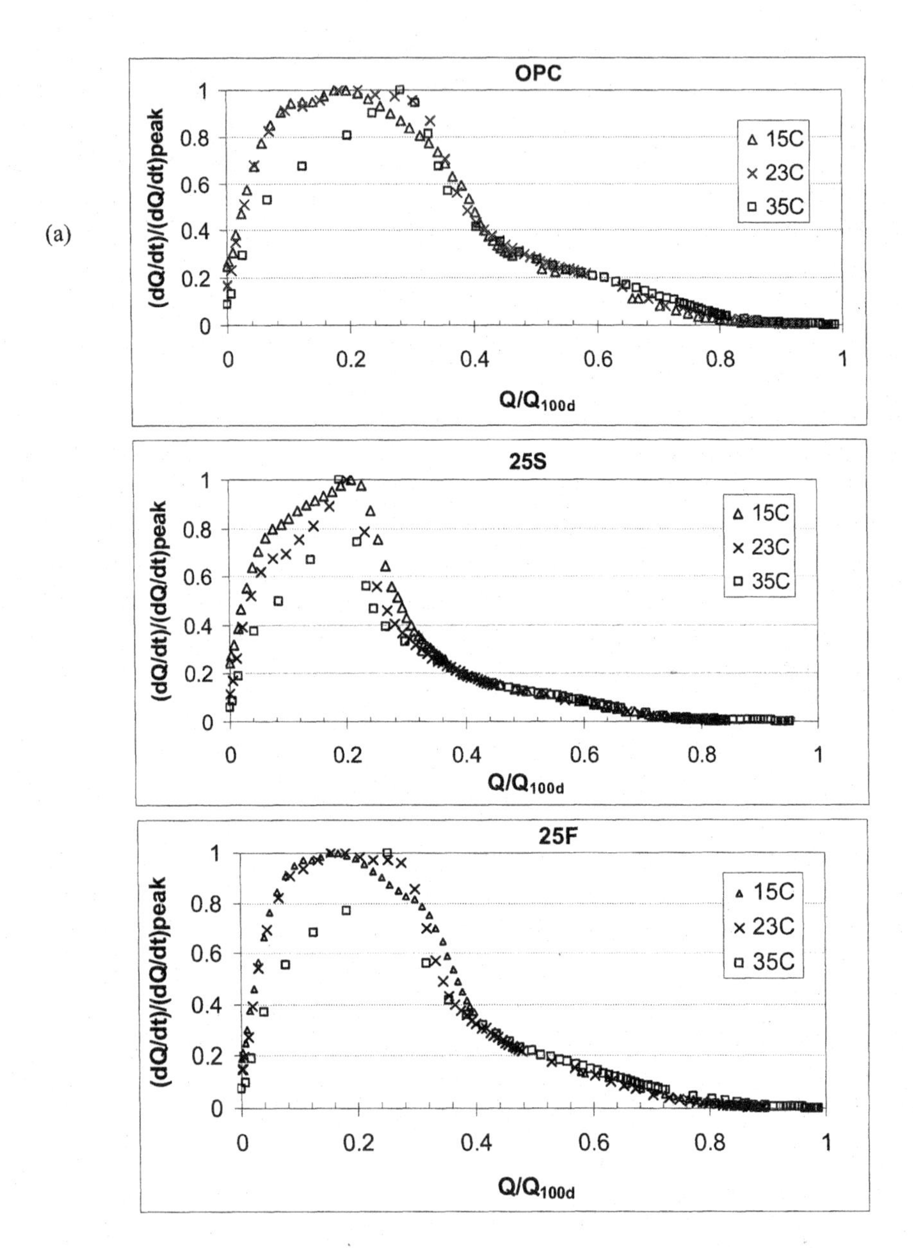

Figure 5. Normalization of hydration curves for (a) 100%OPC (b)75%OPC+25%slag (c) 75%OPC+25%fly fly ash (F)

Now, the *absolute* rate function can be written as a product of two separable functions as seen in Equation 3, wherein, the *absolute* rate constant *k(T)* is numerically equal to the peak rate at each temperature. When the relative heat development $\hat{\alpha}$ is used, the effect of water binder ratio $f(w/b)$ and absolute degree of hydration α are incorporated into $g(\hat{\alpha})$.

$$\frac{dQ}{dt} = k(T)g(\hat{\alpha}) = (\frac{dQ}{dt})_{peak} g(\hat{\alpha}) \tag{3}$$

where $k(T) = (\frac{dQ}{dt})_{peak}$ (the *absolute* rate constant)

and $g(\hat{\alpha}) = g(\alpha, w/b)$.

The *absolute* peak rate constants *k(T)* are listed in Table 3. It is seen that the slag systems have slightly lower *k(T)* than OPC at low temperature and higher *k(T)* at higher temperatures, e.g. it is more temperature sensitive. The two slag concentrations show the same rate constants at the same temperatures, though, there is inadequate data to indicate the concentration needed for this change. On the other hand, the fly ash systems always develop a lower rate constant than OPC with a nearly linear decreasing rate constant as a function of the concentration of fly ash. This indicates that the fly ash systems have lower reactivity than the slag systems. It is obvious that the slag systems are thermally activated and that the hydration of fly ash systems is slower than OPC at all temperatures.

The variation of *k(T)* with temperature obeys an Arrhenius type law, see Equation 4:

$$k(T) = A\exp[-(\frac{E_a}{RT})] \tag{4}$$

where E_a = the activation energy,
R = the gas constant
and A = a pre-exponential factor.

The activation energy of SCM containing systems can be obtained by treating the blended cement as a single *substance*, e.g. using a lumped parameter approach. Peak rate values, listed in Table 3, were used to determine the activation energy for each SCM-containing system. From Table 4, it is seen that SCM reactions have much higher activation energies than the OPC, and the activation energy increases with the percentage of SCMs. The higher activation energies of SCM reactions suggest that they are more temperature sensitive than OPC hydration. Such higher activation energies cannot be attributed at this time to any one mechanism, i.e. does not imply that the activation energy for neat slag hydration is greater than for neat OPC hydration. In fact, others have reported the activation energy for slag hydration, for example, to be considerably lower than that for OPC even at very early age [10]. One possible explanation, however, is that dilution with slow reacting SCMs extending the high activation energy nucleation and growth period of OPC hydration, thereby, resulting in an increase in the apparent lumped activation energy. Other explanations may also be plausible and must be investigated experimentally before putting forth a theory. Furthermore, while the net rate of reaction for the SCM-blended cement may be slower, as indicated by heat evolution as well as strength development, this does not imply a specific change in activation energy. The magnitude of the activation energy is only an indicator of temperature sensitivity, while the absolute value of the rate constant is an indicator of rate at some constant extent of reaction.

In order to explore how slag and fly ash affect hydration, the cumulative heat curves are plotted for the measured OPC system, and compared to the measured OPC-SCM system by scaling the OPC curves for the SCM replacement content. In the following analysis, the individual heats of hydration for clinker, slag and fly ash are assumed to be the same on a mass basis. Furthermore, one must assume that the overall stoichiometry of reaction on a mass basis is the same. Neither of these assumptions is strictly or necessarily true, however, at this time the thermodynamic data (heats of reaction) are not known for either slag or fly ash hydration. Given these assumptions, then, if the OPC-SCM system is below the scaled OPC curve, the SCM is reducing the reactivity of the OPC, while if above this curve the SCM is contributing to heat of hydration either through accelerating the OPC hydration and/or providing heat from the hydraulic/pozzolanic reaction. These results are shown in Figure 6.

Table 3. Absolute peak-rate constant *k(T)* (kJ/kg/hr)

	15 °C	23 °C	35 °C
OPC	7.02	11	32.58
25S	6.66	12.42	35.7
50S	6.7	12.43	34.65
25F	5.26	8.05	25.6
50F	3.72	6	16.33

Table 4. Activation energy (kJ/mol)

	Ea
OPC	57
25S	66
50S	71
25F	75
50F	85

An example is the effect of 25% cement substitution by fly ash. In addition to the measured OPC curve, a scaled OPC curve the 0.75×OPC is found, assuming that the 25% replacement is inert filler and does not contribute to hydration. The difference ΔQ between 25F and 0.75×OPC is due to the effect of the fly ash (Figure 6(b)). $\Delta Q<0$ means the reaction is retarded by the SCM, and $\Delta Q>0$ shows that hydration is accelerated and/or the pozzolanic reaction is contributing. The same scaling can be done for all types of SCMs at different replacement levels. A similar scaling has been done in [11]

The differences in heat ΔQ at three temperatures are shown in Figure 7 for the different OPC-SCM systems. For each temperature, the figures are shown for the first 100 hours and for the first 500 hours. The full time scale gives a general idea of the effect of SCMs, and the first 100 hours illustrates more clearly the details during early age hydration.

For both slag and fly ash containing systems, the contribution of SCM increases with temperature. This suggests that the latent hydraulic (slag reaction) and pozzolanic reaction are activated by temperature. At the three temperatures, the slag takes a more active role in the reaction than fly ash, as suggested by the greater values of ΔQ; again recalling that this interpretation assumes that heats of hydration are the same for all components, clinker, slag and fly ash. From the detailed figure of the early age, it is seen that the slag almost starts to contribute to hydration from the beginning, whereas the fly ash retards the hydration first, then accelerates the reaction. The transition times between retardation and acceleration for fly ash systems are 40, 30, and 20 hours for 15°C, 23°C, and 35°C respectively. For the slag system more hydraulic reaction is found with increasing slag content, while the fly ash systems develop increasing retardation at early ages with increasing fly ash content. However, later age acceleration of OPC increases. These results may be used to determine the optimal percentage of SCMs.

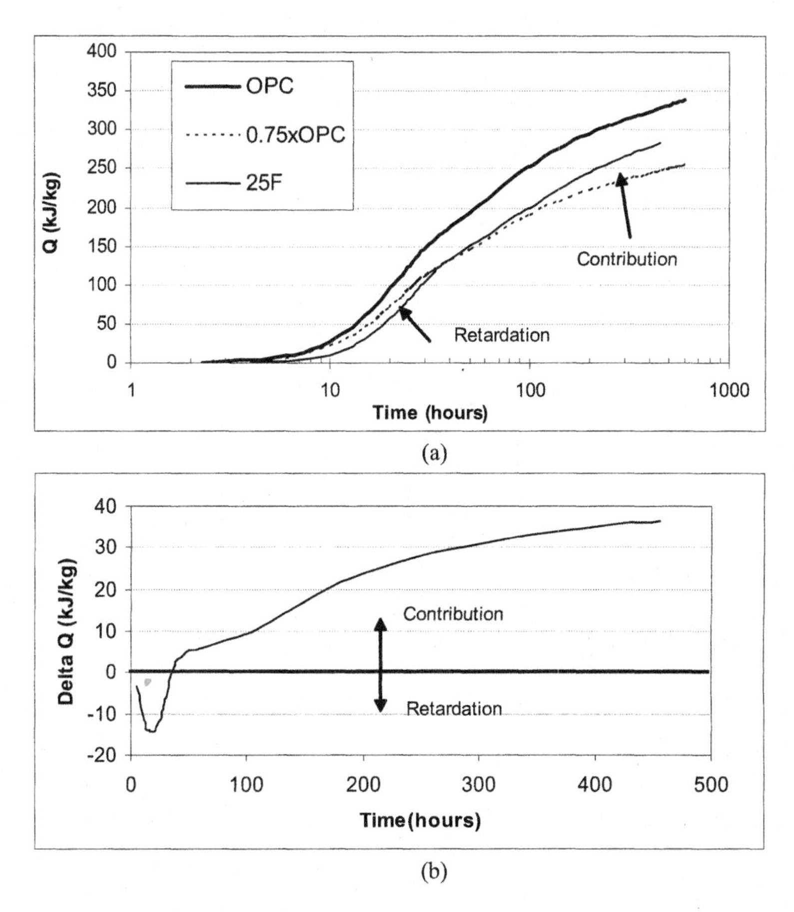

Figure 6. (a) Total heat of hydration for OPC and OPC with SCM. In (b) the effect of SCM on hydration is found from the difference in cumulative heat of the OPC-SCM system and the scaled OPC system.

The CH contents determined by TGA were used to further quantify the effect of SCMs on hydration. Since CH is produced only by cement hydration, the CH content was scaled by the percentage of OPC present in a given OPC-SCM system. These results are shown in Figure 8 (a). The slag system developed greater CH content than OPC during early ages up to 72 hours, and had reduced CH content afterwards. This is evidence that the slag accelerates hydration of the OPC in an OPC-slag system, consistent with the calorimetry results. After 72 hours, the effect of the slag reaction becomes apparent as indicated by a decrease in CH content relative to neat OPC. Therefore, the contribution of slag to hydration of the OPC-slag system includes two parts: (1) acceleration of OPC hydration and (2) reaction of slag with water and the cement hydration products.

The fly ash systems were found to have an initial retarding effect followed by an acceleration. This agrees with the finding in the literature [12]. From Figure 8(b), the CH content of fly ash systems is lower at the beginning for up to 18 hours, followed by a period with higher CH content till around 28 days, then lower again after that. This indicates that the fly ash retards, and then accelerates the hydration of OPC. The initiation of the pozzolanic reaction is, apparently, not happening until much later (> 1000 hours).

CONCLUSIONS

Hydration kinetics was investigated for OPC (Type I) paste containing SCMs based on isothermal calorimetry and TGA results. The SCMs were ground granulated blast furnace slag and fly ash (class F). Cement replacement levels were 25% and 50% by weight. The major findings are:

- Hydration kinetics of the OPC with SCM varies with SCM type (slag versus fly ash). Class F fly ash initially retards hydration with retardation more pronounced with decreasing temperature. This retardation is followed by an acceleration of hydration attributed to the OPC. Pozzolanic reaction is very slow and was found to develop only after about 1000 hours at room temperature. Slag, on the other hand was found to accelerate OPC hydration with the acceleration increasing with slag-cement replacement level. Hydraulic reactivity of the slag becomes apparent after about 72 hours at room temperature.

- The cement-slag system had slightly higher overall hydration rate than the OPC system above room temperature, whereas the cement-fly ash system had lower rate than the OPC system at any temperature.

- The methodology used in this study is well suited as a suite of tools for analyzing the hydration kinetics of OPC-SCM systems and the factors controlling compatibility.

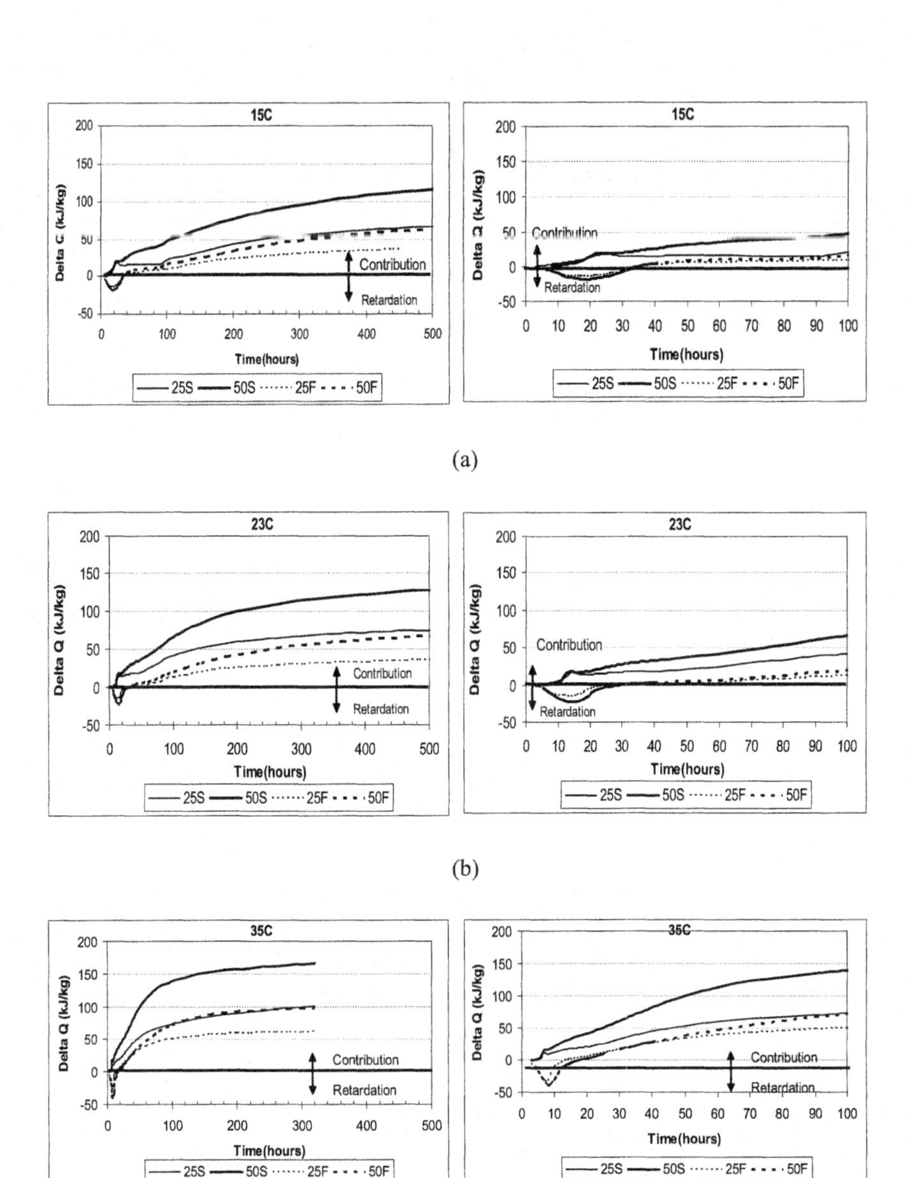

(c)

Figure 7. Effect of SCM on hydration (a) 15°C (b) 23°C (c)35°C

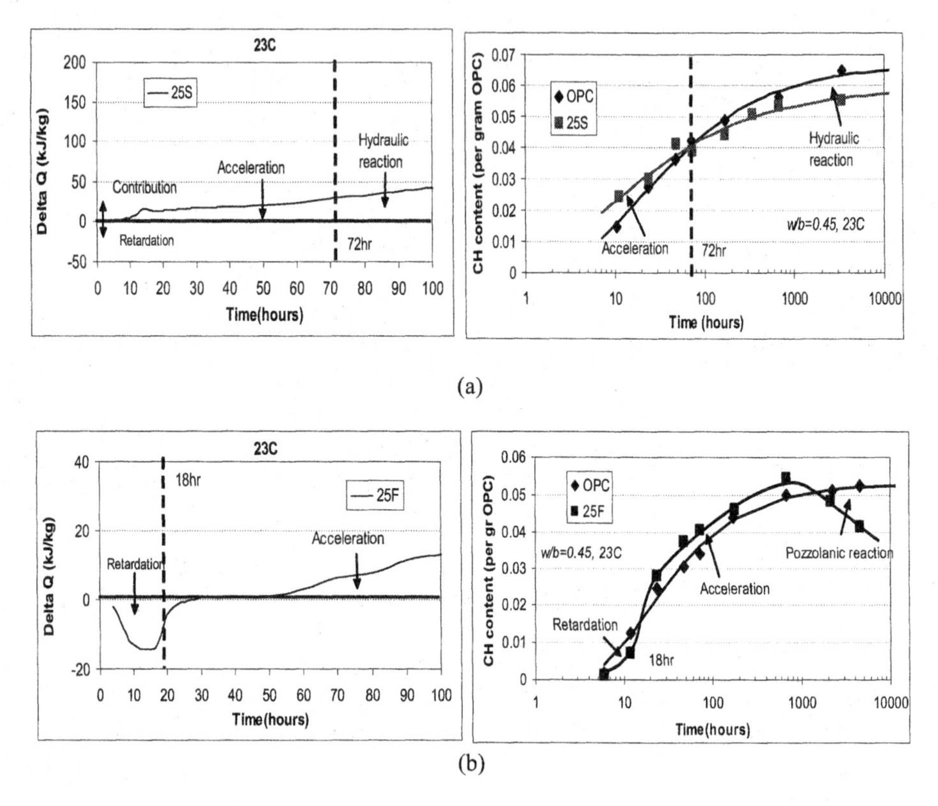

Figure 8. Correlation of heat data and TGA results: (a) for slag system and (b) for fly ash system.

REFERENCES

[1]J.F. Young, "Hydration of Portland Cement in Instructional Modules in Cement Science," Ed. D. M. Roy, Materials Education Council, MRL, University Park, PA, USA 1985

[2]Y.A. Abdel-Jawad, "The Relationships of Cement Hydration and Concrete Compressive Strength to Maturity," Ph.D. Thesis, Univ. of Michigan, 1988

[3]J.K. Kim, Y.H. Moon and S.H. Eo, "Compressive Strength Development of Concrete with Different Curing Time and Temperature," *Cement and Concrete Research*, 28 (12) 1998, pp 1761-1773

[4]D. Li, Y. Chen, J. Shen, J. Su and X. Wu, "The Influence of Alkalinity on Activation and Microstructure of Fly Ash," *Cement and Concrete Research*, 30 (2000), 881-886

[5]P.K. Mehta, "Pozzolanic and Cementitious By-Products in Concrete—Another Look, Fly Ash, Silica Fume, Slag, and Natural Pozzolans in Concrete," Proceedings of Third International Conference, Trondheim, Norway, 1989

[6]A.L.A. Fraay, J.M. Bejen, *Cement and Concrete Research* 19 (1989) 235-246

[7]S. Brunauer, M. Yudenfreund, I. Odler and J. Skalny, "Hardened Portland Cement Pastes of Low Porosity VI. Mechanism of the Hydration Process," *Cement and Concrete Research* 3, 1973, 129-147

[8]H.W. Reinhardt, J. Blaauwendraad and J. Jongendijk, "Temperature Development in Concrete Structures Taking Account of State Dependent Properties," RILEM International Conference on Concrete of Early Ages, 1982, Paris

[9]G. De Schutter, "Hydration and Temperature development of Concrete Made with Blast-furnace Slag Cement," *Cement and Concrete Research* 29 (1999) 143-149

[10]J.J. Biernacki, J.M. Richardson, P.E. Stutzman and D.P. Bentz, "Kinetics of Slag Hydration in the Presence of Calcium Hydroxide," *J. Am. Ceram. Soc.*, 85 (9) 2261-67 (2002)

[11]W. Ma, D. Sample, R. Martin, and P.W. Brown, "Calorimetric Study of Cement Blends Containing Fly Ash, Silica Fume, and Slag at Elevated Temperatures", *Cement, Concrete, and Aggregates*, Vol. 16, No. 2, 1994

[12]P.C. Hewlett (editor), *Lea's Chemistry of Cement and Concrete*, fourth edition, John Wiley & Sons Inc., 1998

NEXT-GENERATION CEMENT BLENDS FOR HIGH PERFORMANCE CONCRETE

A. K. Jain
Grasim Industries Ltd (Cement Business)
Ahura Center, 1st floor, Mahakali Caves Road
Andheri East
Mumbai – 400 093
(India)

ABSTRACT

The compressive strength and other properties of concrete have significantly improved over the years making much higher strength to weight ratios possible. These developments have enabled concrete to be used for a variety of new applications.

The emergence of high-performance concrete and self-compacting concrete and their wide spread use have enabled cement manufacturers to redesign their products to meet the new requirements. Some cementitious products promise even further enhancement of the properties of concrete and subsequent durability of structures. In this paper some of these products have been examined which have potential for mass production. It is expected that the traditional Portland cement and its other derivatives will be replaced by these new-generation cement blends in the coming years.

INTRODUCTION

The world cement consumption for 2003-2004 is estimated at 1760 million tons. It is expected that production could increase to 1835 million tons by the year 2005[1]. Portland cement has been the main hydraulic binder of the 20th century because it offered setting and strength development appropriate to modern construction needs. However, one ton of Portland cement requires about 1.5 tons of limestone and 3500 kJ energy, and releases 1 ton of CO_2, besides SO_2 and NOx emissions to the atmosphere. Declining sources of good quality raw materials and the large amount of pollutants generated during production will significantly influence the course of the cement industry in future. The overriding environmental considerations will ultimately force a change and ways will be sought to minimize over all consumption of raw materials, energy, pollution and production of wastes in general.

The Earth Summit in Rio de Janeiro in 1992 has also stressed sustainable economic development with respect to the earth's eco-systems. Future cementitious materials have to fulfill this basic premise. One aspect of the problem is global warming. There are different views as to how rapidly this is happening but there is no doubt about the role of carbon dioxide in promoting it. Cement production is not the largest source of carbon dioxide emission, but it is a significant one due to decarbonation of limestone and use of fossil fuels. It is estimated that cement production contributes about 7% of the total CO_2 emissions into the environment. Other aspects of the problem are development of fuel-efficient technologies, utilization of industrial by-products and waste materials, and demand on cementitious materials for higher performance and durability. Future innovation in the cement industry must address these considerations and their solution will decide the type of raw materials and the resulting cementitious binders for the construction industry in this century.

The future of the cement industry in India and elsewhere will be mainly influenced by (a) environmental issues (b) energy conservation (c) utilization of industrial – byproducts and waste materials, and (d) development and use of new cementitious materials to meet the needs of the construction industry for higher performance and durability.

On the other hand ordinary portland cement has been widely used in the past and some of its performance problems can be identified (a) portland cement based concretes suffer from poor toughness or crack resistance on impact (b) it has the tendency to crack or shrink on cooling and drying (c) it has high permeability to fluids, which is a concern, particularly in aggressive environments where durability is affected, and (d) it has problems at low temperatures (i.e. freeze/thaw).

The areas of application of cement and concrete have been diversified greatly in the 20th century. The emergence of concrete formulations for specific applications such as self compacting concrete, polymer impregnated concrete, roller compacted concrete, fiber reinforced concrete, high strength and high performance concrete have necessitated the development and use of new–generation cement blends.

Silica Fume Blended Cements

Portland cement and silica fume (12 to 15%) blends homogenized in high efficiency mixers in cement plants is very suitable for producing high strength concrete. In the portland cement system, the concrete strength is less than the paste matrix and is related to lower bond strength between the paste and the aggregate. When the weak interfacial bond is eliminated by adding silica fume, the concrete strength is no longer less than the paste matrix strength. In such systems the main role of the silica fume is in enhancing concrete strength by providing improved bond, whereas the paste matrix strength is practically unchanged when silica fume replaces part of the cement.

In normal concrete, a thin film (duplex film)[2] is deposited on the aggregate surface, beyond which is a zone rich in oriented $Ca(OH)_2$ and ettringite with porosity greater than the bulk paste matrix. The thickness of this zone is normally 20 to 40 μm. The causes for the formation of this structure are related to the formation of water-filled, low density, spaces around the aggregates at early age, wherein there is greater tendency for $Ca(OH)_2$ and ettringite to grow in such spaces. The high initial porosity in this zone remains even after prolonged hydration. There are two weak zones in the vicinity of the aggregate. The first one is at the actual interface and the second one is in the porous transition zone, away from the duplex film.

It is interesting to note from Fig. 1 that the influence of reduction in water to cement (W/C) ratio on the increase in bond strength is much smaller than the influence of W/C ratio on matrix strength. This suggests that factors other than W/C ratio play a major role in controlling bond and therefore design of the mix composition to optimize for bond may follow different rules than those applicable for achieving high paste strength.

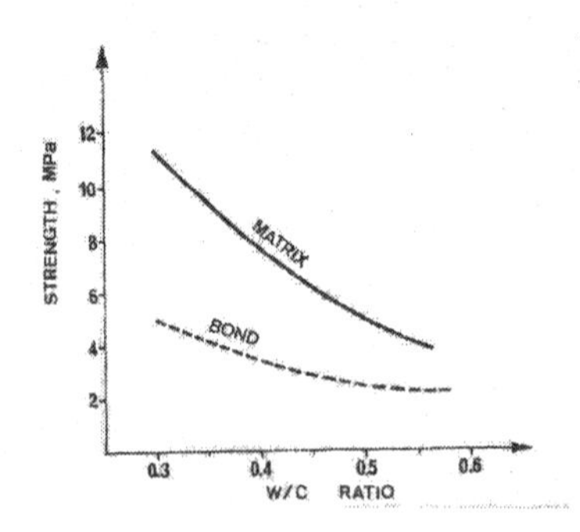

Fig 1 Effect of W/C ratio on aggregate matrix bond and matrix flexural strength (after Giaccio[3])

The improved bond strength has significant effect on concrete strength. Fig. 1 indicates that reduction in W/C ratio improves the strength of the matrix to a much greater extent than the bond strength, and the difference between the two increases, as the W/C is reduced[3]. This implies that in high strength systems (i.e. low W/C ratio) the bond is becoming the "bottle neck" in achieving high strength. Therefore, for producing high strength concretes it is not sufficient to produce low W/C ratio mixes alone, there is also a need to use pozzolans and filler materials such as silica fume, which enhance bond through modification of the interfacial microstructure. This is demonstrated in Fig. 2 in which the strength of the paste matrix and concrete of similar compositions are compared. In the portland cement system, the concrete strength is less than the paste matrix, which can be related to lower bond strength effect. In silica fume – Portland cement system, the bond is strengthened and concrete strength is no longer smaller than the paste matrix strength. The main role of the silica fume is in enhancing concrete strength by providing improved bond, where as the paste strength is practically unchanged by silica fume additions[4,5].

Silica fume has two types of effects, physical effect (i.e. filler effect due to particle size and shape) and chemical effect (i.e. pozzolanic reaction). The filler effect plays a major role in controlling bond[6], by modifying the properties of fresh concrete to reduce internal bleeding, voids and transition zone.

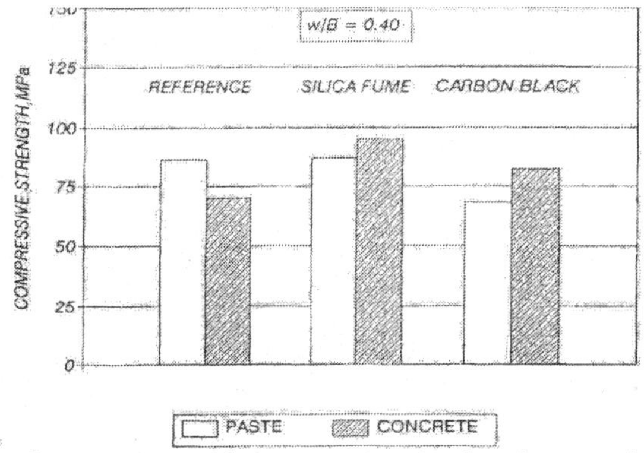

Fig. 2 Comparison of cement paste and concrete strength values in cementitious systems with the same water/binder ratio in which OPC is substituted with 15% of silica fume or 15% carbon black (after Goldman and Bentur)[4].

Fiber Reinforced Cements

The properties of fiber-reinforced cements are largely dependent on the nature and content of the fibers. Numerous studies of fiber-reinforced cements have focused on fibre–matrix bonding. In such studies the following assumptions are generally made:

a) The fibres are well defined geometrically, and uniformly dispersed in the matrix.
b) The matrix is uniform up to the fibre surface and debonding occurs at the actual fibre-matrix interface.

The structure and dispersion of the fibres can be quite complex and the matrix microstructure at the transition zone is different from that of the bulk[(7)]. In case of fibres of well-defined geometry, such as steel fibres, a transition zone is developed around the fibres with a microstructure, which is similar to that which develops around aggregates. The surface treatments of steel fibres to improve bond are not always effective since they affect the interface, which remains a weak zone where debonding can take place. In steel fibres improvement in bond is usually achieved by mechanical anchoring, using fibres of complex geometry.

In many systems, the fibres are not disbursed uniformly and clusters of filaments remain together. This is characteristic of many types of fibres, which are composed of thin filaments bundled together. A multifilament structure of this kind has been observed in glass fibres, asbestos fibre, and in some fibrillated polypropylene fibres. In such systems, the microstucture at the transition zone becomes very complex and non-uniform. The fibres in the bundle are not bonded uniformly and a common effect in bundle type reinforcement is changes in the properties with age. The inner portions of the bundle become filled with hydration products, improving bond. The resulting increase in anchoring causes fibre fracture instead of fibre pullout.

In fibre-reinforced cements, a crack is first initiated in the weak brittle matrix, and as it propogates towards the fibre it will first come across the porous region in the transition zone[8]. Since this is a weak zone, the crack will be deflected into it and the debonding will not occur at the actual interface. This sequence is advantageous, since it provides a crack arrest mechanism and prevents brittle behavior. During further loading of the composite, cracks are bridged by the fibres. Thus fibre cements provide much better resistance to cracks and improves the ductility and toughness of the composite.

Activated Belite-rich Cement

Significant research work has been done on reactive belite cements[9]. In this type of cement the alite phase (C_3S) has been restricted to a minimal level and belite phase (C_2S) is increased to 55-60%, with a partially glassy matrix containing a low proportion of C_3A and C_4AF. Numbers of studies have been conducted with regards to stabilization of phases through conventional burning and rapid quenching. In addition to this, certain dopants like K_2O, Na_2O, SO_3, B_2O_3, Cr_2O_3 were also tried [2].

In the commercially available portland cement clinkers, belite (C_2S) occurs primarily in its β modification, minor quantities also occurring in the ά and ά[1] modifications. Unlike alite, belite develops considerably lower strengths at earlier ages. It is only after longer periods (about one year) that it develops strengths comparable to those of alite. An improvement of the hydraulic qualities of belite can be achieved by stabilizing the hydraulically more active high-temperature modifications ά and ά[1]. This stabilization is obtained either by quick cooling of the clinkers or by integrating Na_2O, or SO_3 in the cement clinkers, or by doing both things at the same time.

The need of reactive belite cement is often required to conserve energy or to minimize limestone consumption. Chatterjee[10] reports that activated belite cement mortars and concrete follow the behaviour of alitic cements with respect to strength development. Furthermore, belitic cements show lower carbonation and have capillary porosity of 7-14% as compared to 15-24% in alitic cements. He is also of the opinion that belitic cements can be used quite effectively to produce blended cements. The comparative behaviour of mortars and concretes prepared from ordinary portland cement, moderate heat cement and belite cement is presented in Table-I[11]. This data shows that improvement in later age strength with lower heat of hydration is achieved for the belite cement.

The present investigations of activated belite-rich cements have shown that it is possible to improve essential cement qualities such as strength development, carbonation behavior, and porosity by using Na_2O or Na_2O and SO_3 as phase stabilizers. The belite-rich cements have low capillary porosity and relatively high density of the hardened cement paste that contributes to better durability of concrete.

Table :I Properties of concrete prepared from ordinary, moderate heat and belite cements [11]

Properties	Unit amount of Concrete (kg/m^3)				Slump (cm)	Compressive strength of Concrete (Mpa)			Adiabatic temp. Rise of concrete (°C)
	W	C	S	G		7d	28d	91d	
OPC	158	300	803	1045	10.5	20.3	39.1	45.6	46
MPC	158	300	803	1045	11.2	13.8	35.5	46.8	38.5
BPC	158	300	803	1045	12.7	8.5	32.6	58.9	29.5

OPC: Ordinary Portland Cement; MPC: Moderate Heat Cement; BPC: Belite cement.

High Silica Modulus Cement for High–Performance Concrete

High performance concrete (HPC) is gaining such rapid acceptance among designers and owners that concrete producers are now faced with the problem of producing and delivering very low (0.25 to 0.35) water/cementitious concretes. One of the greatest challenges facing today's HPC producer is finding the optimum superplasticiser/cement balance. The ideal concrete mix should have moderate slump loss over a period of one hour to ensure easy delivery and placing, yet safeguard against undue retardation.

Finding a portland cement for high-strength applications in not very difficult. A number of precast plants produce high-strength structural element, but casting should be done within a very short time otherwise fluid concrete can loose its plasticity. In some cases, plasticity is not recoverable, even by redosing the superplasticiser. Very few fundamental studies have been reported on the reactivity of portland cement at very low W/C ratios with high dosage of superplasticiser. These, however, report that the slump loss (or the loss of viscosity in grout) associated with super plasticiser use is related to the amount and reactivity of C_3A, the fineness of cement and its soluble alkali content [11].

Therefore, a high silica modulus portland cement also called HTS cement with very low C_3A and alkali contents appears promising for making HPC. HTS cement has considerably more silica (24.2%) compared to about 21% in normal Portland cement. It also has the lowest Al_2O_3 and Fe_2O_3 and relatively low alkali content. As indicated by the Bouge composition, this cement has a very high total silicate content 84% compared to 75% in normal OPC. Its interstitial phase content is the lowest (10.5%), composed of only 3.6% C_3A and 6.9% C_4AF. The LSF is the lowest but the SM is appreciably high. The alite and belite crystals are well formed, small in size, and relatively pure, that is, substitution of foreign ions is low. Very few impurities are present in this clinker[12].

In achieving the same initial and final slumps, HTS cement is far more economical in terms of the amount of super plasticiser needed to reach the same level of slump. Type I cement (due to its higher C_3A content) requires an additional 67% superplasticiser[12]. The compressive strength at 28 days of type I cement was marginally higher than HTS cement but one-day strength was lower. The lower one-day strength of type I cement can be a consequence of somewhat higher dosage superplasticiser (when compared to HTS). The exceptionally low interstitial phase and alkali content, and very high silica modulus HTS cement was made from a clinker composed of well-formed small alite and belite crystals, coarse crystalline C_4AF, and minute C_3A crystals. The slump loss study and compressive strength results obtained on two series of concretes made with different types of cement and at different W/C ratios clearly illustrate the advantage of using a coarse cement with low C_3A and alkali content for making high-performance concrete[12].

This type of cement makes it possible to produce flowing concrete with W/C ranging from 0.30 to 0.20, yet maintains a slump ranging form 200 to 250 mm with as little as to 115 l/m^3 of water and using about 10 to 20 l/m^3 of superplasticiser. Thus, this type of cement is highly economical in terms of superplasticiser dosage.

CONCLUSION

This brief review indicates that commercially produced cementitious materials will be available in the market, and are ready to use for production of high performance concrete. The present system of intermixing different mineral and chemical admixtures with basic cement will pave the way for factory-produced products for better consistency and performance. The cementitious materials of the future will be complex composites and their performance will be largely controlled by the quality of basic ingredients and production technology.

REFERENCES

[1]P.K. Mehta, "Role of Pozzolanic and cementitious materials in sustainable development of the concrete industry," *Proceeding 6th CANMET / ACI International conference* pp.178, Vol I, 1 – 20,1998

[2]S. Diamond, "The microstructure of cement paste in concrete," pp. 122-127 in Vol. I, *Proc. 8th International Congress Chemistry of Cement*, Rio de Janeiro, 1986

[3]Gi Giaccio, "Factors Affecting Cement Paste – Aggregate Bond," pp 331–333, Vol VI, *Proc, 8th International Congress Chemistry of Cement*, Rio de Janeiro, 1986

[4]A. Goldman and A. Bentur, "Bond Effects in High Strength Silica Fume Concretes," *American concrete Institutes Materials* J., **6**[5] 440-447 (1989)

[5]P.C. Aitcin "From Gigapascals to Nanometers," pp. 105-130 in Advances in Cement Manufacture and use. E. Gartner, ed. *Engineering Foundation Conference*, The Engineering Foundation, 1989

[6]P. K. Mehta, "Mechanism by which Mineral Admixtures improve concrete Behaviour," *International workshop on condensed Silica Fume in concrete*" Paper No. 8, Montreal, 1987.

[7]A. Bentur, "Micro Structure, Interfacial Effects, and Micro mechanics of cementitious Composites," pp. 523-549 in Vol 16, Ceramic Transactions, *American Ceramic Society*, Inc. Westerville, Ohio, 1990

[8]S.P. Shah and Y.S. Jenq, "Fracture Mechanics of Interfaces," pp. 205-216 in bonding in Cementitious Composites. *Materials Research Symposium Proceedings*, Vol. 114, Materials research Society, Pittsburg, 1988

[9]K.J. Star, A. Muller, R. Schrader, and K. Rampler, "Existence conditions of hydraulically active belite cement," *Zement-Kalk-Gips*, 34, pp 476-481 (1981)

[10]A.K. Chatterjee, "Special cements, Structure and performance of cements," pp.106-232. 2nd ed. Edited by J. Bensted and P. Barnes, 2002

[11]H. Uchi Kawa, "Characterization and Material design of high-strength concrete with superior workability," PAC RIM meeting, SII Cement, *Technol. Symp*; Honululu, Hawai(1993)

[12]T. Nawa, H. Eguchi and Y. Fukaya, "Effects of Alkali sulfate on the Rheological Behavior of Cement Paste containing a Superplasticizer," pp. 405-424 in *Proc. 3rd Canmet/ACI conf.* on Superplastizers and Other Chemical Admixtures in Concrete. Ottawa, 1989

[13]P.C. Aitan, Shondeep L. Sarkar, Roger Ranc and Christophe Levy, "A high Silica Modulus Cement for High-Performance Concrete," pp. 103-121 in Vol. 16, *Ceramic Transactions, American Ceramic Society*, Inc. Westerville, Ohio, 1990

FLY ASH BASED HIGH PERFORMANCE CEMENTITIOUS COMPOSITES

K.G. Babu
Professor
Department of Ocean Engineering
IIT, Chennai -600036

P. Dinakar
PhD Scholar
Department of Ocean Engineering
IIT, Chennai -600036

ABSTRACT

Fly ash, due to its pozzolanic nature and abundant availability, could be a great asset for the modern construction needs, particularly because fly ash concretes are not only economical but can be of high performance, if appropriately designed. The use of fly ash as a cementitious material as well as a fine filler is being increasingly advocated for the production of High Performance Concrete (HPC), Roller Compacted Concrete (RCC) and Self Compacting Concrete (SCC), etc. However, for obtaining the required high performance in any of these concrete composites, fly ash should be properly proportioned so that the resulting concrete would satisfy both the strength and performance criteria requirements of the structure. The present paper is an effort towards presenting a brief status on the various aspects related.

INTRODUCTION

The substantial increase in power requirements of modern day living has led to an increase in thermal power production and consequently an enormous quantity of fly ash is produced throughout the world, making its disposal an unsurmountable problem in recent years. The quantity of fly ash produced in India is reaching 100 million tons/year, while its utilization is almost negligible. The enormous quantity of fly ash is mostly disposed of in the slurry form in ash ponds requiring large areas of land causing many environmental and ecological imbalances. While many suggest several ways of utilizing fly ash in land reclamation, land filling, soil stabilization etc., the full potential of this very fine powder pozzolan can only be realized in the building industry.

In the building industry, fly ash is used for making building blocks and sintered aggregates and as a pozzolanic additive to cement or as an admixture in concrete, etc. The use of fly ash as a mineral admixture in concrete is not just to save the expensive pyro-processed cement, but to add to the durability of the final concrete, making it a value added product. In particular, the incorporation of fly ash has led to the development of several of the relatively recent concrete composites including High Performance Concrete (HPC), Roller Compacted Concrete (RCC) and Self Compacting Concrete (SCC), etc., which have all been proved to be excellent materials for several innovative applications. Furthermore, most of the fly ash produced is of the low calcium bearing Class F type, which is best suited for improving the durability of these concrete composites.

FLY ASH IN CONCRETE

Over the years a lot of research effort has gone into study of fly ash in concrete. However, one should be reminded of a paper by Popovics [1], initiating the discussion regarding our lack of a quantitative understanding of the role of fly ash in Portland cement concrete. After almost 20 years since this publication, we are certainly better placed, but regrettably the answers to several of the questions Popovics raised are scattered far and wide amongst the huge quantity of misinformation that is largely associated with the ubiquitous theoretical and experimental investigations on highly limited and sometimes even totally confused research efforts. His original suggestions were to recognize that:

(a) the contribution of fly ash to the quality of concrete is not a constant value determined solely by the physical and chemical characteristics of fly ash but rather it can vary depending on the type of Portland cement and water to cement ratio used; and

(b) it is not known at present what factors maximize the fly ash contribution; therefore, in many cases, less benefit is obtained from the fly ash in concrete than possible.

It was suggested that research should be attempted to achieve a better predictability of the behaviour of fly ash in concrete, paving the way for an optimum utilization under various circumstances. In the light of the above, this paper is an effort to present the various possibilities to achieve high performance cementitious composites, utilizing fly ash. This paper, however, does not address blended cementitious composites which are factory produced and are regulated by established codal provisions for such products.

Moreover, it is possible to achieve high performance through an appropriate utilization of a normal cement, pozzolan (fly ash), an optimal grading of the different aggregates and appropriate water to cementitious materials ratio that will give the required flexibility for realizing a well compacted mass. It is imperative to understand at this stage that the performance requirements could be vastly varying depending on the type of structure (i.e., the size of the members, the reinforcement detailing) and the environment in which it is to perform. It is obvious from this that the strength is not a part of the high performance requirement but is only a structural requirement (i.e., the need for only a fairly low strength but high impermeability for dams). Naturally, such lower strength applications can facilitate the use of significantly high volumes of fly ash without sacrificing the performance requirements of the structure. It is also to be recognized that a well-designed concrete with fairly small quantities of fly ash could achieve strengths far higher than that possible with the cement alone. This obviously shows that the strength of the composite cementitious system is inversely proportional to the amount of pozzolan (fly ash) that can be incorporated into the system. At this stage it is only appropriate to introduce the concept of the cementitious strength efficiency of fly ash in concrete.

Earlier researchers have adopted three different methods for the proportioning of fly ash concrete mixes: (1) a simple replacement, (2) modified replacement and (3) rational methods. In spite of the numerous research efforts however, an exact quantitative understanding of the contribution of fly ash to the strength of the concrete was still elusive. In response, a systematic evaluation of the available results was undertaken. Only the results of recent investigations in which the fly ashes conforming to the minimum characteristics specified by ASTM C 618 were chosen for evaluation.

Efficiency Concepts

The efficiency of fly ash is generally defined in terms of its strength relative to control concrete. The *rational methods* were expected to take into account the characteristics of fly ash, which are known to influence the workability and strength of concrete. Smith [2] was one of the first to propose a factor known as cementing efficiency (k) such that a weight "f" of fly ash would be equivalent to a weight "kf" of cement. The strength and workability of this concrete with fly ash is comparable to that of the normal concrete with a water cement ratio of [w/(c+kf)]. Based on the experimental investigations and the results available, the value of the cementing efficiency factor "k" was assessed to be 0.25. Later investigations [3] have shown that the fly ash efficiency factor was a minimum of 0.3 and this value was also used in the German Concrete Standard DIN 1045 [4]. However, a more recent evaluation of concretes containing different cements and fly ashes [5] has shown this value to be 0.5. It was also observed that the significant differences in the properties of fly ashes used (particularly fineness) influenced the compressive strengths only marginally. It also proposes that the consequent reduction in water cementitious materials ratio (Δw) of fly ash concrete compared to the water cement ratio of the reference concrete to be:

$$\begin{aligned} \Delta w &= (w/c_o) - \{w/(c+f)\} \\ &= \{w/(c+kf)\} - \{w/(c+f)\} \\ &= (w/c)\,[1/\{1 + k(f/c)\} - 1/\{1 + (f/c)\}] \end{aligned} \tag{1}$$

The above formulation clearly shows that this reduction depends not only on the efficiency factor (k), but also on the water cement ratio and more importantly the fly ash content.

Evaluation of Efficiency

The "Δw" concept explained earlier was used in evaluating the efficiency of fly ash. Fig.1 shows the relation between the 28-day compressive strengths of concrete and the water cementitious materials ratio for normal and fly ash replaced concretes. "A" and "B" are the two typical water cementitious material ratios of concretes at the higher and lower percentages of replacement, with that of the control concrete being at "N" for the same strength. The method now tries to bring the [w/(c+f)] ratios nearer to that of the control concrete by applying the cementitious efficiency of fly ash (k). The figure is replotted to check whether a unique value of "k" can help in bringing both "A" and "B" to "N". The value of "k" which is generally applicable for all the replacement percentages is henceforth defined as the general efficiency factor (k_e).

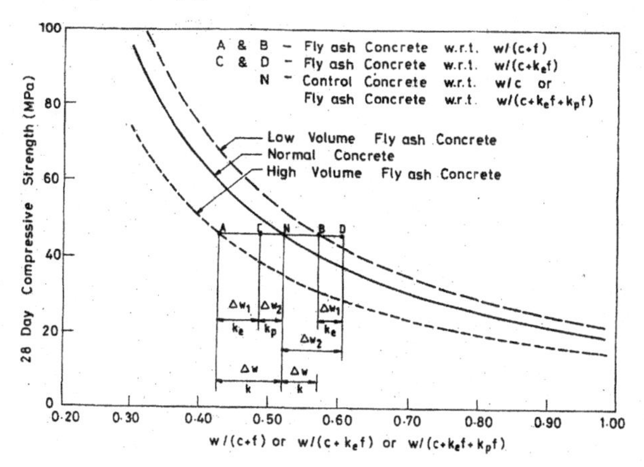

Fig. 1. Conceptual diagram showing the effect of efficiency factors

This means that the points "A" and "B" now shift to their revised locations "C" and "D", by a distance of "Δw_1", with the axis as [$w/(c+k_ef)$]. The revised correction now required (Δw_2) to counteract this effect of the percentage replacement has been evaluated through an additional factor "k_p". These two corrections together will bring the points "A" and "B" to "N" so that the water cement ratio of the control concrete and the water cementitious material ratio [$w/(c+k_ef+k_pf)$] will all be the same for any particular strength.

The results of different investigators from literature for concretes containing ordinary portland cement with fly ash replacements ranging from 0 to 75% were compiled and evaluated earlier [6]. The variation of the 28-day compressive strength with [w/(c+f)] for all the concretes at the different percentages of replacement is presented in Fig.2.

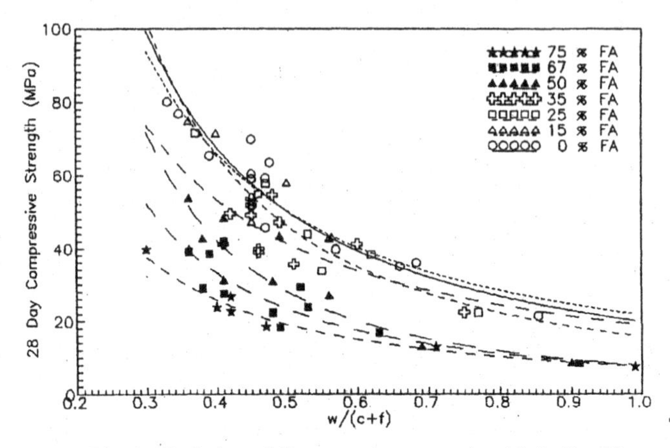

Fig. 2. Variation of Compressive strength with [w/(c+f)]

At a general efficiency factor (k_e) of 0.5 the compressive strengths of fly ash concretes come closest to the control as shown in Fig. 3.

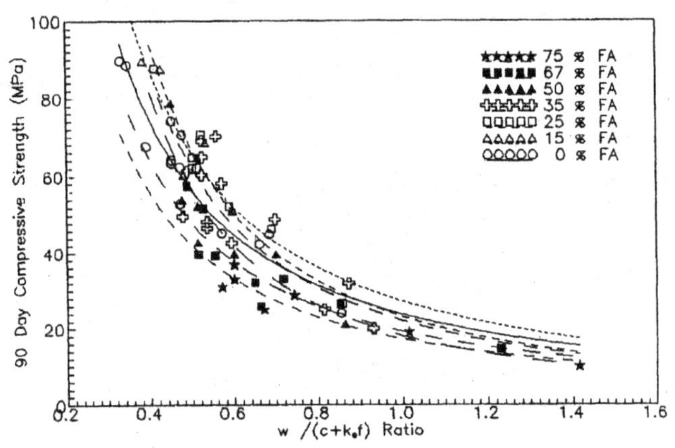

Fig. 3. Variation of strengths after correction for k_e

Now, the differences between the water cement ratio including the effect of general efficiency factor and that of control concrete was computed (Δw_2) for the individual mixes at various water cement ratios and the effect of percentage replacement was calculated through an additional percentage efficiency factor (k_p). Considering the average of the k_p values at different percentages, the variation of compressive strength with $w/(c+k_ef+k_pf)$ is presented in Fig. 4.

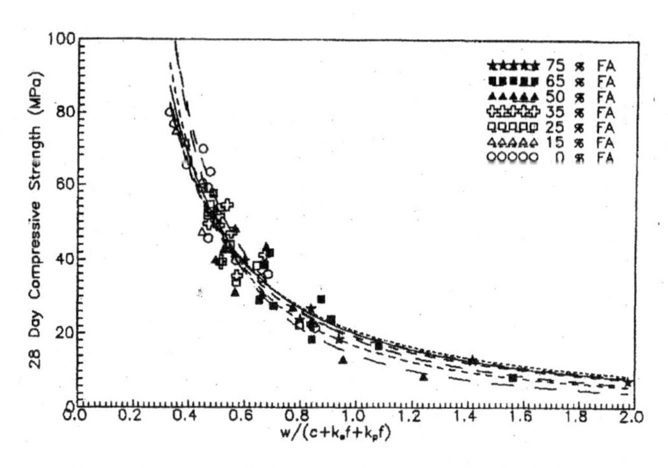

Fig. 4. Variation of strengths after correcting for k_e for k_p

It can be clearly seen that this has resulted in a reasonably close agreement with the control concrete strengths at all the percentages of replacement ranging from 15–75%. It was also observed during these evaluations that some of the specific parameters like curing conditions, addition of admixtures and in certain cases even the type of aggregates and modifications of their proportions have significant effect on the efficiency of fly ash. A detailed presentation of this evaluation was presented earlier [6]. Also, similar evaluation for the k_e and k_p values was also carried out with the water cementitious materials ratio to strength at ages of 7 and 90 days. The details of this part of the assessment were also presented in an earlier paper [7].

The variation of k_p with the percentage replacement and the variation of the overall efficiency factor k ($k = k_e + k_p$) is presented in Fig. 5. This concept of the general and percentage efficiencies for fly ash has also been successfully utilized to establish efficiencies of other pozzolans like silica fume, GGBS, metakaolin, zeolite and rice husk ash. The very high cementitious efficiency of silica fume was clearly established [8] and verified through a systematic experimental investigation.

Mix Design Methodology Using Efficiency Concept

The efficiency approach, being based on the cementing efficiency of the specific pozzolan, will automatically limit the maximum replacement percentage possible. A typical relation between the strength and maximum replacement percentage possible is presented in Fig.6. However, the efficiency approach can easily predict the strengths at any other specific percentage also. This figure also presents an earlier such proposal by Munday et al. [9], which is about 10% lower. Recent experimental results have shown that is possible to replace even higher percentages if one optimizes the aggregate gradings and the filler proportion to minimize the water content needed.

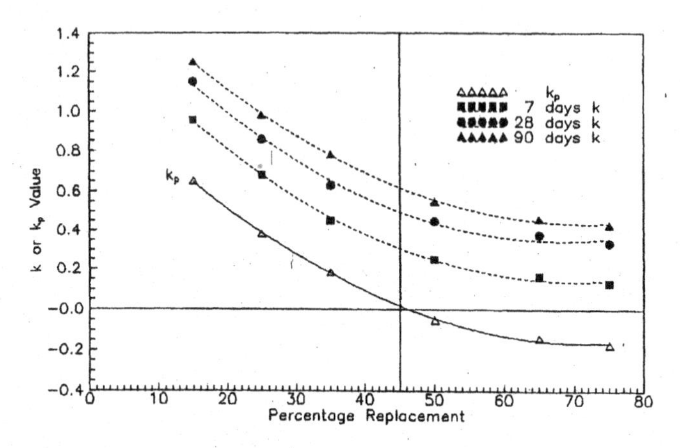

Fig 5. Variation of Efficiency factors k_p and k at different ages

The water cement ratio (w/c_o) is chosen for the strength requirement as per any recognized mix design methodology for normal concrete. The water content (w) required from the workability consideration is also chosen from the same procedure. This basically means that the mix design procedure for fly ash concretes is only a simple, yet significant, extension of the normal concrete mix design. The cement content (c_o) required for normal concrete is also estimated. The percentage replacement level required and the corresponding efficiency factor (k) is taken from Fig. 6.

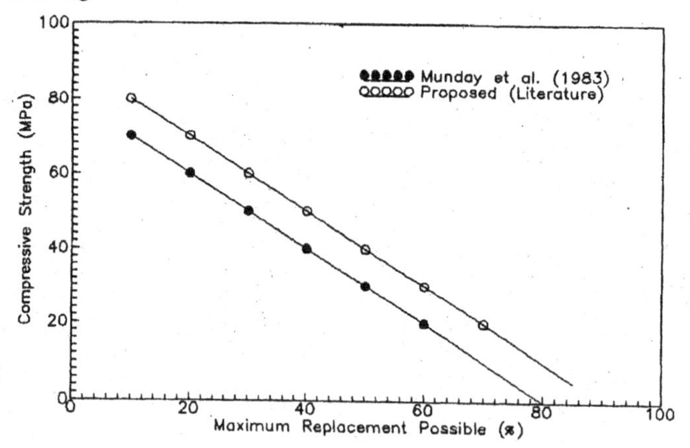

Fig. 6. Maximum possible percentage replacement Vs Strength

The quantities of cement (c) and fly ash (f), to have a cementitious efficiency equivalent of c_o, is calculated. The total aggregate content is then determined by the absolute volume method. The aggregates are then proportioned as in the normal mix design, considering the slump, size and grading of aggregates.

At higher percentages of fly ash replacement, in spite of the absolute volume method accounting for the increased fines in the matrix, may still require modifications in the aggregates. This is to be attempted keeping in view the percentage replacement, to account for the additional fines in the matrix. A good approximation to start with may be to replace an amount of fine aggregate, equivalent to half the quantity of the total cementitious materials in excess of c_o, with the coarse aggregate. This method was found to result in concretes with the desired quality of high performance in limited laboratory experimentation to date [10].

HIGH PERFORMANCE CONCRETE

American Concrete Institute (ACI) defines HPC as "Concrete which meets special performance and uniformity requirements that cannot always be achieved routinely" [11]. Mehta and Aitcin [12] propose that the term HPC should be applied to concrete mixtures possessing high workability, high strength and high durability. HPC was also defined by Zia et al. [13] for transportation applications as one having a water cement ratio below 0.35 along with the correspondingly higher strength of 70 MPa and a minimum durability factor of 80%. However, it can be easily observed that this high performance need not be limited to only the high strength concretes. In general, a "high performance concrete" can be defined as that concrete which has the highest durability for any given strength, of course with the understanding that it should be produced at an economical cost. The improvements in the performance of concrete can be achieved through the following - change in microstructure through pozzolanic additions (particularly at the transition zone) along with reduced voids and micro cracks, increased workability through chemical and mineral admixtures, and more importantly an appropriate grading of the aggregates. It is also important to critically assess the amount of fines through cement, pozzolans and other fillers if need be to ensure the least water content that can achieve the highest workability. Having the composition in order it is now possible to achieve the higher strength, if required, through a modification of the water cementitious materials ratios using workability agents. It is also important to ensure that the concrete produced is appropriately compacted and cured to ensure higher durability. At this stage it may be appropriate to look at the durability aspects critically, to ensure high performance in the different environmental exposure zones.

DURABILITY OF CONCRETE STRUCTURES

The inadequate performance of structures during the past necessitated a better understanding of the parameters influencing durability. Initial efforts to overcome the durability problems through higher strength concretes did not prove to be very effective. This has lead to a reassessment and reformulation of the criteria for ensuring durability by the different authorities and standards related. At this stage, it is essential to note that high performance can only be achieved, even in a particular structure, only if one attempts to consider all the relevant parameters from the conceptual stage of the construction. This obviously means that one should look at several parameters that are highly interrelated and should consider their effects appropriately at all stages. To illustrate the different parameters and their relationships with durability, the authors present various causes, the parameters related, their effects and possible remedial measure, Table I.

Table I. Durability of concrete structures

CAUSES	MATERIALS	CONCRETE MAKING	CONCRETE STRENGTH	LOADING	DETERIORATION OF CONCRETE	CORROSION OF STEEL IN CONCRETE
PARAMETER	Cement Water Aggregates Admixtures	Design Mixing Compaction Curing Temperature	Water content Cement content Admixtures	Early age Later age Over loading	Pore size Micro-cracking Porosity ASR	Type of steel Creep Relaxation Stress Bond
EFFECTS	Pla. settlement Pla. shrinkage Heat of hydration ASR	Density Segregation Bleeding Drying shrinkage	Water cement ratio Porosity Micro-cracking	Micro-cracking Cracking	Erosion, Abrasion Acid attack Sulfate attack Temp. effect	Carbonation Chloride ion ingress Cracking
REMEADIAL MEASURES	Agg.grading Cement content Admixtures	Agg. grading Cooling / Heating Curing comp. Admixtures	Min. cement content Min. strength Admixtures Agg. grading	Age of stripping Age of loading	Cement type Cement content Max. w/c Pozzolans	Admixtures Cover Coatings Inhibitors CP systems

CODAL PROVISIONS

Having understood the various aspects related to durability it is important to understand how the codal provisions, typically the CEB-FIP International code [14], attempt to achieve this. Table II presents a detailed summary of the durability norms as per CEB-FIP and Euro norms. The maximum limitations on the water cement ratio as given by the ACI are also given, which are significantly lower than the Euro norms. This lower water to cement ratio is essentially to ensure discontinuity of pores resulting in impermeable concretes, particularly for seawater environment. Based on these limitations of water to cement ratio and cement content the design strength of concrete in various environments is also given in this table. Recognized the general possibilities and recommendations it is felt that with the available methods i.e., ready mixing, it is now possible to control the aggregate grading, the type of cement and its content (along with the pozzolans), the water cementitious materials ratio and the chemical admixture contents to a high degree of accuracy so as to achieve the required strength as well as durability.

APPLICATIONS

The first major practical application of fly ash in concrete was in the construction of the Hungry Horse Dam (USA) in 1948. Later the use of fly ash up to 30% of cement in concrete became common in construction, mostly in the European countries and USA. Many examples were listed by Dunstan [15] where high volume fly ash concretes of 50 to 80% replacements were used in mass concretes for dam and road constructions. Fly ash has also been used in high strength concrete structures. It was reported that 30% class 'C' fly ash produced 83 MPa strength at 56 days and was used in a 72-story office building [16]. Fly ash has a special place in the production of Roller Compacted Concrete (RCC) and Self Compacting Concrete (SCC), as discussed in the following paragraphs.

Roller Compacted Concrete is a relatively stiff mixture of aggregates, cementitious materials and water that is generally placed by asphalt paving equipment and compacted by vibratory rollers. RCC is placed without forms, finishing, and surface texturing and does not require joints, dowels, reinforcing steel, or form work. Therefore, relatively large quantities of RCC can be placed rapidly with minimal labour and equipment, resulting in speedy construction of dams and pavements. Because of the low water to cementitious materials ratio, RCC mixtures typically exhibit strengths equivalent to, or even greater than, conventional concrete pavements. The use of fly ash as a replacement for cement in RCC increases the amount of fine materials in the mixture. It also decreases the water requirement, improves consistency, and contributes to the strength development due to improved microstructure resulting from the pozzolanic reactions of fly ash. It also decreases the heat of hydration. Literature available shows that the higher replacements (up to 80%) were used in dam constructions but for pavements this replacement was confined to only about 35%. Table III presents a very brief overview of the RCC mixes containing fly ash and the corresponding strengths obtained [17, 18]. It also shows that the use of superplasticizers can improve the situation significantly [18]. The table also contains information regarding some of the mixes that were developed in this laboratory.

Table II. Durability recommendations related to environmental exposure

Env Class	Environmental Conditions	Liquids pH	Liquids CO_2 *mg/l*	Liquids SO_4^{-2} *mg/l*	Soil SO_4^{-2} *mg/l*	Strength *MPa*	Maximum w/c ratio: Normal	Maximum w/c ratio: with $SO_4^{-2\#}$	Min. cement *kg/m³*: Normal	Min. cement *kg/m³*: with $SO_4^{-2\#}$	Max. water pen. ♦ *mm*	Cov *mm*
1	Dry env.	-	-	-	-	≥C16/20 (25)	0.65	-	270	-	-	15
2 A	Humid env., no frost	-	-	-	-	≥C20/25 (30)	0.6	-	300	-	-	30
b	Humid env. + frost	-	-	-	-	≥C20/25 (30)	0.55 (0.45)	- (0.5)	300	-	50	30
3	Humid env. + frost & deicing	-	-	-	-	≥C20/25 (30)	0.55	-	300	-	50	40
4 A	Sea water Env. + no frost	-	-	-	-	≥C25/30 (35)	0.55 (0.4)	- (0.45)	300	-	30	40
B	Sea water Env., + frost	-	-	-	-	≥C25/30 (35)	0.5	-	300	-	30	40
5 A	Slightly aggressive chem.. env.	6.5 - 5.5	15-30	200-600	2000 6000	≥C20/25 (35)	0.55	0.55 0.6@	300	300	50	--
B	Moderately aggressive chem.. env.	5.5 - 4.5	30-60	600-3000	6000-12000	≥C25/30 (40)	0.5	0.50 0.5@	300	330	30	--
C	Highly aggressive chem. env.	4.5 - 4.0	60-100	3000-6000	12000	≥C30/35 (40)	0.45	0.45 0.4@	300	370	50	--

\# Aggressive environments in which sulfate resistant cement is used; ♦ Maximum water penetration.
@ Aggressive environments in which high sulfate resistant cement is used.

It is well recognized that fines play an important role in the behaviour of SCC. It is felt that the powder content (cement + fillers + fines from the aggregate) should be in the range of 500 to 600 kg/m³ for SCC. The vast availability of fly ash at low cost makes it an ideal powder for use in self-compacting concrete [19]. Fly ash can improve packing density, reduce inter-particle friction and viscosity, improve deformability, self-compactabiltiy, and stability, and finally improves the performance of SCC. Use of fly ash reduces the dosage of super plasticizer in SCC needed to obtain similar slump/flow compared to concrete made with portland cement only. The well known advantages of using fly ash in concrete such as improved rheological properties and reduce cracking of concrete due to the reduced heat of hydration of concrete can also be useful to SCC. The impermeability of SCC with fly ash was also found to be remarkable [19]. As in the case of RCC, Table III presents a brief set of SCC mixes along with the compressive strengths achieved in concretes with high volume fly ash replacements [19, 20]. The table also contains a set of mixtures used in this laboratory which confirm to the self-compacting norms.

Table III. Mix proportions of different concrete composites

Type of concrete.	Reference	Cement	w/(c+f)	f/(c+f)	Compressive Strength(MPa)		SP
		kg/m^3		%	28 d	90 d	kg/m^3
Roller compacted concretes	Ragan* [17] (field concretes)	331	0.35	0	43	-	---
		565	0.46	18	36	-	---
		207	0.32	34	48	-	---
		181	0.33	35	34	-	---
	Cheng [18]	210	0.34	39	55	67	2.42
		150	0.33	60	40	46	2.63
	Babu & Amar [21]	175	0.39	50	18	23	---
		138	0.30	70	---	16	---
		224	0.41	30	---	26	---
		85	0.62	50	---	10	---
Self compacting concretes	Bouzouba [19]	336	0.50	0	27	35	0
		247	0.45	40	21	36	1.2
		200	0.40	50	19	35	1.7
		169	0.45	60	16	30	0
	Khurana [20]	300	0.60	0	---	40	3
		240	0.41	45	---	48	5.9
		200	0.36	60	---	47	6.5
	Babu & Dinakar [21]	360	0.34	40	53	65	8.76
		360	0.37	35	52	68	5.61
		360	0.40	30	55	72	3.24
		462	0.29	40	72	93	9.24

DISCUSSIONS

It is clear from the above discussion that fly ash has abundant use as a replacement material from the production of ordinary concretes to the production of concrete composites like RCC and SCC. It is also often said that the characteristics of fly ash and that of the cement will have a significant bearing on the behaviour of fly ash cementitious composites. In particular a lot is talked about classifying the fly ash to ensure the reasonable fineness. However, with the adoption of pulverized fuel and high efficiency burners in the thermal power stations, the resulting fly ash even from the first field of an electrostatic precipitator, which accounts for most of the fly ash produced, was found to be eminently suitable in the laboratory as well as field for almost all the applications listed earlier.

Looking at the possible options for fine materials even to compensate as fillers in a concrete matrix, not even looking for the pozzolanic activity, the possible options are firstly fly ash, GGBS, silica fume, metakaolin, zeolite and rice husk ash. Of these both GGBS and rice husk ash need grinding (an expensive operation adding to cost). Metakaolin and zeolite are naturally available fine powders (but may need to be calcined to be pozzolanic). However, fly ash as well as silica fume (the most efficient but expensive pozzolan available) are generated in the process as fine powders.

In recent times silica fume is being recommended for many applications due to its high reactivity. As already reported the authors have been involved in a very large experimental investigation on the development and performance evaluation of silica fume in concrete and thought it be prudent to write the following few lines about its potential. These discussions are only to project the limitations of silica fume, but certainly there are applications where silica fume could be used to realize certain advantages. While it is advantageous in applications needing high strength and impermeability (probably for under water structures), the effectiveness of silica fume in the long term is highly suspect. The high reactivity of silica fume is associated with a higher heat of hydration, naturally leading to higher shrinkage and temperature effects. Moreover, the self-desiccation effects in the long term have proved detrimental to the behaviour of several structures. In the long term the creep and shrinkage effects on the total structure may also need a specific understanding. Lastly, it is well known and there have been reported failures of silica fume concretes when subjected to fire and high temperature effects. Silica fume when added to concrete, due to its high surface area, will result in a significant increase in the volume of fines, which will result in a dense and impermeable concrete. This concrete when subjected to high temperatures as in nuclear power applications or when exposed to fire will not allow the internal water to come out of the pores. An enormous pressure thus builds up within the concrete resulting in heavy spalling. Such heavy spalling was reported in the case of a fire in a tunnel in literature [22]. These few lines indicate that the use of silica fume for internal lining and for thermal and nuclear power structures is not advisable. An alternative that is always suggested is to use silica fume along with other pozzolans like fly ash and GGBS. However, with the experience the authors have gained in the laboratory it is obvious that the strength requirements can easily be met through the use of fly ash or even GGBS. For higher strength applications probably simpler and more amiable materials like metakaolin or zeolite along with fly ash may be more suitable.

CONCLUSIONS

The paper is an attempt to bring together the various aspects related to an effective utilization of fly ash. In specific the methodology for obtaining concretes of a specific strength at a specific replacement level (below the maximum permissible for that strength) was presented. The methodology of proportioning, compensating for the additional fly ash fines was also discussed. This method appears to result in a high performance concrete at that strength grade for that replacement level. The various parameters that influence performance and the corresponding codal recommendations were also discussed. The paper also highlighted the importance of fly ash as a cementitious component for supplementing the fines required in both RCC and SCC composites. A brief review of the fly ash levels in such mixes was also presented.

REFERENCES

[1]S. Popovics, "What Do We Know About the Contribution of Fly Ash to the Strength of Concrete?" *Proceedings of Fly Ash, Silica Fume, Slag and Natural Pozzolans in Concrete*, ACI, SP-91, **1**, 313-331 (1986)

[2]I.A. Smith, "The Design of Fly Ash Concretes," *Proceedings of the Institution of Civil Engineers*, **36**, London, 769-790 (1967)

[3]K. Wesche, "Fly Ash in Concrete," RILEM to 67 – FAB, State of The Art Report (1990)

[4]----DIN 1045, "Beton und Stahlbeton", Beton Verlag GMBH, Koln, 1994

[5]P. Schiessl and R. Hardtl, "Efficiency of Fly Ash in Concrete – Evaluation of Ibac Test Results," *Technical Report of Institute fur Bauforschung*, RWTH, Aachen, 1-31 (1991)

[6]K. Ganesh Babu and G. Siva Nageswara Rao, "Efficiency of Fly Ash in Concrete," *Journal of Cement and Concrete Composites*, **15**, 223-229 (1993)

[7]K. Ganesh Babu and G. Siva Nageswara Rao, "Efficiency of Fly Ash in Concrete with Age," *Journal of Cement and Concrete Research*, **26** [3], 465-474 (1996)

[8]K. Ganesh Babu and P.V. Surya Prakash, "Efficiency of Silica Fume in Concrete," *Journal of Cement and Concrete Research*, **25** [6], 1273-1283 (1995)

[9] J.G.L. Munday, L.T. Ong and R. K. Dhir, "Mix Proportioning of Concrete with PFA: Critical Review," *Proceedings of Fly Ash, Silica Fume, Slag and Other Mineral by Products in Concrete,* ACI, SP-79, **1**, 267-288 (1993)
[10] G. Siva Nageshwara Rao, "Effective Utilisation of Fly Ash in Concrete for Aggressive Environment,"*PhD Thesis submitted to IIT Madras*, May (1996)
[11] H.G. Russel, "ACI defines High-Performance Concrete," *Journal of Concrete International*, **21** [2], 56-57 (1999)
[12] P. K. Mehta and P.C. Aitcin, "Principles Underlying the Production of High-Performance Concrete," *Journal of Cement, Concrete and Aggregates*, **12** [2], 70 – 78 (1990)
[13] P. Zia, M.L. Leming and S.H. Ahmad, "High Performance Concretes: A State of the Art Report," *SHRP–C/FR-91-103,* North Carolina State University (1991)
[14] ---CEB-FIP Model Code, "Comite Euro-International du Beton," (1994)
[15] M.R.H.Dunstan, "Fly Ash as the Fourth Ingredient in Concrete Mixtures," *Proceedings of Fly ash, Silica fume, slag and Natural Pozzolans in Concrete,* ACI SP-91, **1**, 171-200 (1986)
[16] J. Cook, "Research and Application of High – Strength Concrete using Class C Fly Ash," *Concrete International*, **4** [7], 72-80 (1982)
[17] S.A. Ragan, "Evaluation of the Frost Resistance of Roller Compacted Concrete Pavements," *Transportation Research Records*, **1062**, 25-32
[18] C. Cheng, W. Sun and H. Qin, "The Analysis on Strength and Fly Ash Effect of Roller Compacted Concrete with High Volume Fly Ash," *Journal of Cement and Concrete Research*, **30,** 71-75 (2000)
[19] N. Bouzoubaa, and M. Lachemi, "Self Compacting Concrete Incorporating High Volumes of Class F Fly ash, Preliminary Results," *Journal of Cement and Concrete Research*, **31** [3],.413-420 (2001)
[20] R. Khurana, and R. Saccone, "Fly Ash in Self-Compacting Concrete," *Proceedings of Fly Ash, Silica Fume, Slag and Natural Pozzolans in Concrete,* ACI SP-199, 259-274 (2001)
[21] K. Ganesh Babu, Y. Amarnath and P. Dinakar, "High Volume Fly Ash Concrete Composites for Pavement Applications," *Proceedings of National Conference on Modern Cement Concrete and Bituminous Roads,* December, Visakhapatnam, 7-13 (2003)
[22] L. T. Phan, "Fire Performance of High Strength Concrete: A Report of the State-of-The-Art," *Building and Fire Research Laboratory, National Institute of Standards and Technology* (1996)

MULTI - COMPONENT CEMENTITIOUS SYSTEMS AND THEIR INFLUENCE ON DURABILITY OF CONCRETE

N. Bhanumathidas
Institute for Solid Waste Research & Ecological balance (INSWAREB)
35, Shri Venkateswara Colony
Visakhapatnam 530 012 (India)

N. Kalidas
Institute for Solid Waste Research & Ecological balance (INSWAREB)
35, Shri Venkateswara Colony
Visakhapatnam 530 012 (India)

G.A.B.Suresh
Ramco Research & Development Centre (RRDC)
11A, Okkiyam, Thoraipakkam
Chennai 600 096 (India)

ABSTRACT

When confidence in Portland cement concrete as the choicest material is at stake considering various durability issues, the need for combining modern practices with sound principles of material science is clear. This orientation led to the development of multi-component cementitious system (MCCS) such as binary, ternary and quaternary cement blends, as a supplement to complementary cementitious materials (CCM) such as fly ash, blast furnace slag, metakaolin, silica fume and rice husk ash.

The issues identified to cause early distress of concrete, when high-grade cements are put to use, could be addressed through the use of CCM. Their key contributions are to impart impermeability to the concrete at early ages with refined microstructure, thus making the mature concrete resistant to chemical attacks and corrosion.

Non-conventional MCCS called FaL-G that contains fly ash as high as 65 to 70%, with the remaining being Ordinary Portland Cement (OPC) and/or hydrated lime and gypsum were found to offer a cementitious media to manufacture cost-effective and durable concrete. By choosing an innovative approach for concrete preparation, FaL-G concrete achieved workability and early impermeability, which otherwise demand the use of chemical admixtures and superpozzolans respectively.

Various comparative engineering and durability characteristics of MCCS concretes are discussed, along with some of the intricate issues of pozzolanic chemistry to substantiate the late-age performance of MCCS.

INTRODUCTION

Do we really have a single-component and stand-alone system in cement history to serve the multiple needs of the construction industry? The answer would be literally no. Lime binders had a limited application in the absence of pozzolanic materials, thus leading to the development of lime-pozzolan binders. The latter demanded addition of gypsum for early strength gain as well as to achieve sulfate resistance for maritime constructions. Thus the application of binary and ternary blends has millennia-old history, which, in fact, laid the platform for the concept of durability. The contribution of these blends to the ancient civilization of many countries, more so in Greek and Roman architecture, has been well established and the legacy of these old structures is the proof for upholding the torch of durability.

Coming to the 19th century, the time of invention of portland cement, it was initially observed that ground portland cement clinker could not yield a workable paste due to quick hydration of calcium aluminate phases. Hence to overcome this deficiency, set regulators such as gypsum was ground along with clinker thus making the product commercially useful. This means complementarity is an in-built mechanism in the very nature of building materials. While blending seems to be an unavoidable phenomenon in cement science, adoption of binary and other MCCS in the advanced era of concrete technology is going to be an irrevocable technical need, something like reinventing the wheel.

NEED FOR BLENDED MATERIALS:

During the transition from pozzolan-lime-gypsum binders to Portland cement the durability was also taken for granted. The need for construction speed brought in a transformation in the chemistry of clinkerisation and, in turn, the rheology and hydration chemistry of cement. While these changes proved as a boon to the fast growing construction industry, the same changes had given way to the adverse impact of strength to durability, thus forcing the concrete industry to identify new approaches for durability enhancement.

At this juncture, the cement and concrete technologists reviewed the material science of yesteryears to find the reasons for the sound health of those structures vis-à-vis that of the high-grade cement concrete, preparation and applicational practices. In the process it was realized that clinker chemistry was manoeuvred to meet irresistible market needs and thus can no more offer materials to render characteristic soundness to the concrete. Mehta (1) discussed at length various issues in this direction. In this backdrop the need for combining the modern practices with sound principles of material science was found imminent that led to the development of binary blended cement systems. However, to meet the growing demand for high performance concrete with sustainable performance the need for multi-functional concretes has possibly enlarged the scope further to MCCS.

CHEMISTRY OF CEMENT METABOLISM:

Hoover Dam, built in 1936 with low-grade cement concrete, is in sound health with a projected life of over 1000 years. But many structures built with high-grade OPC concrete have given way to distress within a decade or two. This problem was overcome with the development of blended cements. It is necessary to understand the cement chemistry before comprehending the role of CCM in association with OPC. With reference to the release of heat and lime, cement chemistry has close analogy to metabolism in life system that consists of two phases (2).

In the first phase, called anabolism or constructive metabolism, the cells combine molecules to assemble new organic matters. This may be identified in cement chemistry as a form of associative metabolism during which the compounds of raw materials associate into anhydrous mineralogy of clinker phases, *entrapping* endothermic heat.

In the second phase called catabolism the cells break down molecules to obtain energy for growth and release heat. In cement this phase may be identified as dissociative metabolism. Water or moisture triggers reactions with anhydrous mineralogy of cement, by which heat is liberated and, simultaneously, compounds are dissociated from the anhydrous mineralogy. The presence of heat, hydrated lime and water helps the compounds to associate into hydrated mineralogy that contributes to strength development in mortar and concrete.

The cells of organisms and particles of cement have close parallels in their functions. A cell is intensely active and carries out life's functions including growth and reproduction. The cell is the source of heat; wherein mitochondria burn the fuel eaten by the organism. Similarly the cement particle is the heat source. The difference is that the intake of food by organisms from time to time works as fuel for metabolism and heat generation on a day to day basis, whereas, in cement the heat is entrapped in anhydrous phases at their formation and are released progressively over ages. In either case the heat results in growth and strength.

In addition to heat, hydrated lime of cement draws parallel with sugar in life system. Both lime and sugar are basic needs and indispensable to their respective systems. When both of them are not engaged in catabolic reactions, they prove to be surplus and cause deleterious effects to their respective systems.

In this background, it is interesting to know about the heat bank and lime bank of cement hydration system, conceptualized by the authors.

The hydraulic cements are associated with two performing features: 1) Heat bank and, 2) Lime Bank. In low-grade cements both heat and lime are available for hydration on a progressive and sustainable basis for healthy performance. In high-grade cements the loci of these very performing-features are disturbed, leading to durability problems.

Heat Bank

Cementitious materials such as OPC, ground granulated blast furnace slag (GGBS) and, for that matter, CCM's such as fly ash, rice husk ash (RHA), silica fume and metakaolin do absorb some heat during the formation of their anhydrous phases as indicated in Table I in terms of heat of solution.

Table I: Heats of solution for different cement materials (studied on typical samples)

	OPC	GGBS	Fly ash	Silica fume	RHA	Metakaolin
Heat Content: (kcal/kg)	612	616	358	562	546	684

This heat, identified as endothermic heat during formation, manifests as exothermic heat during hydration. However, this heat does not get released instantaneously but spans years through progressive formation of hydrated mineralogy by reaction of lime with other compounds such as silica and alumina. In other words, the anhydrous constituents work as a heat bank to release the heat progressively over ages for the formation of lime associated hydrated mineralogy. Thus, heat is one of the essential tools in the cement chemistry, if it is maintained within the threshold levels. Excess heat of hydration of cement causes thermal stresses and microcracks in concrete, leading to a durability crisis. In a way, controlled heat of hydration spells upon the formation of hydrated mineralogy, crystal growth, densification of the matrix and impermeability. Even lime mortar is not an exception to this phenomenon. Sand silica does contain about 60 kcal/kg of heat that gets released very slowly with the reaction of lime. It is the presence of this heat in lime-mortar that results in very slow formation of hydrated mineralogy, where, indeed, the external heat is also availed from the environment.

The pozzolan emergent out of volcanoes also contains a certain quantity of endothermic heat that renders a relatively higher rate of formation of hydrated mineralogy, compared to lime mortars, when associated with lime and moisture. This observation resulted in the development of lime-pozzolan concretes used by Roman architects that proved sustainable even after millennia of age.

With regard to blended cements, the heat values of CCM complement the heat bank of OPC by virtue of very slow and progressive hydration over years and decades.

In a nutshell the following aspects can be summed up:

- The quantum of heat is not the matter but rather the sustenance of that heat liberation (exothermic heat) over prolonged ageing matters for sound performance of cement concrete through microstructure refinement.

- The rate of heat liberation depends on the metabolic reactions of cement chemistry. Despite containing parallel quantities of heat, two cement materials liberate that heat at different rates over different ages. This is evident by the performance of OPC and GGBS, despite containing parallel heat content as shown in Table I.

- Upon scanning the durability issues of ancient pozzolanic and lime concretes with those of modern OPC, it is observed that the slower the rate of hydration-heat, the higher the densification of microstructure; all resulting in impermeability and durability.

Lime Bank

Lime is another principal constituent for the progress of cement chemistry. Lime is a common constituent in age-old lime-pozzolan binders or lime mortars and modern-aged OPC. The saturated lime solution, released upon cement hydration, incites reactions with anhydrous constituents whereby the mineralogical phases are dissociated along with release of heat. The presence of heat accelerates this phenomenon of the formation of hydrated mineralogy through re-association of compounds.

In age-old cements with coarse particle size and higher C_2S content, there was synergy between the release of lime and heat, commensurate to the progress of hydration chemistry, without wasting either of them. In the process, the matrix continuously densified over several years rendering a relatively water impervious mass.

In contrast, in modern concretes with high-grade OPC, the loci of lime and heat are disturbed, inviting numerous woes to the concrete in the durability point of view.

HYDRATION CHEMISTRY OF PORTLAND CEMENT AND BLENDS:

OPC consists of four major mineralogical phases in which calcium silicates are about 75%. In addition to this, the hydration of calcium silicates leaves considerable quantities of surplus hydrated lime as shown below.

$2C_3S$ +	$6H$	$\longrightarrow$	$C_3S_2H_3$	+	$3\ CH$	
456g	108g		342g		222g	(1)
			(61%)		*(39%)*	
$2C_2S$ +	$4H$	$\longrightarrow$	$C_3S_2H_3$	+	CH	
344g	72g		342g		74g	(2)
			(82%)		*(18%)*	

It is evident from reactions (1) and (2) that C_3S hydration contributes a relatively low quantum of strength rendering hydrated mineralogy and high release of $Ca(OH)_2$, whereas the C_2S contribution is high in strength rendering hydrated mineralogy and low in $Ca(OH)_2$. Thus, from the durability point of view, it is essential to have a low C_3S and high C_2S content. However as clinker chemistry cannot be reverted back, the other solution is the addition of CCM that absorb the surplus lime as per the following reaction:

$$OPC + H \xrightarrow{fast} \text{Hydrated primary mineralogy} + CH$$

$$CCM + CH + H \xrightarrow{slow} \text{Hydrated secondary mineralogy} \quad (3)$$

Reactions attain technical significance from three main features:

The reaction is slow and rate of heat liberation is accordingly slow	:	This reduces micro-cracking and improves soundness.
The reaction is lime consuming	:	This leaves little chance for chemical attacks, which otherwise cause deterioration of concrete.
The reaction results in the formation of hydrated secondary mineralogical phases	:	This contributes for the mechanism of pore and grain size refinement resulting in enhanced strength, impermeability and strong transition zone.

TYPES AND MECHANISM OF BLENDS

It is generally agreed that the glass or non-crystalline constituent of CCM's react with $Ca(OH)_2$ or $Ca(OH)_2$ released from hydration of portland cement. Though the hydration chemistry of various CCM's is similar, the rate of hydration varies depending on the characteristic reactivity of each CCM, in addition to other factors such as fineness, particle shape, particle size or particle size distribution within same CCM. Superpozzolanic materials like silica fume and rice husk ash tend to interact with hydrated lime simultaneous to the hydration of OPC, thereby contributing for improvement in early strength. This is possible because of two factors as follows:

1) both these CCM's consist of highly reactive silica available in ready-to-react form, and

2) silica fume has very fine particle at 0.1 - 0.3 μ and RHA has a particle with mean diameter at 5 to7 μ and high surface area at about 38 to 40 m^2/gm.

However, in the case of slag and fly ash, the glass has to get dissolved first, upon which the independent constituents are available for reaction. Slag consists of calcium aluminosilicate glass and ASTM Class F fly ash consists of aluminosilicate glass. However ASTM Class C fly ash consists of glass, that is closer to that of slag but which varies depending on the CaO content of the fly ash. As glass dissolution is the first mechanism in these CCM's their rate of reactions is slow. It has been generally observed that slag takes more than three days and fly ash 7 to 14 days to interact with $Ca(OH)_2$ (3). Hence their contribution to strength is also slower when compared to the superpozzolanic materials. By increasing the fineness the rate of slag reaction is increased. In the case of fly ash, increasing the fineness by grinding does not influence the reactivity to favourable levels. For this purpose, the correct grade of fly ash with high surface area has to be selected for its contribution to early-age reactions. Such fly ash also contributes for improvement in workability of concrete without increasing the water to cementitious material ratio.

The glassy spheroids and smooth surface of fly ash particles have been attributed for the reduction in water for the reasons that they do not absorb water on one hand and contribute for easy mobility of water. Helmuth explains that very fine particles of fly ash, when adsorbed on the oppositely charged surface of cement particles, prevent cement from flocculation (4). In the process, cement particles effectively get dispersed without entrapping large amounts of water. As per him, under these circumstances, the system exhibits an increased water flow. This mechanism is comparable to that of chemical admixtures. The authors studied improvement in characteristics, such as slump, strength and chloride impermeability of fly ash blends, as reported in Table II.

Table II: Influence of fly ash characteristics on concrete characteristics

		Strength MPa				Cl^- Permeability			
OPC: Fa	slump mm	28	90	180	1yr	28d	90d	180d	1 yr
100: ---	14	43	47	51	54	3852	2451	2251	1912
65:35 c	25	29	41	44	52	2361	503	303	253
65:35 f	80	35	54	58	68	2029	313	166	155
55:45 c	35	25	35	42	43	2378	333	195	179
55:45 f	140	29	51	54	60	2058	230	92	64

Cementitious content : 300 kg/m³

c. coarse composite fly ash - Blaine 280 m²/kg < 10 μ - 21%

f. fine fly ash from same source - Blaine 630 m²/kg < 10 μ - 63%

It is noteworthy that the above behaviour is achievable by addition of fly ash alone. As illustrated in Fig.1, this holistic performance of fly ash, as a single component, renders multiple benefits.

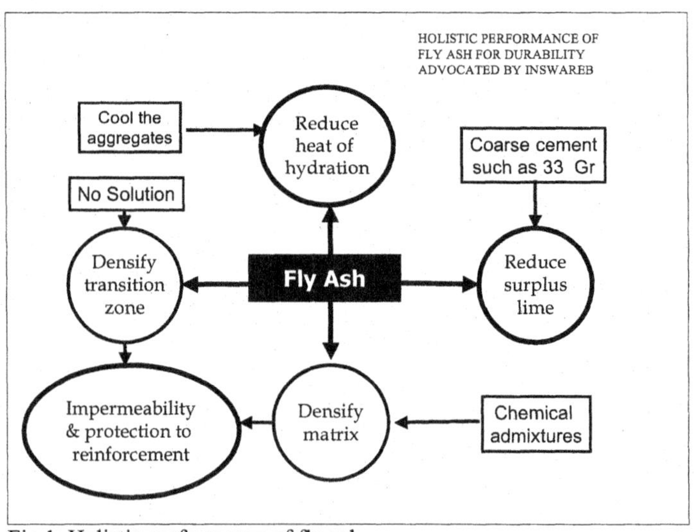

Fig 1: Holistic performance of fly ash

Slag generally does not exhibit such phenomenon because it is a ground material, thus the particles generally have sharp and non-smooth edges. The superfine nature and high surface area are the limiting factors in the case of silica fume and RHA.

The superpozzolans react very fast with $Ca(OH)_2$ released from OPC hydration thereby increasing the content of C-S-H. The total volume of capillary pores in such systems is eventually reduced. The transition zone in silica fume blends is reported to get densified within 48 hours of hydration (3). With these phenomena superpozzolanic materials improve the microstructure and impermeability of concrete at early ages, which, in the case of pozzolanic materials such as fly ash and slag, improve sustainably through late age performance.

Let us conceive a MCCS with different prescriptive inputs based on performance criteria:

- OPC needs to have added CCM's to reduce heat of hydration and utilisation of hydrated lime.
- Fly ash is selected for its specific role as a lime-consuming CCM.
- Hence slag is added as lime-containing CCM, nevertheless with slow release of heat.
- The pozzolanic and slag chemistry being relatively slower, fast early reactions were sought. Hence silica fume or rice husk ash is preferred as superpozzolan.
- To provide an early source of lime, a small portion of $Ca(OH)_2$ is added.
- To alleviate the weakness of calcium aluminate chemistry, gypsum is added by which the former is converted to ettringite as the strength rendering mineralogy.

Upon a review of such issues, it is evident that each CCM has its own advantages and disadvantages. Development of MCCS has been identified as an appropriate approach that helps in tapping the potential of each CCM and at the same time is compensated of its deficiencies by other CCM's. Uchikawa (5) discussed in detail various issues of CCM for up to ternary blends and suggested their effective use in the construction industry in view of their benefits. Consequently, in many high profile constructions and those under aggressive environments deployment of concretes with ternary blends such as OPC with slag + fly ash, fly ash + silica fume, slag + silica fume is observed.

In view of difference in the reactivities and corresponding constituents, the C-S-H of cement paste varies depending on the type of materials used. As calculated by the authors, in slag cement paste the molar ratio of CaO to SiO_2 varies from 1.9 to 1.6, in fly ash blends from 1.7 to 1.0 and in silica fume and RHA blends 1.5 to 0.8 depending on the amount added and condition of hydration. As the CaO to SiO_2 molar ratio decreases, the surface area of such C-S-H increases. With increasing surface area the densification of the matrix improves. With low molar ratio of CaO to SiO_2 the C-S-H becomes more resistant to various chemical attacks. In one of the works of the authors, the comparative characteristics of concretes with different cements under various environments were studied as given in Table III.

Table III: Comparative strengths, MPa, of M 40 concrete with different cements under various environment

Cement	56-day				120-day			
	NC	sulphate	acid	sea water	NC	sulphate	acid	sea water
OPC	68.3	67.7	60.0	56.0	69.1	53.4	58.2	55.8
PPC	65.9	66.5	59.2	61.0	73.1	71.4	65.1	64.8
PSC	62.5	62.7	50.1	59.5	66.8	62.5	63.0	59.2

NC: Normal curing
Base clinker is same for all the above cements

In MCCS the molar ratio of CaO to SiO_2 further comes down due to sharing of released $Ca(OH)_2$ by both superpozzolanic and pozzolanic materials that eventually may lead to superior behaviour of former over that of only binary blends. It is reported in some of MCCS with fly ash, slag and silica fume that the $Ca(OH)_2$ layer around the aggregate disappeared, despite high w/cm, and dense formation of C-S-H appeared.

FaL-G: NON-CONVENTIONAL MCCS

Fal-G's are ternary and quaternary blends containing high volumes of fly ash. Technically christened FaL-G to distinguish from conventional technologies, these cements may be a breakthrough in the fly ash utilisation. FaL-G is the product name of a cementitious binder composed of fly ash (Fa), lime (L) and gypsum (G), that sets, hardens and forms water impervious matrix with hydrated mineralogical phases akin to those in hydrated portland cement. In order to make the technology more flexible, lime is made available either in the form of $Ca(OH)_2$ directly or from the hydration of OPC or both depending on the requirement. The FaL-G technology is based on two principles:

1) the fly ash-lime pozzolanic reaction does not need external heat under tropical temperature condition, and
2) the rheology and strength of fly ash-lime mixtures can be greatly enhanced with the addition of gypsum.

FaL-G technology was developed in India, with an innovative composition, wherein the formation of calcium sulphoaluminate hydrates is attained for early strength development that is achieved through addition of gypsum. In fly ash, the aluminate exists as a dormant phase in association with silica as glass/ amorphous and needs SO_3 ion to dissolve the same and make alumina available for reactions. Hence gypsum is added to accelerate the aluminate reaction. Thus gypsum has as much significant role in FaL-G technology as it has in Portland cement. The major difference is that in FaL-G it facilitates acceleration of aluminate hydration whereas in Portland cement it facilitates retardation or set control of aluminate hydration.

FaL-G is used as a cementitious input both in the production of bricks and concrete. In fly ash brick production FaL-G dispenses the need for heavy-duty press and autoclave equipment, and also makes the process energy-efficient, thus bringing down the capital and production costs. More than 1200 FaL-G brick/block production centers, operating successfully through out India, are the testimony for the acceptance of the technology as well as the product.

The soundness of structures constructed in 1991-92 using reinforced FaL-G concrete has enthused the government to construct check dams and water tanks deploying the same technology. So far about 40 water tanks of different capacities have been constructed, the latest one being with a capacity of 0.25 million liters. Subsequently about 3,000 houses are designed as framed structures with pile foundation keeping in view the black cotton nature of the soil. M20 is the selected grade of concrete for these constructions.

The FaL-G used in various structures as above is a quaternary blend, comprised of 60 to 70% fly ash, 20 to 30% Portland cement and, 10% hydrated lime and gypsum.

The slumps attained for concrete are in the range of 25 to 30 mm. The advantage with FaL-G concrete lies in its yielding good flow when subjected to vibration, despite low slump. This peculiar phenomenon facilitates better compaction and eliminates chances for voids and honeycombing. The super fines of concrete help provide better finish of the structural element provided proper formwork is used. As a quality control measure, specimens are cast everyday using the field concrete and tested at corresponding ages of curing. The range of compressive strengths (MPa) and chloride permeability of concrete specimens, as studied by the authors is given in Table IV.

Table IV: Properties of FaL-G concrete

Age	Compressive strength MPa	Chloride permeability Coulombs
3-day	4.5 to 5.5	-- --
7-day	13.0 to 15.0	-- --
28-day	18.0 to 25.0	1000 -1200
90-day	30.0 to 35.0	400 - 600

The ultrasonic tests conducted on the structures mentioned above, resulted in pulse velocity of 4.0 to 4.5 km/sec, providing an indication of densification, which is also substantiated by low chloride permeability at 400-600 Coulombs for the same concrete. It may be observed that this velocity does not match to the strength of the concrete if it is in terms of OPC. This is where delinking soundness and strength proves the point that has been substantiated with data on chloride permeability.

CONCLUSIONS

With the shift in attention to durability, cements with controlled heat of hydration and moderate levels of lime release become important. When these two aspects are inconsistent with modern high-grade cement chemistry, they can be controlled in association with CCM's. When CCM's are available with independent characteristic contributions towards combined holistic performance, their potential is realized. This culminates in successful implementation of MCCS's.

REFERENCES

[1]P. K. Mehta, "Durability of Concrete - Fifty Years of Progress?" Proceedings of ACI Special Publication, SP-126 pp. 1-31 (1991) ed. V.M. Malhotra

[2]N. Bhanumathidas and N. Kalidas, "Metabolism of Cement Chemistry." Indian Concrete Journal **77** [9] 1304-06 (2003)

[3]P. K. Mehta, "Pozzolanic and Cementitious Byproducts as Mineral Admixtures for

Concrete-A Critical Review," Proceedings of ACI Special Publication SP-79 pp. 1-46 (1983) Ed. V.M.Malhotra

[4]R. Helmuth, "Water Reducing Properties of Fly Ash in Cement paste, Mortars and Concretes: Causes and Test Methods," Proceedings of ACI Special Publications, SP-91 pp 723-740 (1986) ed. V.M. Malhotra

[5]H. Uchikawa and T. Okamura, "Binary and Ternary Components Blended Cement," Mineral Admixtures in Cement and Concrete, Vol 4 pp 1-83 (1993) ed. Shondeep, L. Sarkar and S.N. Ghosh

NEW GENERATION ADMIXTURES FOR ENHANCED PERFORMANCE OF CONCRETE

R. Shridhar
General Manager (South & East) & Head of Marketing
Fosroc Chemicals (India) Pvt Limited
4th Floor, Shankar House
RMV Extension, Mehkri Circle
Bangalore 560 080

ABSTRACT

Until the recent past, the strength of concrete was the principal criterion for the performance of concrete structures. Structures that are exposed to an aggressive environment reveal that high strength of concrete alone cannot guarantee long term performance. Due to the high cost of repair or replacement, concrete structures are required to have service lives of more than a hundred years as against 50 to 75 years of normal service life. The change in the needs of the civil engineering industry has led to the development of high performance concrete.

The author briefly describes the properties and composition of enhanced performance with special reference to self-compacting concrete and polycarboxylate admixture.

INTRODUCTION

Concrete technology has made tremendous strides in the past decade. Concrete is now no longer a material consisting of cement, aggregates, water and admixtures but it is an engineered material with several new constituents. The concrete today can take care of specific requirements under most different exposure conditions. The concrete today is tailor made for specific applications and it contains several different materials like fly ash, ground granulated blast furnace slag, microsilica, metakaolin and several other binders, fillers and pozzolanic materials. The development of specifying the concrete as per performance requirements rather than its constituents and ingredients has opened innumerable opportunities for producers and users of concrete to design concrete as per specific requirements. Self-compacting concrete is one such development in recent past.

There are several definitions of HPC given by different individuals, institutions and organizations. HPC should have at least one out standing property, namely compressive strength, self-compatibility, increased resistance to chemical or mechanical stresses, very low permeability, high durability etc. Depending upon the usage, the concrete should have different properties to suit different project requirements. HPC usually has low water - cement ratio, high workability requirements and in many cases both. The other properties for the sake of placement would be lower loss of slump with respect to time coupled with high early strength. These contradictory properties were difficult to achieve with the earlier generation admixtures. Therefore, the civil engineering community has welcomed the technology of new generation water reducing admixtures. It is possible to obtain higher workable concrete at very low water-cement ratio with this family of new generation admixtures.

SELF-COMPACTING CONCRETE (SCC)

SCC is another dimension of HPC, which can compact itself by its own weight – without any external vibration and compaction. SCC can compact automatically while leveling itself. Hence, SCC should have very high flowability with out bleeding or segregation. SCC should be able to flow in the most complicated formwork, such as those for decorative elements in the precast industry and therefore should not block when congested reinforcement is encountered. Fig .1 shows the flow of SCC.

Fig.1. Flow of self-compacting concrete.

Materials

Aggregates: SCC can be made from most normal concreting aggregates, although grading envelopes may be tighter than those in any international codes. SCC has been produced successfully with coarse aggregate sizes up to 40 mm. The maximum size depends on reinforcement lay out and formwork dimensions in the same way as traditional vibrated concrete. Sand can be finer than normal, as the material smaller than 150 μ m may help increase cohesion, thereby resisting segregation.

Cement and Filler (fines): Cement and in many cases, fillers are required in larger proportions than in traditional concrete for cohesion and stability. These fines can be fine sand or fillers derived from crushed rock or active materials such as ggbfs and pfa. Portland cement is often partially replaced with ggbfs or pfa for technical and economical reasons. Fillers must be assessed for their effect on water demand.

Admixtures: Admixtures are essential in determining flow characteristics and workability retention. Ideally, they should also modify the viscosity to increase cohesion. Newly developed types of High-range water-reducing admixtures, known as Polycarboxylate Ethers (PCEs), are particularly relevant to SCC. They reconcile the apparently conflicting requirement of flow and cohesion, avoiding potential problems with unwanted retardation and excessive air entrainment (generally caused by over dosing), particularly at higher workabilities if the mix design is correct. Additional viscosity modification may not be required. Other elements may indicate a requirement for additional segregation control admixtures, such as ultra-fine silicas, polysaccharide gums, modified polyethers or even simple air -entrainers.

New Generation High-Range Water–Reducing Admixtures Based on Polycarboxylate Ether

With the advent of designer molecules in the field of concrete admixtures, a whole new set of admixtures came into existence. New molecular architecture changed the action and performance of high-range water-reducing admixtures greatly. Polycarboxylates (PC) is one such class of new generation high-range water-reducing admixtures that are tailor – made to specific needs of concrete industry. These new molecules offer immense possibilities of designing new mixes that were hitherto not achievable with the conventional high-range water-reducing admixtures.

Polycarboxylate Ether based products are composed of water dispersible comb-polymers with a more complex molecular structure than conventional high-range water-reducing admixtures. The two most important mechanisms attributed to dispersion in polycarboxylate systems are linked to polymer adsorption and steric hindrance caused by the thickness of the adsorbed polymer layer on to the cement particle and electrostatic repulsion through induced electrical charge.

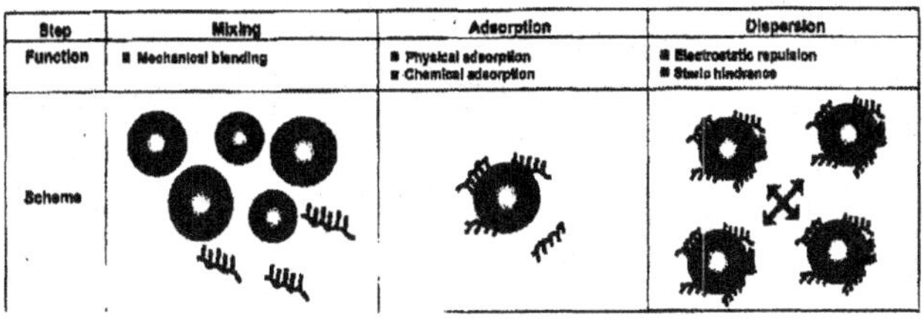

Step	Mixing	Adsorption	Dispersion
Function	■ Mechanical blending	■ Physical adsorption ■ Chemical adsorption	■ Electrostatic repulsion ■ Steric hindrance
Scheme			

Fig.2. Mechanism of dispersion PCE-admixture

Advantages of Polycarboxylate Admixtures

- Significant reduction of the water demand of the mix
- Little loss of consistency
- Short setting time
- High early strength
- Low tendency to segregation
- Self Compaction

Mix Design Consideration: The objective of traditional concrete mix design is to provide the most cost effective selection and proportioning of the materials, while achieving the concrete performance specification. They are based on the inclusion of sufficient cement to achieve strength and durability, and sand content that provides adequate cohesion at the required workability, permitting material variability and minor batching errors.

This approach is unsuitable for SCC, where the main objective is a mix that will flow under its own weight with out blocking or segregation, while satisfying the performance specification when hardening.

To design concrete with these new properties, several principles, some of which are unconventional in relation to current code based mix design terms, should be observed. General criteria are specified in Table 1.

Table I. Design Principles

	Principles	Practical guidelines
Coarse aggregate	Must be limited to reduce inter-particle friction and prevent blocking	Approximately 700-800 kg/m^3
Sand	High sand content is required as it contributes to a part of the total content of fines	Usually > 50% of total aggregate content.
Cement and cement replacement	High binder content, however effect on water demand should be assessed	Typically around 500 kg/m^3
water	High water content will result in segregation and bleeding therefore should be controlled using Viscosity enhancing admixture	Free water < 200 litres/m^3
Water/ (cement + filler) ratio		Less than 0.42 (based on the concrete grade)
Admixtures	Dosage must be determined by trials using materials, mix design and conditions experienced in use.	Trial mixes and manufacturer's advice are essential.
Paste and mortar		Paste >40% by volume of mix. Sand <50% by volume of mortar

Testing of SCC

For determining the flowabilty of SCC, it is necessary to measure the flow spread on a flow table. This flow is measured without imparting external energy to the concrete mix by letting concrete flow under gravity. The filling ability of SCC is the time required for flow of specific quantity from a standard viscosity cone or the time required for free flow up to a specific length is measured. For testing of self-levelling characteristics, there are existing procedures which measure the height differences at various points under free flow. These heights can be measured at two different levels on the testing apparatus. A further important criteria of for SCC is the resistance against blocking. This property can be measured by using the L & U shaped moulds.

Fig.3 below shows different apparatus for testing the fresh concrete and rheological properties of SCC.

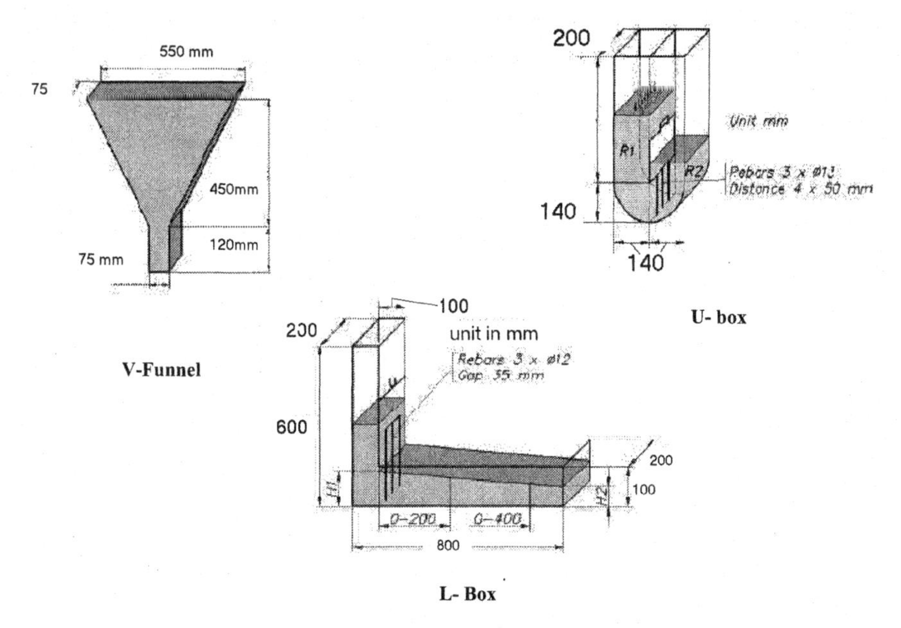

Fig.3 Apparatus for testing the Fresh Concrete and Rheological Properties of SCC.

As far as testing of hardened SCC is concerned, the properties of SCC should be compared with concrete of similar formulation. From the strength point of view, compressive strength, tensile splitting strength and elastic modulus under static pressure is compared. From durability aspect, the shrinkage, the creep, the water absorption and abrasion/mass loss for free & thaw cycle can be compared. Determination of both fresh concrete and set concrete properties can only establish a proper formulation of SCC.

Case Study

The above principles were used to design a SCC mix at KRCL test track Project in Goa for very congested reinforcement piers. The quantity of concrete poured was 1000 m^3 and temperatures were around 32^0C .

Grade of concrete: M45 SCC

1.	Cement	:	450.00 Kg/m^3
2.	Metakaoline	:	34.00 Kg/m^3
3.	Water	:	157.64 Kg/m^3
4.	Sand	:	950.55 Kg/m^3
5.	10 mm	:	427.75 Kg/m^3
6.	20 mm	:	350.00 Kg/m^3

Results obtained when the concrete was in fresh state

1.	Slump flow	:	680 mm
2.	V – funnel	:	7 sec
3.	U- box (H2-H1)	:	30 mms

Compressive strength was 65 N/mm^2

CONCLUSION

The increasing demands of HPC in congested reinforcement will lead to greater use of SCC. The use of SCC will lead to an increase in the usage of ggbfs, pfa, metakaolin, etc. and new generation high range water reducing admixtures. A new generation high range water reducing admixtures would be necessary considering the stringent requirement of high workability, self-compaction and long hauling times. The new generation admixtures would have to be of the type that gives very low loss of flow over a period of time with out delaying the hydration process. The other major advantage of the new generation high range water reducing admixtures is that it is not affected by the mode of addition. The usage of new generation high range water reducing admixtures is currently not high in India, but the technology is expected to pick up soon considering the rate at which infrastructure projects are being launched in the country.

REFERENCES

[1]Japan Concrete Society, "Recommendations for self-compacting concrete," Tokyo Japan 1999

[2]Roger Rixon and Noel Mailvaganam, "Chemical admixture for concrete," *E&FNSpon*, Third Edition UK.

[3]EFNARC (2002) Specifications and guidelines for SCC

[4]Collepardi S., Coppala L., Troli.R. and Collepardi M., "High-range water-reducing admixture: Types Compositions and Properties," pp. 355-366

[5]Mittal Amit, "New Generation High-range water-reducing admixtures for High Performance Concrete," *Construction World*, April 2001

[6]Rob Gaimster and John Gibbs, "Self-Compacting Concrete - Part 1," *The Materials and its Properties Concrete,* July/August 2001

[7]Okamura H. and Ouchi, M., "Creep, Shrinkage and Elastic Modulus of Self-Compacting Concrete," *Proceedings of the First international RILEM Symposium,*. Stockholm 1999

INFLUENCE OF FINE AGGREGATE LITHOLOGY ON DELAYED ETTRINGITE FORMATION IN HIGH EARLY STRENGTH CONCRETE

A.M. Amde
Department of Civil Engineering
University of Maryland
College Park, MD 20742

R.A. Livingston
Federal Highway Administration
6300 Georgetown Pike
McLean, VA 22101

K. Williams
Department of Civil Engineering
Cornell University
Ithaca, NY 14853

ABSTRACT

The study investigated the influence of Maryland fine aggregates on delayed ettringite formation (DEF). The main variable in the study was MDSHA expansion test rating of the fine aggregates, which is used as a surrogate for ASR reactivity. The scope of the study was limited to three fine aggregates currently being used in Maryland State Highway projects: Millville, Brandywine, and Silver Hill. Only one coarse aggregate, non-reactive Frederick limestone was used throughout the study and all other concrete mix factors were kept constant. Two replicate sets of prisms were involved; one set stored under water, the other in 100% RH. Also, as a control, a third set of Millville specimens was stored under dry, i.e. ambient RH, conditions. The Duggan test as modified by UMD/FHWA, SEM with EDXRD, and X-ray Computed Tomography were used to investigate DEF as the possible damage mechanism. In addition, compressive strength was measured at 28 days and 270 days.

This study concludes that fine aggregate type can significantly predispose a concrete sample to DEF and associated damaging expansion, while the exposure condition may exacerbate the effect. For example, the samples made from the sand with the highest MDSHA rating, Silver Hill, and stored under water exhibited the most ettringite and had the most severe expansion in terms of rate, duration, and amount. Additionally, water exposure seemed to enable DEF.

The expansions observed agreed with the ranking obtained in the MDSHA test: Silver Hill was the highest, Brandywine intermediate and Millville, the crushed limestone, the lowest. However, for the two siliceous aggregate types the expansion measurements did not predict the correct rank order of damage as measured by compressive strength at 270 days. The kinetics of the expansion varied among the aggregates. It is important to examine the expansion rate curve, i.e. the time derivative of expansion, to characterize the process rather than relying solely on the measured expansion at some fixed point in time. The various aggregates showed different responses to fog vs. submerged conditions. This suggests that the effect of each type of aggregate on the microstructure and hence the absorptivity of the cement paste needs to be considered in addition to its potential for ASR in evaluating expansion test data.

Ettringite was observed in most of the specimens. The morphology of the crystals may be an indication of the progress of damage. However, there is no quantitative way of correlating DEF with expansion measurements at the present time. Finally, the results show that the measured expansions come from a combination of moisture absorption, DEF and ASR. Consequently, expansion tests do not measure ASR effects only.

INTRODUCTION

A potentially serious deterioration problem in concrete is associated with delayed ettringite formation (DEF). Many countries have reported deterioration of concrete structures where the main cause of distress has been identified as DEF. In 2000, it was reported that 375 out of 860 precast elements of Brewer Stadium, North Carolina, in the United States showed severe cracking and spalling. In addition, 100 precast elements were slightly deteriorated. The cracks and spalling started only after eight years of construction and the main mechanism of deterioration identified was distress caused by DEF. The reconstruction of the stadium was chosen as the most economical solution. In 1995 the Texas Department of Transportation found significant cracks in 56 precast concrete beams waiting installation in highway bridges. Analysis indicated that the damage appeared to be caused by DEF. The beams were written off at a cost of U.S. $250,000. Other countries that have reported damage include South Africa, Egypt, United Kingdom, Germany and Scandinavia.

Cement Hydration, Ettringite and DEF

Cement hydration is driven by the reactions of water molecules with the cement's constituents (C_3S, C_2S, C_4A, C_4AF). These reactions commonly proceed in the presence of gypsum and proprietary additives designed to alter the properties of the resulting concrete. Generally, the reactions proceed in sequence. For example, tricalcium aluminate (C_3A) hydrates first with gypsum to produce ettringite and heat. Pure ettringite has the chemical formula:

$$3CaO\ Al_2O_3\ 3CaSO_4\ 32H_2O \quad (1)$$

or in cement chemists notation,

$$C_6\ A\ \hat{S}_3\ H_{32} \quad (2)$$

The resulting ettringite then may participate in subsequent reactions to produce monosulfoaluminate. In a recent study, Stark, Moser and Bellmann (1) dispute this hydration model. Using Environmental Scanning Electron Microscope with Field Emission Gun (ESEM-FEG) they found that ettringite does not become unstable and does not gradually convert to monosulfate or tetracalciumsulfoaluminatehydrate. They conclude that both ettringite and tetracalciumsulfoaluminatehydrate are the final products of hydration of ordinary Portland cement.

The primary products of cement hydration are calcium silicate hydrate (C–S–H), calcium hydroxide (CH), and ettringite. Ettringite naturally forms during the early hydration process of cement, while the concrete is still plastic. Its formation greatly influences many properties of concretes made from Portland cement, such as setting of cement, strength, durability, etc. However, the early type of ettringite is harmless and does not produce any damage. The early ettringite decomposes, but in some cases, it may reform at a later stage, after months or years, a process, which is called, delayed ettringite formation (DEF). Sometimes it is also referred to as secondary ettringite formation. It is generally believed that DEF formation can cause expansion and deterioration of concrete. DEF damage is typically viewed as the result of an expansive process within the material similar to that of alkali-silicate reaction (ASR), but many aspects of this process remain controversial.

After curing, concrete remains a highly reactive material and it continues to change chemically and physically as its constituents interact with the external environment. Often these changes degrade the overall integrity of the concrete. The extent and severity of any physical or chemical change is influenced by the exposure conditions of the concrete. Some common mechanisms enabling the deterioration of concrete are reinforcement corrosion and expansion, weathering, alkali-silica reaction (ASR), and sulfate attack. Delayed ettringite formation is considered to be the result of an internal sulfate attack since it requires no source of sulfate from the surrounding environment.

Objectives and Scope

The main objective of this research was to examine the influence of fine aggregates on delayed ettringite formation in high early strength (Type III) concrete. The only factor that is allowed to vary is the type of fine aggregate, which has an associated Maryland State Highway Administration (MDSHA) expansion test rating. This is usually assumed to be a measure of ASR reactivity. Thus this research can establish the dependence of DEF on the fine aggregate expansion ratings. Also since ettringite is a hydrous phase of cement material, different exposure conditions including maximum exposure of the concrete sample to water were investigated. The scope of the research has been limited to three fine aggregates with a single unreactive coarse aggregate. The fine aggregates selected for the study vary in geographic place of origin, MDSHA expansion test rating, method of production, and petrographic makeup (2).

MATERIALS

Portland Cement

The cement used for all batching was Portland cement Type III, which is commonly used to produce high early strength concrete. This type of cement was chosen because it is suspected of being particularly susceptible to DEF. In comparison to other cement types, Type III cement has a higher sulfate content, which would favor the crystallization of ettringite (3). To eliminate possible variation in the cement sample, all cement used originated from the same batch produced in Allentown, PA. Though the cement was purchased pre-packaged, all bags were thoroughly blended together in a drum mixer to achieve a uniform and consistent cement sample.

Fine Aggregates

The scope of this research project included only fine aggregates used in Maryland state highway projects. Three fine aggregates were selected from each possible MDSHA expansion test rating: low (0 – 0.09%), medium (0.10% – 0.19%), and high (0.20% - 0.29%). Fine aggregates with ratings above this range would usually not be acceptable in state highway projects. Aggregate Industries, Inc. currently distributes each of the fine aggregates selected for this study throughout the state of Maryland. Two of the three fine aggregates are produced in Maryland (Brandywine and Silver Hill) and one is produced in West Virginia (Millville). All applicable ASTM test procedures were followed during the collection of the sands to ensure a sufficiently random sample (ASTM C-702, D-3665, and D-75).

Millville Fine Aggregate: Aggregate Industries, Inc. currently produces highly uniform and non-expansive sand in Millville, West Virginia. Railroad cars bring the sand to the company's Bladensburg plant where it is distributed throughout the state of Maryland. The sand is milled from quarried limestone gravel. Since its ASR rating is very small (0.028%), this sand was considered to be a control sample.

Brandywine Fine Aggregate: Brandywine is produced from alluvial gravel collected from creeks and streambeds. By nature it is highly heterogeneous. According to the Maryland Geological Survey (Geologic Map of Maryland, 1968), this aggregate is composed of orange-brown siliceous gravel, locally limonite-cement in the Brandywine, and minor silt with red, white, or gray clay. The sand is also very coarse relative to the other two sands and it has a mild MDSHA expansiveness (0.14%). The sand is part of an upper loam layer found in southern Maryland. The layer has a depth of up to fifty feet from the surface.

Silver Hill Fine Aggregate: Silver Hill is highly expansive natural sand. It is also very uniform in terms of color, morphology, and grading considering it was naturally produced. However, it contains a small amount of a deleterious substance (chalcedonic chert). The sand is pale brown but weathers to white or pale gray. It is part of a fine-grained argillaceous sand layer found in southern Maryland with a depth of up to one hundred and fifty feet from the surface.

Coarse Aggregate

To eliminate possible variables, only one coarse aggregate was used for all three concrete batches. The gravel is dolomite limestone with a maximum diameter of one inch. The gravel originates in Frederick, MD and has a specific gravity of 2.72 and 3% absorption.

Potassium Carbonate as Admixture

The only admixture allowed under this research project was anhydrous granular reagent grade potassium carbonate (K_2CO_3). Its effect on the pore water chemistry is the same as naturally occurring potassium in unhydrated Portland cement. Potassium in cement primarily occurs as potassium sulfate (K_2SO_4). During hydration, the potassium sulfate exchanges ions with calcium and hydroxide to produce a potassium hydroxide solution and a calcium sulfate precipitate:

$$K_2\,SO_4 + Ca^{2+} + 2OH^- \rightarrow 2K^+ + 2OH^- + CaSO_4 \quad (3)$$

The potassium carbonate admixture hydrates as follows. The potassium carbonate exchanges ions with the calcium and hydroxide to produce a potassium hydroxide solution and a calcium carbonate precipitate:

$$K_2\,CO_3 + Ca^{2+} + 2OH^- \rightarrow 2K^+ + 2OH^- + CaCO_3 \quad (4)$$

Since the number of hydroxide moles on both sides of equations 3 and 4 remain constant, neither reaction affects the pH of the mix. The advantage of adding potassium carbonate rather than potassium sulfate is that the former does not increase the sulfate content, which also favors DEF. The potassium carbonate was added directly to the mix water to achieve a target potassium level of 1.50% by weight of cement or 1.41% by weight of the mix water. The target potassium level included the added potassium plus the natural potassium in the cement, in this case potassium sulfate.

Sample Preparation and Treatment

Casting: All concrete batches made under this research program were proportioned with a water-to-cement ratio of 0.5 and high potassium content (1.50% by weight of cement), which produced a very dry mix with low workability and reduced set time. These factors made adequate consolidation in the molds a challenge. Since the slump was less than 25 mm, a table vibrator was used to consolidate the concrete in the molds per ASTM C192. The table vibrator used had a surface area of approximately 0.37 m^2 and accommodated up to four cylinders and two prisms at one time. The molds were held down as close to the center of the tabletop as possible to insure maximum energy transfer from the vibrating surface to the molds. Casting began by placing the empty molds on the vibrating table, which was set to maximum (approximately 60 Hz). The molds were then filled incrementally with two equal layers of concrete. The molds were left on the vibrating surface for up to ten seconds. This technique yielded excellent consolidation despite the difficult workability issues. The molds were stripped after 24 ± 8 hours per ASTM C192. Very little honeycombing, a symptom of under consolidation was discovered in the freshly stripped samples. No segregation of the coarse aggregate, as a result of over consolidation, was found.

High Temperature Curing: All high temperature samples were cured almost immediately after casting (usually within 1 hour). The goal was to simulate commercial steam curing using the conventional ovens available in the lab. The samples were placed in the oven with pans or bowls full of water. The oven was then preset to 85° C and left to run for four hours.

Low temperature Curing: One day after casting, the samples were submerged in limewater at room temperature (21° C) and allowed to continue curing for seven days. The lime or calcium hydroxide [$Ca(OH)_2$] prevents leaching between samples and is a requirement for water storage per ASTM C192 .The samples where then heat-treated using the Duggan Cycle. After sitting at room temperature for two days in a dry container, the samples were finally placed in controlled moisture environments for long-term storage.

Exposure/Storage Conditions: Samples were stored in one of three exposure conditions: dry (21^0C w/ 50% humidity), fog (21^0C w/ 100% humidity) or submerged (limewater). ASTM C192M-95 requires concrete samples to be stored in limewater as opposed to a plain water bath to stimulate pH conditions in actual concrete.

Modified Sample Preparation Method: The specimens used in this research differ from the core samples that are recommended for use in the Duggan test. Duggan suggested the use of cores 25 mm in diameter and 50 mm in length with parallel and smooth ends. The use of cores allows for sampling from a general source of concrete that may be of interest including existing structures. However, when laboratory specimens are used, this method of obtaining test samples is unduly arduous. Therefore, under this research program, a modified Duggan Sample Preparation method, proposed by UMD/FHWA, was used (4). The modified sampling technique uses prisms (75 mm x 75 mm x 275 mm) cast from concrete mixed in a laboratory instead of the cores Duggan suggested. The prisms are made with end steel inserts for linear length change measurements following ASTM C192-88 and ASTM C1293. Before heat treatment, the initial lengths of the prisms are recorded.

Duggan Heating Cycle: The Duggan test heating cycle is an eight-day heating regimen with prescribed intervals of heating and cooling at specific temperatures and duration. The heating temperature is eighty-two degrees Celsius and the cooling temperature is twenty-one degrees Celsius. The first two cycles are identical and include one day of heating followed by one day of cooling. The samples are cooled in a limewater bath. The third and final cycle involves three days of heating followed by a day of cooling. It is here noted that water-cooling can have its drawbacks especially when dealing with a large temperature gradient, in this case from 82° C to 21° C. Water is an excellent conductor of heat and has the potential of rapidly cooling concrete samples, particularly those of the size used in this experiment, thus causing excessively large cracks in the sample. To avoid this potential problem, the samples where allowed to cool in air for no more than one hour before placement in the water bath.

Preparation of Fracture Surfaces for SEM: Specimens are broken from 75 mm x 75 mm x 275 mm concrete prism by means of hammer blows. To collect the fractured surfaces, the prisms where sliced into three equal segments each approximately four inches in thickness. The slices where placed in a plastic bag and broken using a hammer to expose fresh surfaces. Fragments were then sampled from the bag and sorted into groups of "internal" or "external" chips. External fragments were defined as a fragment with a surface in common with the outside surface of the prism. The outside surface made the external fragments readily identifiable. Generally, ideal fragments were between 2-10 mm in size, flat, and included at least one cement paste-aggregate interface or water void. The chips were mounted with carbon paint to 13 mm diameter carbon stubs and dried at 55° C in a vacuum oven. The samples were then coated with a film of carbon approximately 100 Angstroms thick (1 Angstrom = 10^{-6} cm) to insure electrical conductivity and thus prevent charging effects in the microscope.

DISCUSSION OF RESULTS

The influence of ASR rating on expansion and damage is summarized in Table I. The expansions at 200 days qualitatively agreed with the ratings from the MDSHA ASR test. Silver Hill with highest ASR test rating had the greatest expansion (Figures 1 and 2) while Millville with the lowest ASR test rating had the lowest expansion, and Brandywine fell in between. This

Table I. Influence of ASR Rating on Expansion and Damage

	Batch #1 Silver Hill Agg.		Batch #2 Millville Agg.		Batch #3 Brandywine Agg.	
ASR Rating	0.28		0.028		0.14	
Exposure Condition	Sub-merged	Fog	Sub-merged	Fog	Sub-merged	Fog
Expansion @ 200 days (%)	0.64	0.48	0.125	0.05	0.49	0.26
Compressive Strength @ 270 days (N/m^2)	$5.2x10^6 \pm 550$	$9x10^6 \pm 3450$	$7x10^6 \pm 690$	$27.6x10^6 \pm 570$	$4.1x10^6 \pm 900$	$4.1x10^6 \pm 900$
Peak Expansion Rate(mm/day)	0.2325	0.575	0.0875	0.0225	0.350	0.1150
Time for Peak Expansion Rate (Days)	60	90	90	110	98	100

ranking was the same for both submerged and fog moisture conditions. Conversely, Millville had the highest compressive strength for both cases. However, Brandywine, instead of Silver

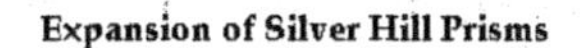

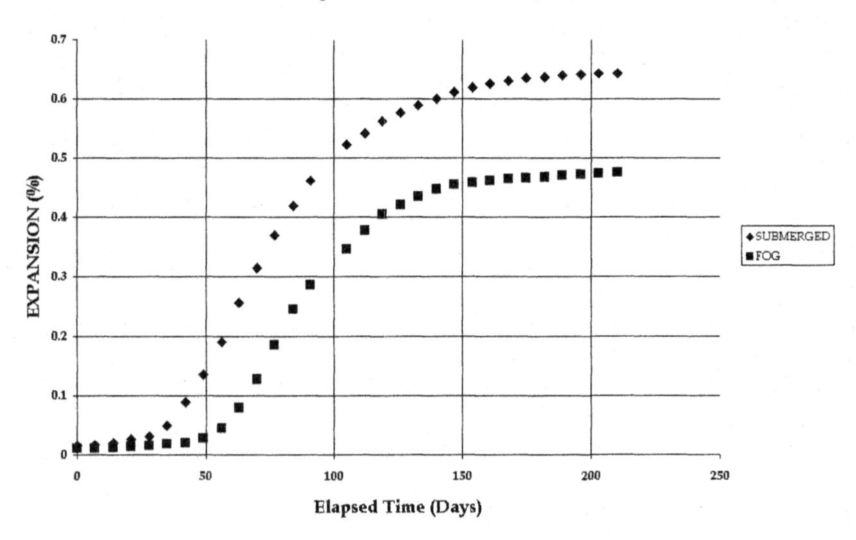

Figure 1. Summary of Expansion of Silver Hill Prisms

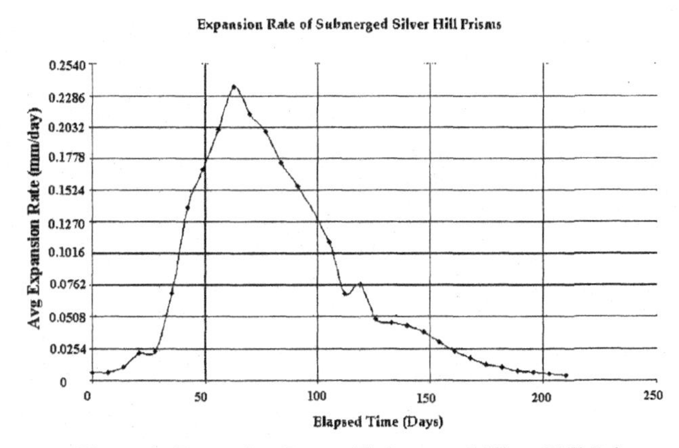

Figure 2. Expansion Rate of Submerged Silver Hill Prisms

Hill, had the lowest compressive strengths. It should be noted that in the submerged case, the difference between the two might not be statistically significant. These results indicate that expansion tests alone are not an infallible predictor of actual concrete durability as measured by long term compressive strength.

Considering the kinetics of expansion, there was a clearly identifiable main peak in each case, except for Brandywine stored in fog. This peak occurred significantly earlier in Silver Hill than the others for both moisture conditions. Since the timing of the peak depends mainly on the kinetic rate constant, this implies that the Silver Hill aggregate is more reactive. The peak in Silver Hill occurs at 60 days. This also suggests an alternative method of defining the expansiveness ratings of aggregates. Instead of using the amount of expansion at an arbitrary fixed time, for example 16 days in ASTM C1260, one could use the time of the peak, which is physically more meaningful.

The shape of the expansion curves usually exhibits a very slow expansion initially. The curve for the Millville prisms stored in dry air shows significant shrinkage during this early period. There are at least three components to shrinkage: drying, thermal and autogenous or chemical. Given the time delay between casting and the start of initial measurement, thermal shrinkage would be negligible in these data. Thus the observed shrinkage is a combination of drying and chemical shrinkage. Both of these concern the demand for water in the curing concrete. Since the shrinkage curve levels out just when the moist storage curves take off, this suggests that in the latter the chemical shrinkage dominates over the expansive processes in the consumption of water from external sources. It is only when the chemical shrinkage is largely completed, that the expansive processes take over. Only chemical shrinkage is important here, since drying shrinkage by definition could not happen in submerged or 100% RH conditions.

Contrasting the submerged and the fog-stored cases reveals some interesting differences. Both Millville and Silver Hill showed much reduced compressive strengths for the submerged cases compared to the fog case. However, Brandywine, which had the lowest compressive strengths, showed no statistically significant difference. To understand this, it is necessary to consider the processes that control the rate of reaction of the water. First, it must physically pass from the outside into the concrete, which is a physical mass transfer, or diffusion-limited, process. Then it must chemically react to generate expansive products. The diffusion rate into the concrete should be much lower for the fog case since the surface concentration of water molecules is orders of magnitude lower. The fact that Brandywine compressive strength is insensitive to this difference indicates that in it the processes are controlled by chemical reactions, or in essence the rate of reaction is slower than the rate of delivery of water. Conversely, the compressive strengths of the other two mixes show considerable sensitivity to mass transfer conditions; hence the chemical reactions must be faster than the diffusion processes. This is also borne out in the expansion rate data. The time of the peak shifts significantly for the Silver Hill and Millville mixes, but not for the Brandywine.

Note that these differences among the aggregate types involve both physical and chemical aspects. The overall damage process involves several steps. The water has to physically transfer into the concrete, a diffusion-controlled process. Then it has to chemically react with the aggregate, creating expansive stresses, and finally the stresses produce an expansive strain of the material depending on its constitutive law (5). The diffusion rate is mainly determined by the microstructure of the cement paste in terms of pore size distribution and connectivity. These in turn depend upon the grain-size distribution of the fine aggregate and its reactivity with the Portland cement. The chemical reaction rate depends upon the specific surface of the fine aggregate and its reaction rate with water molecules.

The shapes of the expansion curves also show differences between the two storage conditions. For Silver Hill, the two curves have effectively the same shape up to 100 days, but the fog curve appears to be lagging in time, which would be a reflection of the slower water transport. The Millville mix also seems to show a small time lag for the fog case, but the slopes of the two curves differ. However, the Brandywine curves show no time lag. The two curves coincide up to 75 days.

A major and unavoidable problem in DEF research is the lack of a reliable analytical method for quantifying the amount of ettringite present. This arises because the mineral form of ettringite, as distinct from monosulfate or gypsum, must be explicitly identified in the cement paste regions that are randomly distributed around fine and coarse aggregate particles. The available methods are powder X-ray diffraction, thermogravimetric analysis (TGA) and scanning electron microscopy (SEM) with energy dispersive X-ray diffraction (EDXRD). These all are limited to very small volume of sample, on the order of 10 mm^3, that may not be representative of the overall volume of the prism. Moreover, quantitative X-ray diffraction requires extensive and painstaking sample presentation, but even then may not be sensitive to minerals that are present in amounts of less than 1% (6). TGA is a more robust method, but the ettringite may be interfered by the presence of other phases. The SEM method relies upon visual identification of the characteristic ettringite crystal shape in the image. However, the image area is only on the order of 100 by 100 μm, which makes it essentially impossible to carry out an exhaustive examination of the sample. Consequently, in the following discussion it should be understood that the results are qualitative.

Ettringite was found by SEM fracture surface analysis (Figure 3) in all three batches that were stored in water at 9 months. For the submerged cases, there appears to be a trend toward more expansion as the frequency of ettringite observations increases. Several morphologies of ettringite were observed, with lamellar structures developing at later times. This is consistent with observations from the cores taken from Maryland bridges in which the lamellar structure developed with more advanced damage (7).

Figure 3. Ettringite from Interior Fragment of Silver Hill Prism at Nine Months

Finally, although ettringite was frequently observed, there was only one SEM sample that contained ASR gel. The implication is that expansion tests do not measure ASR effects only, but rather the combined effects of ASR, DEF and moisture absorption.

CONCLUSIONS

The expansions observed in this research program agreed with the ranking obtained in the MDSHA test: Silver Hill was the highest, Brandywine intermediate and Millville, the crushed limestone, the lowest. However, for the two siliceous aggregate types the expansion measurements did not predict the correct rank order of damage as measured by compressive strength at 270 days. The kinetics of the expansion varied among the aggregates. It is important to examine the expansion rate curve, i.e. the time derivative of expansion, to characterize the process rather than relying solely on the measured expansion at some fixed point in time. The various aggregates showed different responses to fog vs. submerged conditions. This suggests that the effect of each type of aggregate on the microstructure and hence the absorptivity of the cement paste needs to be considered in addition to its potential for ASR in evaluating expansion test data.

Ettringite was observed in most of the specimens. The morphology of the crystals may be an indication of the progress of damage. However, there is no quantitative way of correlating DEF with expansion measurements at the present time. Finally, the results show that the measured expansions come from a combination of moisture absorption, DEF and ASR. Consequently, expansion tests do not measure ASR effects only.

ACKNOWLEDGEMENTS

The research reported herein was sponsored by the Maryland State Highway Administration (MDSHA). Sincere thanks are due to Mr. Paul Finnerty, Office of Materials and Technology, Precast/Prestressed Concrete Division, MDSHA. Special thanks to: Mr. Earle S. Freedman, Mr. Peter Stephanos, Mr. Jeffrey H. Smith, Ms. Barbara Adkins and Ms. Vicki Stewart of MDSHA; and Dr. W. Clayton Ormsby and Mr. T. Tessema of FHWA. Appreciations are also due to Mr. Gary M. Mullings and Mr. Solimon Ben-Barka of NRMCA; Dr. Amal Azzam of University of Maryland; and Mr. Indra Kumar of Aggregate Industries Lab, Mid – Atlantic region.

REFERENCES

[1]J. Stark, B. Moser and F. Bellmann, "New Approaches to Cement Hydration in Early Hardening Stage," pp. 261-277 in Proc. of the 11th Int. Congress on the Chemistry of Cement (ICCC), *Cement's Contribution to the Development in the 21st Century,* Edited by G. Grieve and G. Owens, Document Transformation Technologies, Durban, South Africa, May 2003

[2]A.M. Amde, K. Williams and R.A. Livingston, "Influence of Fine Aggregate Lithology on Delayed Ettringite Formation in High Early Strength Concrete," Maryland SHA, *Report No. MD-04-SP107B4U*, July 2004

[3]H.F.W. Taylor, "Delayed Ettringite Formation," pp. 121-131 in Advances *in Cement and Concrete*, New Hampshire Conference, American Society of Civil Engineers, New York, 1994

[4]E. Ramadan, A.M. Amde and R.A. Livingston, "The Effect of Potassium and Curing Conditions on Delayed Ettringite Formation," *Journal of Materials*, ASCE (under review), 2004

[5]R.A. Livingston, "The Damage Function Concept in the Deterioration Science of Concrete" *Materials and Construction*, (in press), 2004

[6]P.E. Stutzman, "Guide for X-ray Powder Diffraction Analysis of Portland Cement and Clinker," *NISTIR 5755*, NIST, 1996

[7]A.M. Amde, M. Ceary, R.A. Livingston and N. McMorris, "Pilot Field Survey of Maryland Bridges for DEF Damage," Maryland SHA, *Report No. MD-04-SP107B4U*, July 2004

POTENTIAL USE OF BENEFICIATED FLY ASH IN HIGH PERFORMANCE CONCRETE

K.A. Riding and M.C. Garci Juenger
Department of Civil Engineering
The University of Texas at Austin
Austin, TX 78712

ABSTRACT

Fly ash is commonly used as a supplementary cementing material in concrete to improve performance and reduce cost and environmental impact. Some fly ash suppliers have begun marketing "ultra-fine" or "beneficiated" fly ash which has been processed to have a particle size less than about 45 μm. This material boasts many improvements over traditional fly ash, including higher reactivity, and has the potential to serve as an alternate or supplement to silica fume in the production of high performance concrete. In this paper we discuss the physical and chemical characteristics of beneficiated fly ash and critically review data in the literature concerning its impact on concrete performance.

INTRODUCTION

Supplementary cementing materials (SCM) are being used in increasing amounts in high performance concrete because of their potential to improve concrete durability and sustainability. Traditionally, high performance concrete contains silica fume at cement replacement levels of 5-15%. Because of its small size and pozzolanic reactivity, silica fume increases concrete strength by reducing the porosity of the interfacial transition zone (ITZ) and creating C-S-H at the expense of CH. Fly ash, ground granulated blast furnace slag, and other SCM have been used less frequently in high performance concrete because their particle sizes are generally too large to significantly improve the ITZ and, therefore, concrete strength. Furthermore, their reaction rates can be slow, which negatively impacts early strength. However, in the interest of improving concrete performance and life cycle-costs, and not simply strength, these other SCM can impart many benefits.

There is an interest in exploring alternative SCM to silica fume in high performance concrete. Silica fume has limited availability in some parts of the world and also decreases workability of the fresh concrete, limiting the maximum replacement level and necessitating the extensive use of superplasticizers. A material has recently emerged that can potentially challenge the dominance of silica fume as the SCM of choice in high performance concrete, namely beneficiated fly ash. This is fly ash that has been processed to reduce its particle size, increasing its reactivity and its ability to fill in ITZ spaces and thereby increase strength. Processed fly ash has been shown to increase concrete strength without decreasing workability [1], making it an attractive material for use in high performance concrete. Furthermore, fly ash is readily available, with over 450 million metric tons produced annually worldwide [2]. The objective of this paper is to critically review and compile data in the literature concerning the physical and chemical properties of beneficiated fly ash along with its effects on the properties of fresh and hardened concrete. The intention is to demonstrate the potential of beneficiated fly ash for use in high performance concrete.

FLY ASH PROCESSING

There are several available methods of reducing the particle size of fly ash including grinding, air-classification, magnetic separation, and sieving. The most commonly used methods are grinding and air-classification, with the latter being the most successful. Air-classification is usually done using a cyclone separator. Fly ash is introduced into the top of a separator and whirled about the inside by either blowing the fly ash into the separator off-center or by using fans or rotating discs. The coarse fraction of the ash is forced to the outside wall of the separator, as in a centrifuge. When reaching the bottom of the separator, the air carrying the fine particles is forced upward, leaving the coarse fraction behind. The fine particles are then removed from the air with bag filters. The air flow, speed, and other separator properties can be adjusted to separate different sized particles [3].

PHYSICAL CHARACTERISTICS OF BENEFICIATED FLY ASH

Unprocessed fly ash particles are generally spherical, with some irregular shaped particles, hollow spheres (cenospheres) and hollow spheres filled with smaller spheres (plerospheres). Scanning electron microscopy has shown ground fly ash particles to be angular and irregularly shaped [3,4]. Air-classified fly ash has been shown by scanning electron microscopy and shape factor analysis to be made up of mostly spherical particles [3-5]. Hemmings and others [6] found that classified fractions have a greater proportion of spherical particles than the parent fly ash. This is because the irregularly shaped particles tend to be in the larger size fraction and are removed during classification.

Grinding and air-classification decrease the particle sizes in fly ash. Laser diffraction techniques give accurate size distributions for small particles, but the equipment is expensive and not readily available. Therefore, it is rarely used for characterizing cements and SCM. Particle fineness is more commonly assessed using surface area, which increases as particle size decreases. Mehta [7] found that surface area as measured by Blaine fineness [8] was a better indicator of pozzolanic activity than is surface area as measured by nitrogen adsorption for fly ash because small sulfate particles covering the spherical glassy particles inflate the values of the latter. The Blaine surface area value will be used in this paper as an indicator of particle size, or the extent of beneficiation, and will be correlated to other fly ash characteristics and concrete performance indices. Figure 1 shows the median particle size as measured using laser diffraction techniques versus Blaine fineness for several fly ashes studied by different researchers. The relationship is non-linear because a small decrease in particle diameter greatly increases the surface area.

Beneficiation also increases the bulk density of fly ash because the altered particle size distribution increases packing efficiency. This effect is enhanced when the ash is ground because cenospheres (hollow fly ash particles) are broken up, increasing the bulk specific gravity [9]. Figure 2 shows the relative increase in bulk density (ratio of processed fly ash value to unprocessed fly ash value) versus the relative increase in Blaine fineness (ratio of processed to unprocessed). The trend is linear; increasing the Blaine fineness raises the bulk density.

CHEMICAL CHARACTERISTICS OF BENEFICIATED FLY ASH

Most of the research on fly ash beneficiation has been done with ASTM Class F [10] fly ashes. The glassy calcium aluminosilicates in ASTM Class C [10] fly ash are already very reactive, giving little incentive for beneficiation [7,11]. The beneficiation of fly ash, whether

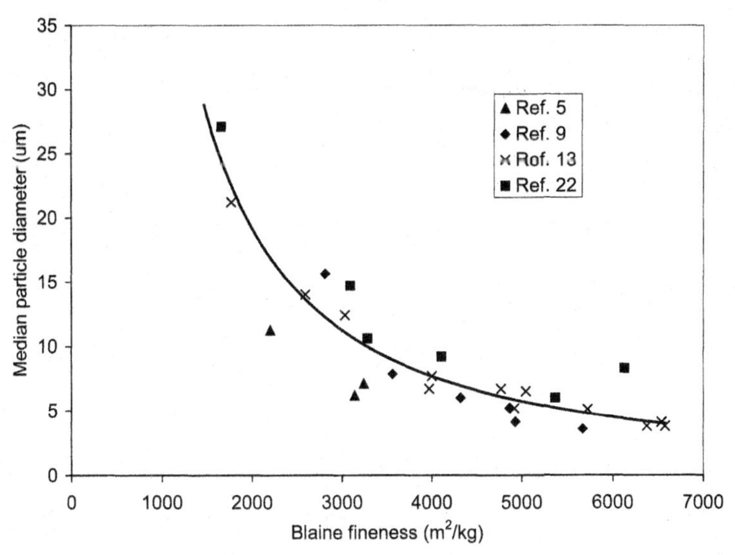

Figure 1 - Median particle diameter vs. Blaine fineness for several beneficiated fly ashes

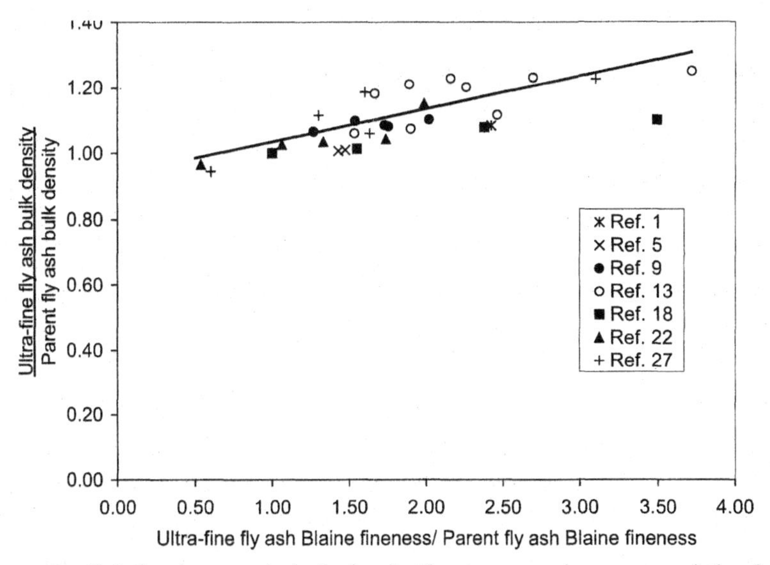

Figure 2 – Relative increase in bulk density due to processing versus relative increase in Blaine fineness due to processing

through grinding or air classification, has been shown to improve reactivity through more than just a reduction in particle size, as will be discussed here.

The chemical compositions and phase distributions in fly ash are complex and variable. Different particles within the same fly ash sample can have different chemical compositions and phases present. Differences can be categorized based on particle size, magnetic properties, and density; some studies have also shown intra-particle variations [12]. Trends in chemical composition can be defined based on particle size, but exceptions are common because of the complicated and varied nature of fly ash. In other words, one cannot easily generalize that larger fly ash particles contain more calcium oxide, for example, than smaller particles.

Researchers working on ground fly ash generally assumed that the particles remain unchanged chemically after grinding. One group found that grinding produces higher calcium carbonate and sulfide levels [9]. However, there is little evidence from other studies to support this conclusion. As discussed earlier, crushed plerospheres and cenospheres can look like broken egg-shells. Solid particles can also break apart, leaving solid, irregularly shaped particles [13,14]. The grinding not only increases the surface area of the particles, but exposes the interior of the spheres that may be more or less glassy and have a different composition than the exterior [12]. In unprocessed fly ash, these interior phases would not participate in early reactions; exposing them may influence early reactivity.

Fly ash reactivity has been shown to be greatly dependent on the glass content of the ash [11,15]. Classified fly ash has an increased amorphous content compared to the coarser fractions. Fine particles are glassier than larger particles because the rate of cooling of the particles is greater with fine particles [12]. Quartz tends to be found in the heavier, larger particles. When the ash is air classified, the larger detrital sand particles are removed; the fine portion is left with a lower percentage of crystalline particles. However, some crystalline material can still exist in the fine fraction as small crystals of mullite and magnetite, for example. The fine fraction has been found to be at least 20% more amorphous than the unclassified ash and at least 40% more than the coarse fraction [1,16].

The composition of the glassy phases can greatly influence the reactivity of the fly ash. In particular, the Al/Si ratio is important in determining the reactivity of the glass. Furthermore, magnetic fractions of fly ash have been found to produce lower strengths than non-magnetic fractions [17]. This indicates a difference in reactivity between the ferro-aluminosilicates and aluminosilicates. Dense iron particles can be rejected as coarse particles by air separators, depleting the fine fraction of iron and increasing the percent of aluminum [16], thereby increasing the reactivity of the fine fraction compared to the parent ash.

Classified ashes tend to have higher sulfur contents than unprocessed ash [1,3,18]. Alkali sulfates are found in higher percentages in the ultra-fine particles and tend to be concentrated on the surface of the particles. Figure 3 shows the sulfur content versus the mean particle diameter for several fly ashes. The sulfur content is normalized against the sulfur content of the parent ash. The increase in sulfur due to classification varies widely, but the trend is for sulfur content to increase as mean particle diameter decreases. There is dispute as to how and why these elements are more concentrated on the outside of the smaller particles [6,12].

EFFECTS OF BENEFICIATED FLY ASH ON CONCRETE PROPERTIES

Workability

The workability of concrete can be improved by the addition of air-classified ultra-fine fly ash. Several theories have been given as to why air-classified fly ash improves workability. The most common explanation is that the increased number of spherical particles act like ball bearings, decreasing the inter-particle friction [19-21]. Another theory suggests that ultra-fine particles help release water trapped by bulked particles [19]; like aggregate gradation, there exists an optimum cementitious gradation for workability. Fly ash particles also adsorb less water per unit surface area than cement. This benefit is said to be counteracted somewhat by a large increase in surface area with ultra fine particles [20].

The workability of mortars with ground fly ash decreases with increased grinding of fly ash. This can be attributed to the fact that ground particles have a higher surface area than spherical particles [14]. The sphericity of the particles also decreases with increased grinding, eliminating the beneficial ball bearing effect.

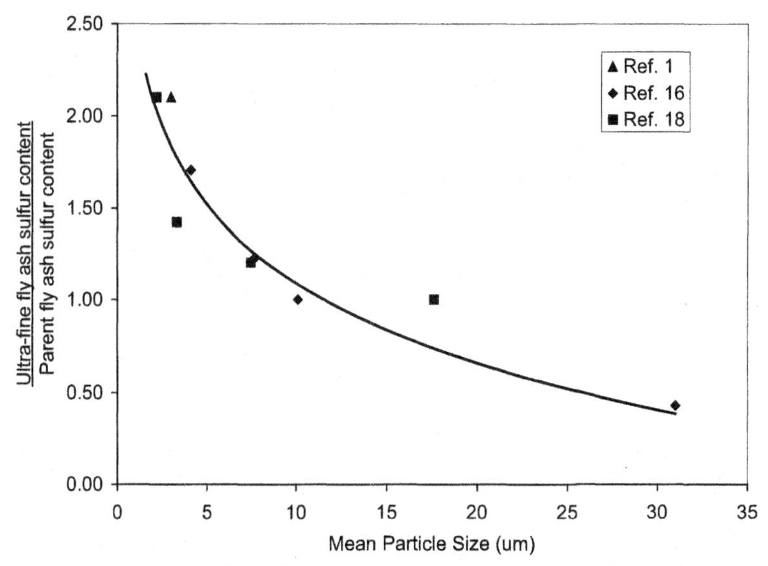

Figure 3 – Sulfur content (beneficiated/parent ash) versus mean particle size of beneficiated ash

Compressive Strength

Beneficiated fly ash is more reactive than unprocessed fly ash, leading to higher compressive strengths in ultra-fine fly ash concrete compared to traditional fly ash concrete. The increased reactivity of ultra-fine fly ash is evidenced by the increase in heat evolved during hydration [18]. Furthermore, 3-day chemically bound water tests show an increase in C-S-H production from control samples, evidence of early fly ash reaction [16]. As discussed previously, the high surface area of beneficiated fly ash increases its reactivity, as does its increased glass content. In addition to increased reactivity, ultra-fine fly ash increases concrete strength by filling in voids and the interfacial transition zone between paste and aggregates [22].

Figure 4 shows the 28 day strength activity index [23] versus Blaine fineness [8] for several fly ashes. Strength activity index [23] is a comparison between the compressive strength of a mortar containing 20% fly ash and that of a control mortar without fly ash. The values shown in Fig. 4 are taken either directly from the literature or are calculated based on concrete strengths reported in the literature. Strength activity index data from the work of Bouzoubaâ and others [13] are for samples containing raw and ground ashes. The data from Obla and others [1] are for an air-classified ash and its parent ash. The data from Mehta [7] are from several raw ashes. Blaine fineness is a general indication of the degree of beneficiation of the fly ash. Fig. 4 shows that the strength activity index increases as the pozzolan's surface area increases. The considerable scatter could be due to the different chemical and physical characteristics of the different ashes; the correlation is much tighter within a given fly ash than for all of the points together.

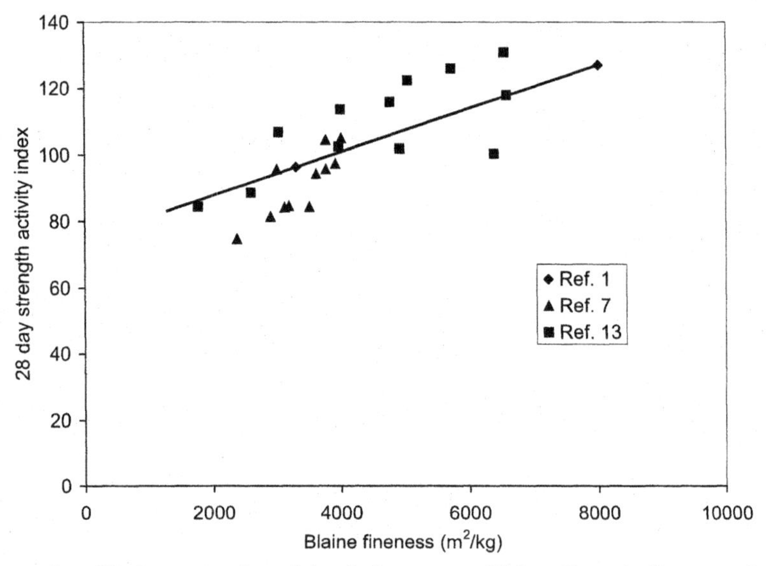

Figure 4 – 28 day strength activity index versus Blaine fineness for several fly ashes (processed and unprocessed)

When investigating the dependence of concrete strength on fly ash particle size, researchers have attempted to correlate strength with the percent of particles smaller than a critical value, either 10 or 45 μm. Some studies have found that that the percent of particles under 10 μm is a critical parameter governing the reactivity of the fly ash, and thus the strength development of fly ash concrete [7,24]. Some researchers argue that there exists a correlation between strength and percent of fly ash particles smaller than 45 μm [7]. But others have found no correlation between the percent of particles smaller than 45 μm and strength [25]. In order to address this controversy, we have plotted the relative compressive strength of mortars and concretes (30% fly ash/0% fly ash) from several sources versus the percent of fly ash particles smaller than 10 μm in Figure 5. The samples tested by Berry and others [16] used raw and air-classified fly ash. The data from Hassan and Cabrera [5] come from samples made with unprocessed fly ash. Payá and others [17] used both ground and raw fly ash. There does indeed appear to be a weak positive correlation between the percent of particles smaller than 10 μm and strength.

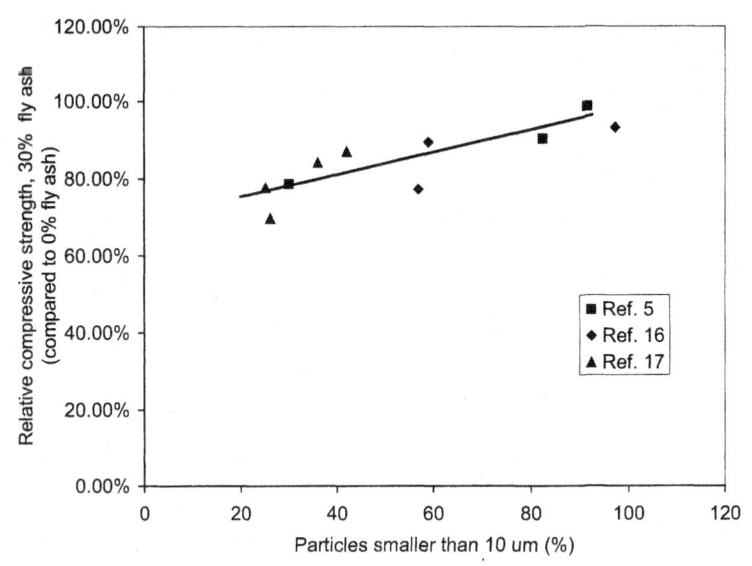

Figure 5 – Compressive strength of concrete and mortar with 30% fly ash (expressed as a percentage of control with 0% fly ash) versus percentage of particles less than 10 μm

Durability

It is well known that many durability aspects of concrete can be improved by the replacement of cement with fly ash. Fly ash, however, also has some shortcomings that can be reduced with processing. For example, fly ash has a negative effect on abrasion resistance of concrete. Abrasion resistance has been correlated to compressive strength [18,26]. Since ultra-fine fly ash increases compressive strength relative to unprocessed ash, it is likely that ultra-fine fly ash may produce better abrasion resistance [18].

Studies have shown that the water content is an important factor controlling drying shrinkage [18,27]. Because beneficiated fly ash reduces the water demand, it helps lower the total paste content, decreasing drying shrinkage [18,27]. In shrinkage ring tests, ultra-fine fly ash has shown lower autogenous shrinkage and higher resistance to shrinkage cracking than silica fume concrete and ordinary portland cement concrete without SCM [1].

Ultra-fine fly ash has been shown to be more effective in controlling alkali-silica reaction (ASR) than unprocessed fly ash [1,18,28]. More research is needed to determine the amount of ultra-fine fly ash needed to control alkali-silica reaction. More research is also needed into the mechanisms of ultra-fine fly ash for suppressing ASR. The improvement could be related to the increased reactivity of ultra-fine fly ash. It has also been suggested that the alkali sulfates in ultra-fine fly ash are less soluble than in unprocessed fly ash [16].

The use of beneficiated fly ash can dramatically reduce the permeability of concrete and the chloride penetrability [1,16,18,28,29]. It has been hypothesized that ultra-fine fly ash reduces permeability by the following mechanisms: 1) fills in voids in matrix, creating a denser matrix, 2) reduces the size of interfacial transition zone by more efficient particle packing and consumption of calcium hydroxide by the pozzolanic reaction, and 3) creates a more discontinuous pore structure. Porosity data show that with ultra-fine fly ash, total pore area increases as the median pore size decreases [5,16,22]. This is consistent with a denser, more discontinuous matrix.

Resistance to sulfate attack improves with the use of beneficiated fly ash compared to unprocessed fly ash [1,27,28]. The increased resistance to sulfates can be principally attributed to the decrease in concrete permeability [29].

CONCLUSIONS

Beneficiated fly ash shows potential for use in high performance concrete because it increases concrete compressive strength, workability, and durability compared to unprocessed fly ash. Ultra-fine fly ash has been shown to have an increased reactivity compared to unprocessed fly ash due to its higher surface area and glass content. Air-classified fly ash is more spherical than unprocessed ash, reducing water demand. Ultra-fine fly ash also improves particle packing, as evidenced by an increased bulk density. It decreases permeability and drying shrinkage, and increases resistance to alkali-silica reaction and sulfate attack.

REFERENCES

[1]K.H. Obla, R.L. Hill, M.D.A. Thomas, S.G. Shashiprakash and O. Perebatova, "Properties of Concrete Containing Ultra-Fine Fly Ash." *ACI Materials Journal*, **100** [5] 426-433 (2003)

[2]O.E. Manz, "Worldwide production of coal ash and utilization in concrete and other products," *Fuel*, **76** [8] 691-696 (1997)

[3]M.M.Th. Eymael and H.A. Cornelissen, "Processed Pulverized Fuel Ash for High-Performance Concrete," *Waste Management*, **16** 237-242 (1996)

[4]H.A.W. Cornelissen, R.E. Hellewaard and J.L.J. Vissers, "Processed Fly Ash for High Performance Concrete," pp. 67-80 in *Fly Ash, Silica Fume, Slag, and Natural Pozzolans in Concrete, Proceedings of the Fifth CANMET/ACI International Conference, SP-153, V.1*, Edited by V.M. Malhotra, American Concrete Institute, Farmington Hills, Michigan, 1995

[5]K.E. Hassan and J.G. Cabrera, "The Use of Classified Fly Ash to Produce High Performance Concrete," pp. 21-36 in *Fly Ash, Silica Fume, Slag, and Natural Pozzolans in Concrete, Proceedings from the Sixth CANMET/ACI International Conference, Bangkok, Thailand, SP-178, V.1*, Edited by V.M. Malhotra, American Concrete Institute, Farmington Hills, Michigan, USA 1998

[6]R.T. Hemmings, E.E. Berry, B.J. Cornelius and B.E. Scheetz, "Speciation in Size and Density Fractionated Fly Ash II. Characterization of a Low-Calcium, High-Iron Fly Ash," pp. 81-98 in *Fly Ash and Coal Conversion: Characterization, Utilization and Disposal III, V.86*, Edited by S. Diamond, Materials Research Society, Warrendale, Pennsylvania, USA, 1986

[7]P.K. Mehta, "Influence of Fly Ash Characteristics on the Strength of Portland-Fly Ash Mixtures," *Cement and Concrete Research*, **15** [4] 669-674 (1985)

[8]ASTM C 208, Standard Test Method for Fineness of Hydraulic Cement by Air-Permeability Apparatus, Annual Book of ASTM Standards, ASTM International, West Conshohocken, Pennsylvania, USA, 2003

[25]J.G. Cabrera, C.J. Hopkins, G.R. Wooley, R.E. Lee, J. Shaw, C. Plowman and H. Fox, "Evaluation of the Properties of British Pulverized Fuel Ashes and Their Influence on the Strength of Concrete," pp. 115-144 in *Fly Ash, Silica Fume, Slag, and Natural Pozzolans in Concrete, Proceedings from the Second CANMET/ACI International Conference, Madrid, Spain, SP-91, V.1*, Edited by V.M. Malhotra, American Concrete Institute, Farmington Hills, Michigan, USA, 1986

[26]S.H. Gebler and P. Klieger, "Effect of Fly Ash on Physical Properties of Concrete," pp. 1-50 in *Fly Ash, Silica Fume, Slag, and Natural Pozzolans in Concrete, Proceedings from the Second CANMET/ACI International Conference, Madrid, Spain, SP-91, V.1*, Edited by V.M. Malhotra, American Concrete Institute, Farmington Hills, Michigan, USA, 1986

[27]P. Chindaprasirt, S. Homwuttiwong and V. Sirivivatnanon, "Influence of fly ash fineness on strength, drying shrinkage and sulfate resistance of blended cement mortar," *Cement and Concrete Research*, **34** [7] 1087-1092 (2004)

[28]K.D. Copeland, K.H. Obla, R.L. Hill and M.D.A. Thomas, "Ultra Fine Fly ash for High Performance Concrete," pp. 166-175 in *Construction and Materials Issues 2001*, American Society of Civil Engineers, Reston, Virginia, USA, 2001

[29]B.P. Hughes and M.N.A Al-Ani, "PFA fineness and its use in concrete," *Magazine of Concrete Research*, **41** [147] 99-105 (1989)

[30]J. Payá, J. Monzó, M.V. Borrachero, E. Peris-Mora and F. Amahjour, "Mechanical Treatment of Fly Ashes. Part IV: Strength Development of Ground Fly Ash-Cement Mortars Cured at Different Temperatures," *Cement and Concrete Research*, **30** 543-551 (2000)

Tools for Modern and Next Generation High Performance Research

COMPARISON OF DIFFERENT METHODS FOR CHARACTERIZATION OF CEMENT-BASED MATERIALS SUBJECTED TO SULFATE ATTACK

K.E. Kurtis* and A.C. Jupe
School of Civil & Environmental Engineering,
Georgia Institute of Technology
Atlanta, Georgia USA

N.N. Naik
National Academy of Construction
Hyderabad, India

S.R. Stock
Institute for Bioengineering and
Nanoscience in Advanced Medicine,
Northwestern University
Chicago, Illinois USA

P. Stutzman
National Institute for Standards
and Technology
Gaithersburg, Maryland USA

ABSTRACT

Routine and specialized characterization of cement-based materials can be performed with a variety of tools including those considered here: stereo microscopy, laser scanning confocal microscopy, x-ray microtomography, and electron microscopy with x-ray imaging. The advantages and disadvantages associated with each of these techniques have been described. To illustrate their use and facilitate comparison among the different techniques, the application of each of these techniques in the characterization of sulfate resistance of cement-based materials has been presented.

INTRODUCTION

A variety of tools are available for both routine and specialized characterization of cement-based materials. These range from relatively low resolution optical imaging methods to higher resolution electron microscopy. A combination of backscattered electron imaging and x-ray imaging can provide further information by spatially relating the structure of a polished sample to its elemental composition. X-ray tomography provides an additional advantage of through-depth or volumetric characterization of materials, at high resolution in smaller samples. Quantitative evaluation of a sample can be performed with a variety of image analysis techniques and commercial software. These characterization methods are important components for research and often serve to complement mechanical and physical testing or provide additional understanding not available through mechanical and physical measures.

The objective of this paper is to demonstrate how various characterization techniques may be applied to cement-based materials. Optical microscopy (including both stereomicroscopy and laser scanning confocal microscopy), x-ray microtomography, and electron microscopy with x-ray imaging will be considered. To illustrate their use and to facilitate comparisons between the different techniques, each of these techniques will be described in context of their use during recent investigations of sulfate resistance of cement-based materials. In this context, the advantages and disadvantages of each technique will be described.

* Corresponding author: Dr. K.E. Kurtis, 790 Atlantic Dr., Atlanta, GA 30332-0355; phone: 404-385-0825; fax: 404-894-0211; email: kkurtis@ce.gatech.edu

EXPERIMENTAL PROGRAM

In an investigation of the fundamental relationships between materials and mixture characteristics and sulfate resistance in a range of sulfate environments (i.e., varying solution concentration and associated cation), the influence of sulfate exposure on strength, length change (expansion), structure, and chemical composition were monitored over time. Samples were cast from ASTM Type I and V cements* at w/c of 0.485 and 0.435 and were exposed at room temperature to sodium sulfate (Na_2SO_4) and magnesium sulfate ($MgSO_4$) solutions at 0.1% (1000 ppm), 1.0% (10,000 ppm), and 3.38% (33,800 ppm) sulfate ion concentration. The ratio of solution volume to sample surface area was kept constant at $2.4 cm^3/cm^2$ for all sample types, and solutions were changed weekly.

Cylindrical cement paste samples approximately 40 mm long and 12 mm in diameter were cast in plastic vials from mixtures of Type I or Type V cement and de-ionized water for characterization by microtomography and energy dispersive x-ray diffraction (EDXRD). Scanning electron microscopy with x-ray imaging, confocal microscopy, and stereo microscopy have also been performed on these specimens.

Based upon the sulfate resistance test method described by Mehta [1975] and previously implemented by Kurtis *et al.* [2001], 12.7 mm (0.5 in) cement paste cubes were cast from the same two cements. Compressive strength measurements were performed on the cement paste cubes using a 9980 kg-capacity (22,000 lb), screw driven universal testing machine with a load rate of 272 kg/min (600 lb/min). Compressive strength measurements were made on six replicate samples for each condition.

In addition to strength measurements, measurements of length change were made on mortar bar samples cast in 25 mm x 25 mm x 285 mm (1 in x 1in x 11 in) brass molds from the same two cements, with *w/c* of 0.485 and 0.435 and sand-to-cement ratio of 2.75 by mass. Length change measurements were conducted weekly generally on six replicates, but not on fewer than three replicate samples (as some samples failed during exposure).

All samples (i.e., portland cement-based cylinders, cement paste cubes and mortar bars) were demolded after 1 d of accelerated (ASTM C1012 sulfate attack conditions) or normal curing and were subsequently cured for an additional 2 d in limewater at room temperature. The three-day curing period was determined as described in ASTM C 1012, where the curing period is determined by the time necessary for the compressive strength of 50.8 mm (2 in) mortar cubes to reach 20.0 MPa. After curing, the samples of each type were placed in sulfate solutions, while some samples remained in saturated limewater baths (referred to as unexposed samples) or were stored in sealed containers at room temperature to serve as controls. Further details regarding the results of these tests can be found in [Naik, 2003; Stock *et al.*, 2002; Naik *et al.*, 2004; Jupe *et al.*, accepted; Naik *et al.*, in review].

OPTICAL MICROSCOPY

Stereo Microscopy

Stereo microscopy is probably the most commonly employed form of optical microscopy for routine examination of cement-based materials. Standard practices for petrographic examination of hardened concrete, including examination by stereo microscopy, are described by ASTM C 856. In addition, a good reference for concrete petrography can be found in St. John, Poole, and Sims [1998].

---

* Cement compositions are reported in [Naik *et al.*, in review] and [Naik, 2003]

Stereo microscopy can be performed at a range of magnifications (generally not higher than 80-100x) on polished or unpolished samples, offering much flexibility in terms of the types of samples that may be characterized. The lower magnification of this technique, relative to SEM for instance, is useful for examining larger regions of interest. In addition, images can be acquired in color, which represents an advantage for distinguishing between phases.

However, the rather low magnification available can represent a limitation when attempting to resolve features smaller than ~0.1 mm. Another disadvantage occurs when attempting to image curved or rough surfaces, such as within cracks or on fracture surfaces. With these samples, often certain portions of the sample will be in focus, while other portions remain out-of-focus. This limitation is overcome by confocal microscopy (see below).

Figure 1 shows some images of cement pastes obtained with a Leica WILD MZ6 stereo microscope.* This particular instrument has a 6.3 x to 80 x magnification range and is fitted with a photo/video camera. Images in Figure 1 (a) and (b) were obtained at relatively low magnification. Figure 1 (a) shows the cracked surface of one of the cylindrical cement paste specimens which were also examined by microtomography. The image clearly shows a pattern of surface cracking and some discoloration on the surface. In color images of this sample, the regions which appear lighter here in grayscale are white and are clearly distinguished from the rest of the surface; the whitish color suggests that these are likely products of efflorescence. Also apparent in Figure 1 (a) is the loss of focus at the edges of this curved sample. A higher magnification of a companion sample, Figure 1 (b), shows the cracking and some apparent surface product formation, but also demonstrates the difficulties with focus encountered when imaging rough surfaces by stereo microscopy.

Figure 1 (c) shows spalling (or material loss at the edges and corners) of this cement paste cube exposed to sulfate attack. By varying the sample tilt or by varying the location of the light source, long, needlelike products can be made to reflect light. The morphology, given the exposure conditions suggests that these needles are likely ettringite crystals. Such manipulation of the sample and lighting is an advantage of stereo microscopy for routine examination of samples.

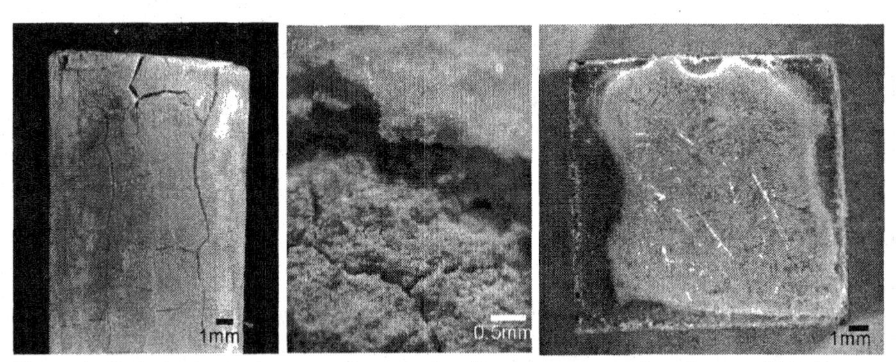

Figure 1. Optical stereo microscopy images showing (a) a pattern of surface cracking in a cylindrical sample of Type V cement paste (w/c=0.485) after 78 weeks exposure to 3.38% (33,800 ppm) sulfate in Na_2SO_4 solution, (b) the cracking and some surface product formation at higher magnification in a companion sample, and (c) some evidence of spalling and apparently crystalline products at the surface of a 12.7 mm cement paste cube.

Confocal Microscopy

Confocal microscopy is a form of optical microscopy, and, hence, like stereo microscopy, also requires little sample preparation prior to imaging. That is, polished or rough surfaces of small and large samples may be imaged at atmospheric pressure. Confocal microscopy's advantage is the ability to generate sharp images of relatively rough surfaces and to image through a sample depth, as described below. Most confocal microscopes have been configured for greater magnification than stereo microscopes, some offering magnification up to ~2000 x. Confocal images are generally obtained in gray scale.

With confocal microscopy only one plane in the sample is imaged at a time. A point light source (laser) is typically used along with an aperture for limiting the beam size. A second aperture is placed in a conjugate focal plane of the objective lens in order to reject all light except that from the in-focus plane in the sample. The sample is scanned across the beam to build up an image of the plane of the sample. This imaging scheme improves resolution as compared to ordinary optical imaging methods, and images collected at a series of focal planes can be reconstructed into a single three-dimensional volume, a projection image, a stereo image, or a video providing perspective. Further details regarding the working of confocal microscopes can be found in [Corle and Kino, 1996; Sheppard and Shotton, 1997; Clarke and Eberhardt, 2002].

For cement-based samples, this "non-destructive sectioning" allows for reconstructions of rough surfaces or for representations of the volumetric structure of transparent (i.e., thin sectioned samples) and semi-transparent materials. Surface characterization is probably the most common use of confocal microscopy in cement-based materials. Confocal microscopy has been used to image concrete, mortar, cement paste, and fiber-cement fracture surfaces. Quantifications of surface roughness derived from the confocal images have been related to mechanical behavior and fracture processes [Lange *et al.*, 1993(a),(b); Zampini *et al.*, 1995; Abell *et al.*, 1998; Kurtis *et al.*, 2003]. Confocal microscopy has also been used to volumetrically characterize pozzolanic reaction products [Kurtis *et al.*, 2003].

A confocal microscopy technique has also been developed to allow "through-aggregate" imaging. This overcomes a key challenge for the *in situ* study of cement-based materials – the inherent difficulty in imagining the aggregate-paste interface without introducing artifacts through sample preparation. By using optically transparent aggregate such as glass or single crystal quartz, through-depth (up to ~2 mm) characterization to the aggregate/paste interface is possible with confocal microscopes using appropriate objectives [Kurtis *et al.*, 2003; Collins *et*

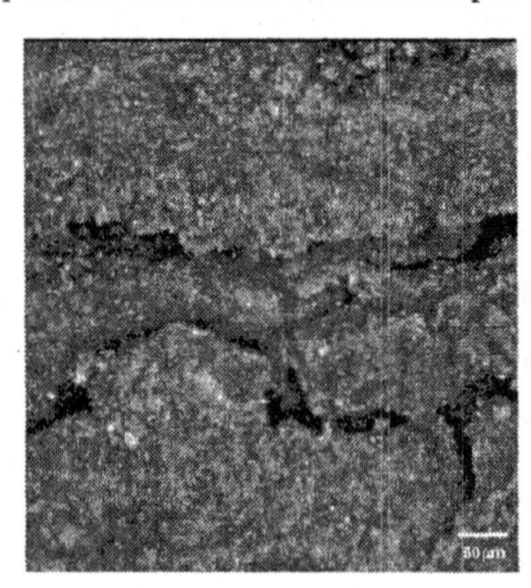

Figure 2. Projection image a Type I cement paste sample with w/c=0.485 exposed to a Na_2SO_4 solution with 1% (10,000 ppm) sulfate ion concentration for 637 days (91 weeks). The image was produced from 125 confocal images spaced 5 μm apart in the z-direction.

al., 2004]. The limitation to this approach is that imaging can occur only through transparent material.

The projection image of a Type I cement paste cylindrical sample in Figure 2 was obtained using an Argon-laser Leica Confocal TCS NT microscope. This is the same sample depicted in Figure 1 (b). Figure 2 clearly shows the improvement in imaging clarity afforded by confocal microscopy. All portions of the region imaged – those both on the surface and within the crack – are in focus. The material at the surface is lighter in color and, as viewed through the eyepiece, apparently crystalline. Although the improved resolution and clarity of confocal imaging is a clear advantage over stereo microscopy, variations in the composition and structure of the sample, which may be apparent during observation or stereo imaging in color, are not as clear in monochrome images.

MICROTOMOGRAPHY

X-ray computed tomography (also known at CAT or CT scanning) is a familiar medical procedure that can also be used advantageously for the characterization of cement-based materials. X-ray microtomography (microCT) may be distinguished from CT scanning by its improved spatial resolution. Most modern medical CT scanners have spatial resolutions on the order of millimeters, while manufacturers of desktop microCT scanners claim resolutions to several micrometers, with newer tools promised to give resolution to 0.5 μm. With desktop scanners, however, sample size in typically limited to the tens of millimeters size range.

X-ray microtomography allows the high-resolution, volumetric characterization of opaque samples. The capacity to characterize a sample through its volume represents a key advantage over conventional microstructural characterization methods where a sample needs to be broken to examine its internal structure. A series of contiguous two-dimensional slices are typically reconstructed in computed tomography and these stacks of slices form the three-dimensional volumetric representations. Microtomography is performed by passing x-rays through the sample at discrete angular increments, with an appropriate detector configuration. Variations in sample composition and density produce variations in x-ray absorption, giving the contrast necessary to discern internal features within the sample. Use of contrast agents can provide further enhancement. By translating the sample through the x-ray source(s), a series of slices can be obtained and can be reconstructed to give a volumetric representation of the sample. Reconstruction time varies with the instrument and software used, resolution, and sample size, but typically can be accomplished in 20 minutes to 2 hours time. Further details regarding the underlying physics and mathematics, as well as details regarding apparatus and data collection, can be found in Stock [1999].

Because microCT allows non-destructive evaluation, *in situ* process monitoring in the same

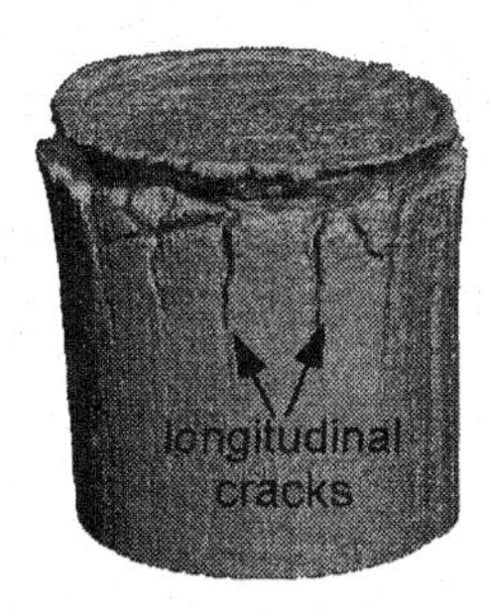

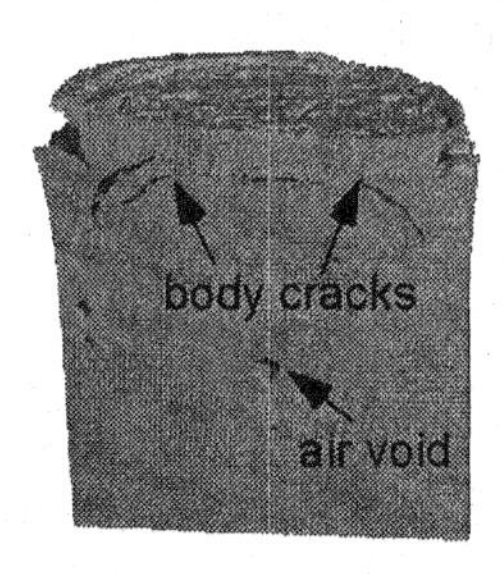

Figure 3. 3-D renderings, from microCT, of a Type I cement paste sample (w/c=0.485) after Na_2SO_4 exposure for 84 days (12 weeks) at a sulfate concentration of 3.38% (33,800 ppm) show (a) evidence of expansion and longitudinal cracking and (b) internal cracking or “body” cracks. Sample diameter is 12 mm.

sample is possible. Potential applications include monitoring of hydration reactions (with very high resolution instruments) and monitoring damage evolution due to environment or loading. For instance, using microCT it is possible to map the development of a crack surface in the same sample with increased loading or exposure. With this data, quantitative measures of damage evolution can be made and perhaps linked to service life predictions.

Figure 3 shows 3-D reconstructions from microtomography of a Type I cement paste cylindrical specimen exposed to sodium sulfate attack. External cracking and expansion are clearly apparent in the full-volume reconstruction (Figure 3 (a)). This surface crack pattern is quite similar to that observed by stereo microscopy in Figure 1 (a). However, a partial reconstruction (Figure 3 (b)) also reveals internal cracking or "body cracks" ~2mm below the sample surface, as well as some pre-existing entrapped air voids, which would not have been apparent with stereo microscopy. The use of microtomography with image processing to qualitatively and quantitatively characterize damage by sulfate attack has been treated extensively in Naik *et al.* [2004]; microtomography combined with energy dispersive x-ray diffraction for compositional analysis of sulfate-damaged cement-based materials has been described in Naik [2003] and Naik *et al.* [in review].

SCANNING ELECTRON MICROSCOPY WITH X-RAY IMAGING

Scanning electron microscopy is a familiar and useful technique for sub-micrometer characterization of cement-based samples. Sample preparation requirements vary with the type of imaging performed (i.e., secondary electron or backscattered electron) and the system used (i.e., traditional SEM performed under vacuum vs. variable-pressure SEM.

Secondary electron imaging may be performed on relatively rough surfaces and is useful for examination of microstructural features at high-magnification; samples may be coated (e.g., with carbon, gold, osmium) for improved conductivity or left uncoated. Epoxy-impregnated, polished samples are typically used for backscattered electron (BE) imaging. BE imaging is a technique which is useful for highlighting compositional differences within a sample based on differential brightness as related to average atomic number. Highly polished samples may be relatively rapidly characterized by backscattered electron and x-ray imaging. X-ray imaging is a technique which is useful for identifying elements present in a region of interest, determining the relative concentration of those elements present, and for phase identification, when spatial distribution and relative elemental concentration can be measured. Stutzman [2000] provides an excellent review of the working principles of SEM, preparation of cement-based samples for SEM, and examples of practical application of SEM and x-ray imaging.

While SEM is performed under vacuum conditions and requires sample drying, environmental scanning electron microscopy (ESEM) and variable-pressure SEM allow imaging of non-conducting, hydrated samples under higher pressures. Because of the advantages of less complicated specimen preparation and the ability to image hydrated samples, these instruments have gained relatively rapid acceptance for characterization of cement-based materials.* Use of ESEM and variable-pressure SEM to characterize cement-based materials is expected to increase with increasing access to equipment.

---

* For examples of use of ESEM and variable-pressure SEM for characterization of cement-based materials, see Sarkar *et al.* [1992], Meredith *et al.* [1995], Katz *et al.* [1995], Kjellson and Jennings [1996], Su *et al.* [1996], Neubauer and Jennings [1996], Deng *et al.* [2002], and Ye *et al.* [2002].

Figure 4 shows a backscattered electron image and x-ray images of an epoxy-impregnated, polished cylindrical, cement paste sample of the same sort previously examined by stereo microscopy (Fig. 1), confocal microscopy (Fig. 2), and microtomography (Fig. 3). These images in Figure 4 were obtained with a variable-pressure SEM (FEI Quanta 600). At higher magnification, evenly spaced cracks, presumably from shrinkage of the sample during epoxy impregnation under vacuum, were observed in the radial direction (indicated "R" in Fig. 4).

The darkening toward the sample surface in the SEM image reflects a loss of near-surface material with 12 weeks exposure to sodium sulfate solution. This outer 150 μm is comprised of C-S-H and minor amounts of residual cement, principally ferrite. Calcium hydroxide, ettringite and monosulfate are absent, having all been leached from this zone. While an EDS spectrum comparing this region to the relatively unaffected portions of the sample did not bear out a Ca-

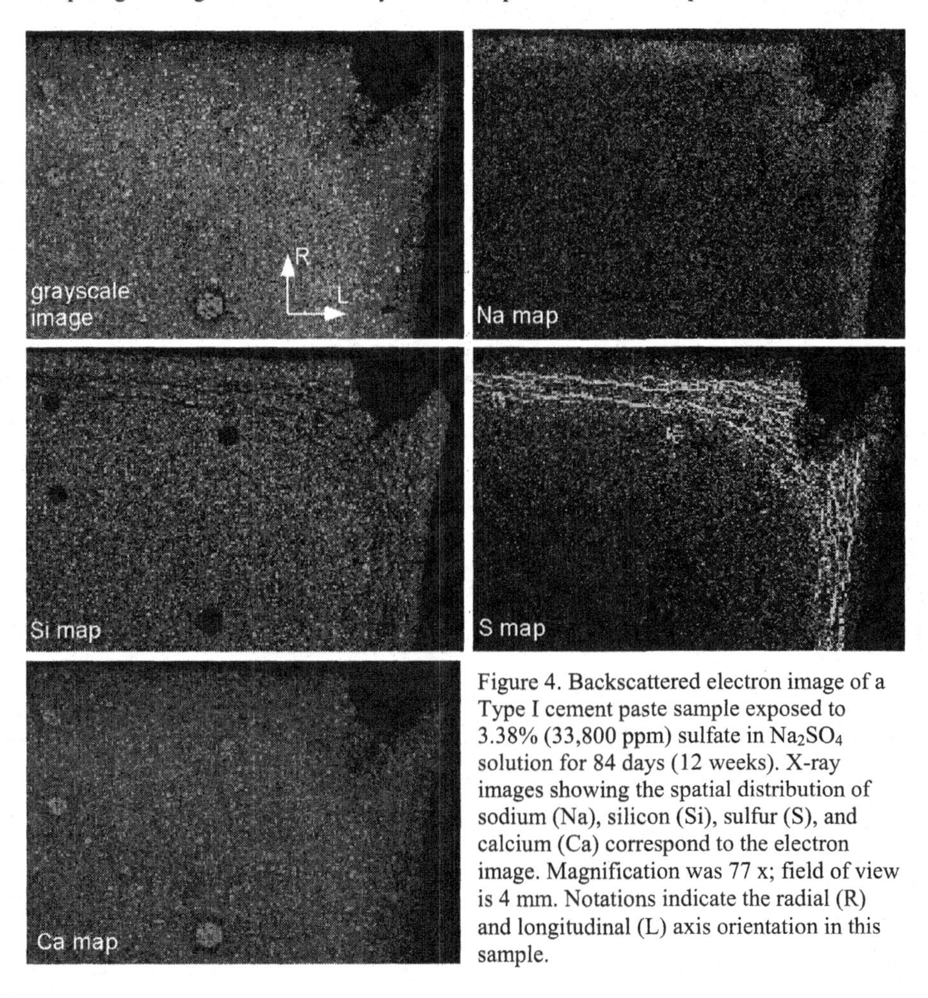

Figure 4. Backscattered electron image of a Type I cement paste sample exposed to 3.38% (33,800 ppm) sulfate in Na_2SO_4 solution for 84 days (12 weeks). X-ray images showing the spatial distribution of sodium (Na), silicon (Si), sulfur (S), and calcium (Ca) correspond to the electron image. Magnification was 77 x; field of view is 4 mm. Notations indicate the radial (R) and longitudinal (L) axis orientation in this sample.

depletion of the C-S-H, a relatively larger carbon peak (from the embedding resin) indicates an increased microporosity of this phase. The sodium and sulfur images demonstrate the ingress of the solution to a depth of at least 2.5 mm. While the Na is present consistently through the near-surface layer, sulfur is absent from the first 150 μm within the highly-leached zone. Also, the sulfur is largely detected in "bands" within the region 151 μm to 450 μm below the surface. In these sulfur-rich banded regions, silicon is not detected, but calcium is, however, detected. Thus, the x-ray data suggests the secondary formation of gypsum in bands which were likely microcracks. Cracks were not detected in this sample at this age by microtomography.

Also of interest is the presence of calcium within each of the four entrapped air voids visible in the SEM image. It is not uncommon to find large calcium hydroxide crystals in these large voids. Silicon and aluminum (data not shown here) are not found in any of these pores, but sulfur is detected in one about 1 mm from the surface. This suggests that calcium hydroxide within the entrapped void has reacted with ingressing sulfate solution to form gypsum in the near-surface void, but has not reacted with those at greater depths at the time of sampling.

IMAGE ANALYSIS

As a footnote to this discussion of imaging methods, it is worthwhile to mention image analysis. Image analysis consists of a powerful and broad class of techniques for manipulation and quantification of images, which may be obtained from any imaging method. Image analysis can allow researchers to obtain more information than may "meet the eye" from an image or series of images. For instance, the size and relative distribution of features such as cracks and voids or various phases in a sample can be quantified. Also, image correlation can be used to map changes, which may not be readily apparent when comparing two or more images of the same region of interest, and to describe these changes in terms of relative displacement or strain.* As previously described, the roughness of fracture surfaces can also be quantified through image analysis and related to mechanical measurements of toughness and fracture toughness. These are but just three examples of the many possible applications of image analysis to cement-based materials.

Much commercial software of varying complexity and cost is available for image analysis. However, when considering the acquisition of image analysis software, it is worth investigating shareware tools, such as those available on line from the National Institute of Health (NIH) (http://rsb.info.nih.gov/nih-image/Default.html). NIH Image (Apple), Image J (a Java version which runs in any environment), and Lispix (Windows, Apple) (found at http://www.cstl.nist.gov/div837/Division/outputs/software.htm), are all available free of charge. Many plugins, also available for free download, have been developed to run with these programs and can be used for a wide variety of image manipulation and analysis.

CONCLUSIONS

A set of cement paste samples subjected to sodium sulfate attack were examined by four characterization methods in order to compare the relative merits of these techniques. Through this examination, the advantages and disadvantages of each can be summarized as:

- Stereo microscopy allows for rapid and simple color imaging, with in situ sample manipulation, making this technique useful for routine examination at relatively low magnification.

* For examples of image correlation applied to cement-based materials, see [Xi et al., 1994; Neubauer *et al.*, 1997; Neubauer *et al.*, 2000; Neubauer and Jennings, 2000; Lopez *et al.*, 2003]

- Confocal microscopy allows for improved focus and resolution when imaging unpolished samples and allows for volumetric or through-depth characterization of transparent or semi-transparent materials; images are typically monochromatic.
- Microtomography allows for non-destructive, volumetric characterization of opaque materials with resolution on the order of several micrometers for relatively small (~10 mm - 20 mm) samples and is quite advantageous for *in situ* process monitoring, although reconstruction times can be lengthy depending upon sample size and imaging resolution.
- Scanning electron microscopy combined with x-ray imaging allows for sub-micrometer characterization of phase and chemical spatial distribution of highly polished samples; when using planar, polished surfaces, the images are amenable to processing and analysis for quantitative microscopy.

ACKNOWLEDGEMENTS

Figure 1 (b) was provided by Ford Burgher, and Figure 2 was provided by Ben Mohr, both in the School of Civil and Environmental Engineering at Georgia Tech. Some of the research presented here was supported by National Science Foundation (NSF) CMS-0084824 and CMS-0074874. The microtomography equipment was acquired under NSF OIA-9977551. Any opinions, findings, and conclusions or recommendations expressed in this material are those of the authors and do not necessarily reflect the views of the sponsor.

REFERENCES

Abell, A.B. and Lange, D.A., "Fracture mechanics modeling using images of fracture surfaces," *Int. J. Solids Struct.*, V35:4025-33, 1998

ASTM C 856, "Standard Practice for Petrographic Examination of Hardened Concrete," Book of Standards, V. 04-02, American Society for Testing and Materials, 2004

Collins, C.L.; Ideker, J.H. and.Kurtis, K.E., "Laser Scanning Confocal Microscopy for *In Situ* Monitoring of Alkali-Silica Reaction," *Journal of Microscopy*, Feb 2004, V.213(2):149-157

Clarke, A.R. and Eberhardt, C.N., Microscopy Techniques for Materials Science, CRC Press, Woodhead Publishing Limited, Cambridge, p.229-302, 2002

Corle, T.R. and Kino, G.S., Confocal Scanning Optical Microscopy and Related Imaging Systems, Academic Press, New York, 1996.

Deng, C.S.; Breen, C.; Yarwood, J.; Habesch, S.; Phipps, J.; Craster, B.; and Maitland, G., "Aging of oilfiled cement at high humidity: A combined FEG-ESEM and Raman microscope investigation," *J. Mat. Chem.*, V12(10):3105-12, 2002

Jupe, A.C.; Stock, S.R.; Lee, P.L.; Naik, N.; Kurtis, K.E. and Wilkinson, A.P., "Phase Composition Depth Profiles using Spatially Resolved EDXRD," *J. Applied Crystallography*, accepted for publication

Katz, A.; Li, V.C.; and Kazmer, A., "Bond properties of carbon fibers in cementitious matrix," *J. Materials in Civ. Eng.*, V7(2):125-8, May 1995

Kjellson, K.O. and Jennings, H.M., "Observations of microcracking in cement paste upon drying and rewetting by environmental scanning electron microscopy," *Adv. Cem. Based Mat.*, V3(1):14-9, January 1996

Kurtis, K.E.; El-Ashkar, N.H.; Collins, C.L. and Naik, N.N., "Examining Cement-Based Materials by Laser Scanning Confocal Microscopy," *Cement & Concrete Composites*, October 2003, V.25(7):695

Kurtis, K.E.; Shomglin, K.; Monteiro, P.J.M.; Harvey, J. and Roesler, J., "Accelerated Test for Measuring Sulfate Resistance of Calcium Sulfoaluminate, Calcium Aluminate, and Portland Cements," *ASCE Journal of Materials in Civil Engineering*, May/June 2001, V13:216-221

Lange, D.A.; Jennings, H.M.; and Shah, S.P., "Analysis of Surface Roughness using Confocal Microscopy," *J. Mater. Sci.*, V28:3879-84, 1993(a)

Lange, D.A.; Jennings, H.M.; and Shah, S.P., "Relationship between Fracture Surface Roughness and Fracture Behavior of Cement Paste and Mortar," *J. Am. Cer. Soc.*, V76(3):589-97, 1993(b)

Lopez, M.; Kurtis, K.E. and Kahn, L.F., "Creep Strain Distributions and Deformation Mechanisms of High Performance Lightweight Concrete," Proc. 9th Int. Conf. on Advances in Cement and Concrete: Volume Changes, Cracking, and Durability, Engineering Conferences International, Copper Mountain, Colorado, Aug. 10-14, 2003

Mehta, P.K. "Evaluation of sulfate-resisting cements by a new test method," *J. ACI*, 72 (1975) 573-575

Meredith, P.; Donald, A.M.; and Luke, K., "Preinduction and induction hydration of tricalcium silicate – An environmental scanning microscopy study," *J. Mater. Sci.*, V30(8):1921-30, April 15, 1995

Naik, N.N., Sulfate attack on portland cement-based samples: Mechanisms of damage and long-term performance, Ph.D. thesis, Georgia Institute of Technology, Atlanta, Georgia, 2003

Naik, N.N.; Jupe, A.C.; Stock, S.R.; Wilkinson, A.P. and Kurtis, K.E.,"Sulfate Attack Monitored by MicroCT and EDXRD: Influence of Cement Type, Water-to-Cement Ratio, and Aggregate," submitted to *Cement and Concrete Research*, June 2004

Naik, N.N.; Stock, S.R.; Wilkinson, A.P. and Kurtis, K.E., "Sulfate deterioration of cement-based materials examined by x-ray microtomography," to be published in Proc. of the SPIE 49th Annual Meeting, Optical Science and Technology: Developments in X-ray Tomography IV, Denver, August 2-6, 2004

Neubauer, C.M. and Jennings, H.M. "The Role of the Environmental Scanning Electron Microscope in the Investigation of Cement-Based Materials," *Scanning*, V18(7):515-21, October 1996

Neubauer, C. M.; Garboczi, E. J.; and Jennings, H. M. (2000). "The Use of Digital Images to Determine Deformation Throughout a Microstructure Part I Deformation Mapping Technique," *Journal of Material Science* **35**: 5741-5749

Neubauer, C. M. and Jennings, H. M. (2000). "The Use of Digital Images to Determine Deformation Throughout a Microstructure Part II Application to Cement Paste," *Journal of Material Science* **35**: 5751-5765

Sarkar, S.L and Xu, A.M., "Preliminary study of very early hydration of superplasticized C_3A+Gypsum by environmental SEM," *Cem. Conc. Res.*, V22(4):605-608, July 1992

Sheppard, C.J.R. and Shotton, D.M., Confocal Laser Scanning Microscopy, Microscopy Handbook 38, BIOS Scientific Publishers with the Royal Microscopy Society, London, 1997

St. John, D.A.; Poole, A.W.; and Sims, I. Concrete Petrography: A Handbook of Investigative Techniques, John Wiley and Sons, New York, 1998

Stock, S.R., "X-ray Microtomography of Materials", *Int. Mater. Rev.,* V44(4):141-64, 1999.

Stock, S.R.; Naik, N.N.; Wilkinson, and K.E. Kurtis, K.E. "X-ray Microtomography (MicroCT) of the Progression of Sulfate Attack of Cement Paste," *Cement and Concrete Research*, October 2002, V32(10):1673-5. Errata, December 2002, V.32(12): 2002

Stutzman, P.E., “Scanning Electron Microscopy in Concrete Petrography,” Mat. Sci. of Concrete Special Volume: Calcium Hydroxide in Concrete, Proc. Workshop on the Role of Calcium Hydroxide in Concrete, J. Skalny, J. Gerbauer, and I. Odler (Eds.), The American Ceramic Society, pp. 59-72, 2001. (also available at http://fire.nist.gov/bfrlpubs/build01/PDF/b01086.pdf)

Su, Z.; Sujata, K.; Bijen, J.M.; Jennings, H.M.; Fraaij, A.L., “The evolution of the microstructure in styrene acrylate polymer-modified cement pastes at the early stage of cement hydration,” *Adv. Cement Based Mat.*, V3(3-4):87-93, April-May 1996

Xi, Y., Bergstrom, T. B., and Jennings, H. M. (1994). "Image Intensity Matching Technique: Application to the Environmental Scanning Electron Microscope," *Computational Material Science* **2**: 249-260

Ye, G.; Hu, J.; van Breugel, K.; Stroeven, P., “Characterization of the development of microstructure and porosity of cement-based materials by numerical simulation and ESEM image analysis,” *Mat. and Struct.*, V35(254):603-13, December 2002

Zampini, D.; Jennings, H.M.; and Shah, S.P., “Characterization of the paste-aggregate interfacial transition zone surface-roughness and its relationship to the fracture-toughness of concrete,” *J. Mater. Sci.*, V30:3139-54, 1995

SUSTAINABLE DEVELOPMENT: APPROACH FOR RESEARCH ON NEXT GENERATION HIGH PERFORMANCE CONCRETE WITH FLY ASH IN INDIA

P.C. Basu
Civil & Structural Engineering Division,
Atomic Energy Regulatory Board,
Mumbai, India

INTRODUCTION

Sustainable development means the development that can be pursued continuously in future. Sustainable development, therefore, should meet the present demands of human needs without compromising the needs of future generation. Concrete is the second most used commodity for human civilization, next to water, is key for sustainable development of India.

The main challenge before the Indian concrete industry now is to meet the demand of economical but efficient construction materials posed by enormous infrastructure needs due to rapid industrialization and urbanization. Again the spree of present development should not jeopardize the environment and also preserve the resources for future. All these call for use of high performance concrete (HPC) with the minimum resources (e.g. limestone, energy, money) and achieving maximization of strength, durability and other intended mechanical properties. Therefore, key to sustainable development of India with respect to concrete is associated with optimal concrete (mix) maintaining environmental balances. With the shrinkage of natural resources to produce ordinary Portland cement (OPC), use of some industrial waste materials having pozzolanic characteristics that can replace cement clinker is one of the efficient ways to meet the present challenge. Fly ash is most suitable cement replacing material for this purpose.

India is a country of wide disparity. Such disparity is prevailing in construction industries too. Concrete structures are being constructed both in organized and unorganized sectors. In the organized sector, construction of concrete structure is carried out following good engineering practices in line with codal requirements, adhering to strict quality control and employing state of the art machineries. Example for such sectors are nuclear and hydal power plant, heavy industries, highways, ports, etc. On the other hand, in unorganized sector, construction is being carried out in a very subjective way. Such sector comprises of construction of small building, infrastructure work of small budget etc. The research work on high performance concrete with fly ash needs to be carried out keeping the above in mind.

FLY ASH CONCRETE FOR SUSTAINABLE DEVELOPMENT

There are a few pozzolanic materials, which are suitable to HPC. Example of such pozzolans is silica fume (SF), high reactivity metakaolin (HRM), rice husk ash (RHA), and fly ash (FA). Silica fume is high reacting pozzolona. This has successfully been used in Indian industry [1]. However, India does not produce this pozzolan, and import of this material makes it comparatively expensive. Moreover, its usage does not address environmental issues prevailing in India. High reactive metakolin has been found to be high reactive pozzolans like silica fume [2]. Basic material – kaolin is to be mined from deposits that are available in abundance in India. Kaoline is also required for other industry. Why not keep it for other industrial uses and for the future. Rice husk ash is also reported to be a good pozzolonic material. Its production is not common in India and also its disposal does not pose environmental problem as that of other industrial waste. On the other hand, use of fly ash has many fold advantages in India; utilization of this industrial waste from thermal power plant in concrete is one of the eco-friendly ways of its disposal, saving of natural resources, and finally, improvement in properties of concrete.

It is estimated that 100 million ton fly ash is annually produced in India from major coal based power plants [3]. This quantity is almost same as that of present cement production in India. Only a fraction, about 15 to 18 percent, of total fly ash is used in India. Increase in usage of fly ash in concrete as cement-replacing material will not only preserve natural resources, but also reduce the emission of CO_2 and problem of its disposal. The positive effects of using fly ash in concrete are summarized as follows,

- Enhancing the quality of concrete mix and durability of concrete structures,
- Saving in energy requirements in the production of OPC,
- Minimizing greenhouse gas emissions associated with the manufacturing of OPC,
- Ecological and economical disposal of millions of tones of fly ash,
- Preservation of lime stone reserve and coal deposits,
- Stoppage of conversion of large area of agricultural land into ash ponds.

The projected availability of fly ash in India is expected to reach 200 million tones in the year 2010. Considering the fact that huge quantity of fly ash is produced in India, use of fly ash in increasing quantity in concrete is definitely the best possible effective utilization of this industrial waste for sustainable development of the country. Fly ash is introduced into concrete in one of the two following ways,

- Blended cement manufactured by mixing fly ash with OPC clinkers,
- Fly ash may be used as an additional ingredient at the concrete mixing stage.

The use of blended cement is easier, since it is free from the complications of batching additional materials at the construction site and may ensure more uniform control. Fly ash blended cement, Portland pozzolana cement (PPC), is gradually becoming a common cementitious material in India. Share of PPC is about 45% of total current cement production in India.

The addition of fly ash at the concrete mixing stage is flexible and allows for more exploitation of fly ash as an ingredient of concrete. Fly ash may be used in concrete in varying percentages. In low volume fly ash concrete, fly ash is added to concrete as a replacement of cement in the range of 10 to 30 % by mass. When the cement replacement level is 50% or above, the class of concrete is termed as high volume fly ash concrete (HVFAC) [4]. This class of concrete has very low water content, which I possible due to use of superplasticizers. Practically no application of HVFA concrete can be found presently in India.

Consistent quality of fly ash for concrete mix could be procured from different thermal power plants. Considering this aspect, the development of proper technology for use of fly ash with Indian cement in concrete both in organized as well as unorganized sectors of concrete construction, is in real sense, the research need for next generation high performance concrete for sustainable development of India.

RESEARCH NEED ON FLY ASH CONCRETE FOR SUSTAINABLE DEVELOPMENT IN INDIA

Use of fly ash as mineral admixture has manifold beneficial effects on the properties of HPC,

- Reduction of CH crystals
- Improvement of transition zone

- Minimization of voids and improved grain packing
- Considerable reduction in permeability
- Discontinuous capillary pore system.
- Less heat of hydration and volumetric change.

These beneficial effects, resulting in considerable improvement in strength as well as durability of concrete against attack due to chemical as well as physical process, has been established in USA, Canada and European countries by means of voluminous data generated from extensive research [4,5,6,7,8,9]. This available data is extremely useful in Indian context. However, there exist certain reasons for further research on fly ash concrete in India. For example, characteristics of Indian cement are different from that of other countries; C_3S content of Indian cement is lower than that of Western countries. Direct application of available data, such as on mix proportioning of fly ash concrete, generated in Western countries may, some times, lead to wrong and confusing outcome. In view of this, further research on fly ash concrete is necessary in certain areas considering specificity of Indian condition. Example of such areas is,

- Characterization of Indian fly ash,
- Mix proportioning,
- Durability,
- Effects of manufacturing process on the behavior of fly ash concrete,
- Curing,
- Utilization of coarser fly ash.

Characterization of fly ash

Most of the fly ashes in India, barring a few exceptions are of low – calcium category. IS3812: 1981 [10] specifies fly ash for use as pozzolona and admixture. The code specifies two grades of fly ash,

Grade I: Fly ash derived from bituminous coal

Grade II: Fly ash derived from lignite coal.

Limiting value of loss of ignition, 12%, specified by the IS Code is rather high, better to be restricted within 6% as per ASTM C618 01. The fly ash grains having size more than 45µ have very negligible pozzolanic property and are non reactive. Although a large quantity of fly ash is available now, all of them may not be suitable for the use in cement or concrete. What urgently required is well-documented information on the characteristics of Indian fly ash. This requires systematic survey, data collection as well as data generation on the fly ash available at different power plants located all over India.

Fly ash concrete mix

Good quantum of effort has been made over the years at different institutions/organizations spread all over India to develop mix proportions of fly ash concrete, especially HVFAC. Tables I, II and III contain the characteristics of fly ash collected from a thermal power plant. This fly ash was used in laboratory trials to develop fly ash concrete mixes having total binder content of 500 Kg per cubic meter of concrete with various cement replacement level [11].

Table I: Physical properties of fly ash

Characteristics	Test Result	IS Code provision
Fineness (m^2/kg) by Blaine's permeability method	675	320
Lime Reactivity (MPa)	6.5	4.0
Compressive strength at 28 days as percentage of corresponding plain cement mortar cubes	92.0	80.0
Drying shrinkage (%)	0.08	0.15
Soundness by Autoclave method (%)	0.18	0.8

Table II: Chemical composition of fly ash

Constituents	Test Result	IS Code provision
SiO_2 (%)	59.27	35.0 (min)
Al_2O_3 (%)	26.84	$SiO_2 + Al_2O_3 + Fe_2O_3 = 70.0$
Fe_2O_3 (%)	5.29	(min)
CaO (%)	2.09	
MgO (%)	0.75	5.0 (max)
Na_2O (%)	0.39	1.5 (max)
K_2O	0.91	
SO_3 (%)	Trace	2.75 (max)
TiO_2 (%)	3.15	
Loss of Ignition (%)	2.50	12.0 (max)

Table III: Grain size analysis of fly ash

Size range (micron)	% Finer
300	100.0
200	99.5
150	97.7
90	80.4
45	34.3
30	33.5
20	25.8
10	22.3
5	18.1

Composition of five trial fly ash concrete mixes is presented in Table-IV. The reference mix, CM, is a control mix in the context that fly ash is not added. Mixes FM1 and FM2 were developed having cement replacement levels 20%, 33.3% by fly ash are of low volume category. The cement replacement level of mixes FM3, FM4 and FM5 are 50%, 60% and 70% respectively, these mixes fall under the category of high volume fly ash concrete (HVFAC) category. Table-V contains the initial slump, 28 days compressive strength ($f_{c,28}$) and split tensile strength ($f_{st,28}$) of the mixes. Fig. 1 contains the variations of slump with time. The variation of compressive strength and split tensile strength with age are given in Fig(s). 2 and 3 respectively for mixes CM, FM1, FM2, and FM3.

Table IV: Mix proportions of fly ash concrete

Mix	Mix Proportions (per m^3)							
	Aggregate (kg)			Binder		w/b ratio	Water (Kg)	Super plasticiser[4] (%)
	Coarse	Fine	Total (Kg)	Cement (Kg)	Fly ash			
CM	1199.2	685.7	500	500	-	0.300	150.0	2.0
FM1	1092.9	673.7	500	400	100 (25^2 : 20^3)	0.300	150.0	0.70
FM2	1080.0	665.7	500	333	167 (50 : 33.33)	0.300	150.0	0.70
FM3	1146.8	655.7	500	250	250 (100 : 50)	0.300	150.0	1.00
FM4	1136.3	649.7	500	200	300 (150 : 60)	0.300	150.0	0.70
FM5	1125.9	643.8	500	150	350 (233.3 : 70)	0.300	150.0	0.70

Note: 1. % by weight of cement.
2. Quantity in terms of addition to cement
3. Quantity in terms of replacement of cement
4. % by weight of cement

Table V: Properties of fly ash concrete mixes

Mix Designation	Properties of fly ash concrete mixes		
	Initial slump (mm)	Compressive strength, $f_{c,28}$ (N/mm^2)	Tensile strength, $f_{st,28}$ (N/mm^2)
CM	150	78.0	5.38
FM1	100	74.7	4.42
FM2	175	69.6	4.31
FM3	170	58.8	3.85
FM4	180	41.1	2.92
FM5	165	28.2	-

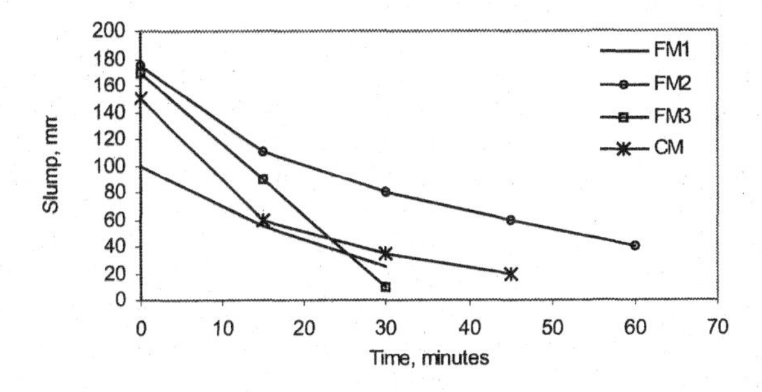

Fig.1: Variation of slump with time

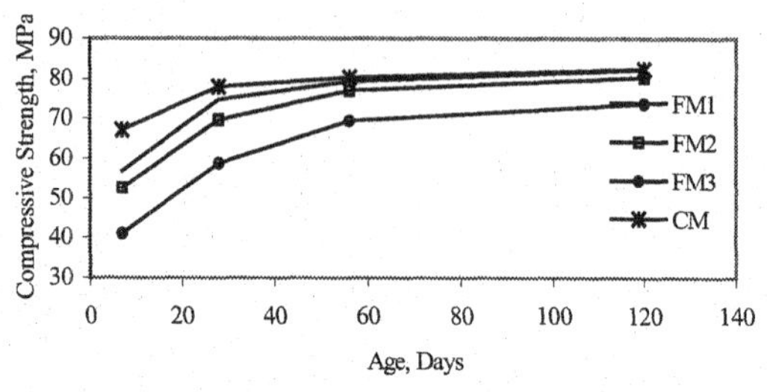

Fig. 2: Variation of compressive strength with age

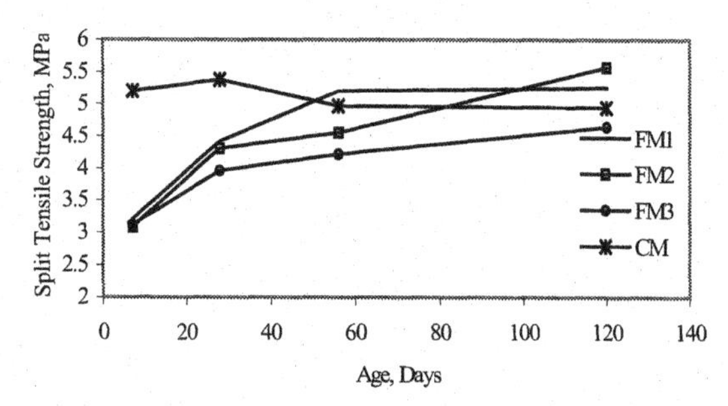

Fig. 3: Variation of split tensile strength with age

Initial slump is maximum for FM4, while it is lowest for FM1. Both the mixes, FM1 and FM2, have same unit water content and superplasticizer content but fly ash of the latter one is more. Initial slump of FM3, FM4 and FM5 are more than FM1 and fly ash content is also higher. The result indicates, in general, that the quantity of fly ash has significant contribution to initial slump. However, Fig. 1 suggests that for a better slump retention, higher unit water content is necessary.

28 days compressive strength is maximum for control concrete mix and decreases with the increase in replacement of cement by fly ash. This is expected because pozzolanic action of fly ash is comparatively slow at initial stage. Compressive strength attains almost same value at 120 days for mixes CM, FM1 and FM2 (Fig. 2), while that of FM3 is about 15% lower than of CM at the same age. The split tensile strength of fly ash concrete is lower than that of control concrete for all five fly ash concrete mixes at 28 days. However, gain in split tensile strength with ages for FM1, FM2, FM3 is considerable. Split tensile strength at 120 days for these mixes is quite comparable to that of CM. Above data clearly indicates that high volume fly ash concrete mix with cement replacement level at least up to 70% can be developed using commercially available Indian cement and the fly ash generated in Indian thermal power plants. Behavior of Indian fly ash concrete is similar to those published in literature [4]. However, further study in this respect is necessary on fly ash collected from different power plants all over the country for HVFAC.

Durability of fly ash concrete

Measure of durability of concrete should, ideally, quantify its ability, in terms of one of its attributes, to resist the effects of deteriorating media maintaining the original form, quality and serviceability of the structure. Therefore, definition of measure of durability involves the identification of the attributes that can represent reasonably well the ability of concrete to resist the deteriorating effects of all phenomena related to durability discussed above, and approach for its quantification. Identification of a single attribute that can cover all aspects of durability seems to be remote considering the complexity and qualitative nature of the problem. As ingress of deleterious fluid is most of the causes for impediment of durability, permeability is generally taken as the measure and assessed by means of rapid chloride penetration test (RCPT) for its qualification. However some engineers still prefer to specify, immediately after the term durability, the type of attack that is involved [12].

Variation of RCPT values with ages for mixes CM, FM1, FM2 and FM3 is shown in Fig. 4. RCPT value of FM1, FM2 and FM3 is higher at 7 days compared to that of CM indicating fly ash concrete has higher permeability at initial age compared to control concrete. But the values decrease with ages for fly ash concrete and such decrease is predominant up to 56 days. The control concrete does not show decrease in RCPT value after 28 days. The results indicate durability of fly ash concrete mix is, in general, higher than the control mix after 28 days and during service life.

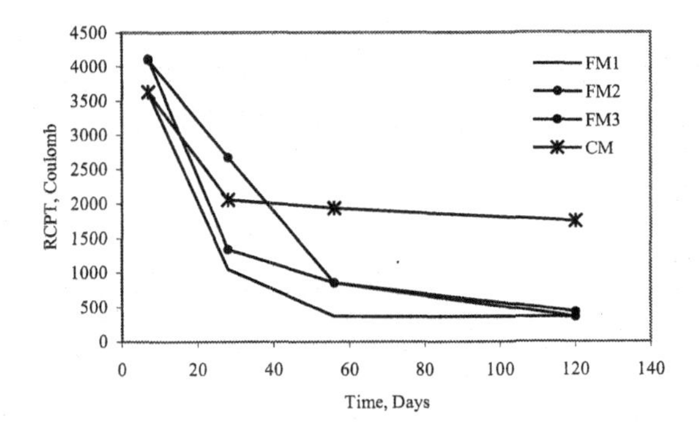

Fig. 4: Variation of RCPT (Coulomb) value with age

It is a common belief that the high durability of HPC composites like that of HVFAC is primarily based on its low permeability achieved due to both grain and pore refinement of hydrated cementious system resulting principally from the secondary hydration. But this aspect needs to be rationalized with feedback data. It is difficult to assess the durability of a new material like HVFAC without having data on the long-term performance of structures with this material in different environmental conditions under various service scenarios. There is a school of thought that durability of the material can be assessed using the accelerated test to meet the immediate need in absence of feed back data of actual structure. Procedures for accelerated tests have been standardized over the years, in which samples are usually exposed to concentrated aggressive solutions under extreme temperatures, load or gradient conditions. It is implicitly assumed that these laboratory conditions represent field conditions fairly well and that only destructive mechanism is accelerated.

No doubt, accelerated test data is useful. It cannot replace the necessity of well-documented field data. It has been observed in number of occasions that behavior of a particular engineered system is quite different from that predicted from accelerated test data. For HVFAC or other types of HPC, well-documented field case studies of its successful or unsuccessful use are not enough. Concentrated effort is necessary in this direction. There is urgent need of creating a database on the durability aspects of actual structure constructed with HVFAC.

Effects of manufacturing process [13]

Limited published literature is available on the effect of manufacturing process on the behavior of concrete mixes. The limited information, whatever is available, indicates manufacturing process, especially the mixing method has noticeable influence on the behavior of the concrete mixes both in fresh and hardened states. Two concrete mixes, one without fly ash (MCM) and another with fly ash (MFA) were manufactured using five different mixing methods. These mixes have total binder content of 500Kg/Mm^3. Tables VI and VII provide the mix proportions and the different mixing methods used for manufacturing these mixes respectively. Five multistage mixing methods were worked out for the study using varying sequence and time of agitation. The mixes thus produced were studied for density, rheology, strength, RCPT and pH value.

Table VI: Mix proportions to study effect of mixing method on fly ash concrete

Mix	Mix Proportions (per m^3)							
	Aggregate (kg)			Binder		w/b ratio	Water (Kg)	Super plasticiser[4] (%)
	Coarse	Fine	Total (Kg)	Cement (Kg)	Fly ash			
MCM	1176.8	672.9	500	500	-	0.325	1625	2.0
MFA	1164.3	665.7	500	333	167 (50 : 33)	0.300	150	0.7

Table VII: Details of different mixing methods

Mixing method	Sequence of mixing
I	CAg+FAg+C+W/3 (60 sec) → W/3+MA (60 sec) → W/3+CA (150 sec) → Total mixing time = 270 sec
II	CAg+FAg+C+ MA + W/3 (60 sec) → (2/3)W+ CA (210 sec) → Total mixing time = 270 sec
III	CAg + FAg + C + MA (60 sec) → W/3 (60 sec) → (2/3)W+ CA (150 sec) → Total mixing time = 270 sec
IV	CAg+FAg+C+ MA + W/3 (30 sec) → (2/3)W+ CA (120 sec) → Total mixing time = 150 sec
V	CAg+FAg+C+ MA + W/3 (90 sec) → (2/3)W+ CA (180 sec) → Total mixing time = 270 sec

Result of the density, rheological observations and pH value are given in Table(s) VIII, IX and X respectively. Strength and RCPT value at 28 days presented in Figs. 5 and 6.

Table VIII: Bulk densities of different mixes

Mixing method	Bulk density (Kg/m^3) MCM	MFA
I	2564	2552
II	2562	2538
III	2567	2515
IV	2586	2520
V	2539	2503

Table IX: Rheological observations on different mixes

Mix	Mixing Method	Visual Observation				Slump	
		Segregation	Bleeding	Stickiness	Cohesion	Initial	1 hr.
MCM	I	Nil	Nil	Not sticky	Cohesive	180	25
	II	Nil	Nil	Not sticky	Cohesive	190	100
	III	Nil	Slight	Not sticky	Cohesive	205	80
	IV	Slight	Slight	Not sticky	Cohesive	160	45
	V	Moderate	Slight	Not sticky	Cohesive	200	165
MFA	I	Nil	Nil	Slightly sticky	Cohesive	80	0
	II	Nil	Nil	Slightly sticky	Cohesive	80	0
	III	Nil	Nil	Slightly sticky	Cohesive	110	0
	IV	Nil	Nil	Slightly sticky	Cohesive	150	0
	V	Nil	Nil	Slightly sticky	Cohesive	110	0

Table X: Variation of pH for different mixes with mixing method

Mixing Method	Mix	
	MCM	MFA
I	12.22	11.96
II	12.01	12.20
III	11.30	12.05
IV	12.38	11.98
V	11.38	11.48

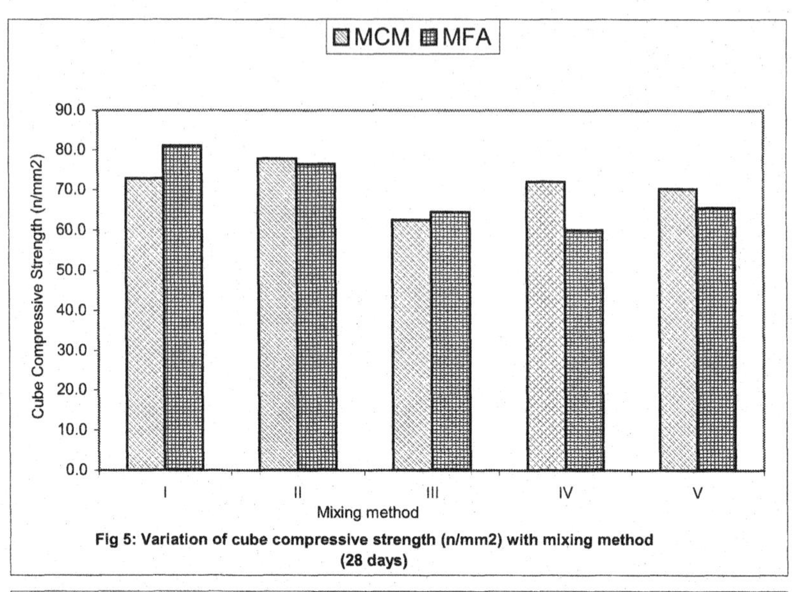

Fig 5: Variation of cube compressive strength (n/mm2) with mixing method (28 days)

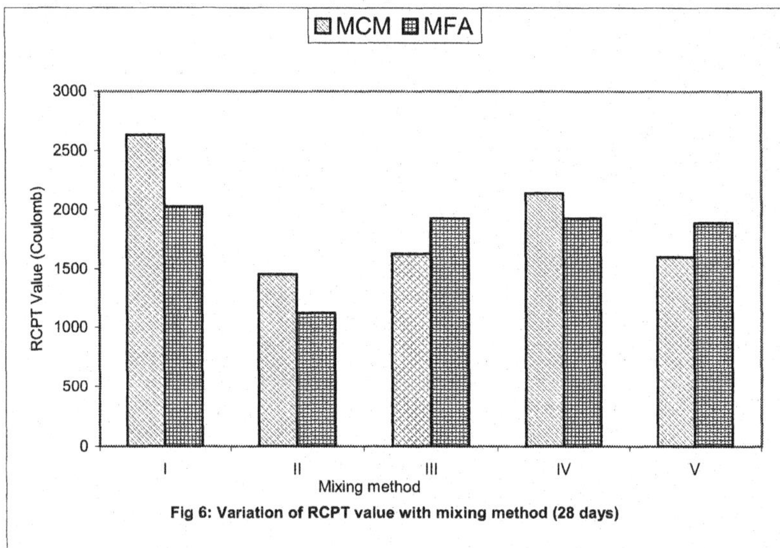

Fig 6: Variation of RCPT value with mixing method (28 days)

Following observations are drawn from the results,

1. Density of fresh concrete appears to be insensitive to mixing methods. Mixing method does not seem to have noticeable influence on bleeding, segregation, stickiness and cohesiveness on both the concrete mixes.
2. Mixing method has influence on the initial slump and slump retention and also on the compressive strength of both control and fly ash concrete mixes.

3. Mixing method has significant influence on the strength and RCPT values of both these mixes. This indicates durability of concrete is also dependent on the mixing method.
4. No noticeable influence of mixing methods was observed on the pH value of both the mixes. An additional observation could be made from this set of results that use of fly ash concrete does not reduce its passive characteristics necessary for corrosion resistance of reinforcements.

It can be reasonably concluded that mixing method has similar type of influence on both control and fly ash concrete mixes. The data presented here is for cement replacement level of 33%. Further work is needed for higher replacement level.

Curing

Various literatures suggest that success of fly ash concrete is greatly dependent on the curing [14]. Curing seems to have even higher influence in hot weather of India. It is suggested that a minimum period of 14 days is required for curing to reap all the benefits of fly ash concrete. However, very limited data is available on this front in India. A systematic research need is felt in this area.

Fly ash as replacement of sand

The fly ash having size of more than 45μ does not possess pozzolanic properties. Use of fly ash having coarser size as a partial replacement of fine aggregates in concrete mix is a matter of interest; especially in the region where availability of natural sand is rare. Pond ash, which is a mix of fly ash and bottom ash, is occupying vast tracts of land around the thermal plants all over the country. The fact that sand supply is diminishing and that the production of pond ash, a waste material is increasing, the research is felt to explore the possibility of using pond ash as a partial replacement to sand.

Since pond ash is much coarser than fly ash, it can be used as a fraction of sand, to modify the characteristics of sand and thus of mortar or concrete both in fresh as well as hardened state. The suitability or otherwise can be estimated only after a thorough study involving characterization of the particular pond ash. This is necessary since the pond ash from different sources can be different and might have different characteristics within a pond at different locations and depths. The economic issues need also to be studied.

Although pond ash is to be added as a partial replacement of sand, which is an inert material, a small amount of pozzolanicity, which is present in pond ash, might add to the later day strength. Benefits like improved workability, better gradation of sand are also possible. The ultimate aim should be to help in codifying the use, achieve economy, reduce consumption of natural resource like sand and increase consumption of waste material.

CONCLUSION

1. Fly ash is the most suitable industrial waste that could be used in concrete for sustainable development of India from four basic counts,
    - Fly ash concrete is better quality concrete in terms of strength, durability, rheology, etc than control concrete mix (i.e. concrete mix without any mineral admixture).
    - It is economical
    - Environmental friendly disposal of fly ash
    - Preservation of natural resources for future.
2. Superiority of fly ash concrete has been established in USA, Canada and European countries [4,5,6,7,8,9] by means of voluminous data generated from extensive research

work. Though this available data is extremely useful for India, further research work is required in this field to take care the specific Indian conditions.

3. Use of fly ash in concrete by means of blended cement; Portland pozzolona cement (PPC) has become common now. A few important structures have been constructed incorporating cement replacement up to about 40% by fly ash.
4. Use of high volume fly ash concrete is almost absent in India till date. Proper utilization of fly ash for sustainable development of India could be achieved through HVFAC. Data available from limited research in this area in India indicates that HVFAC having requisite properties can be developed with at least up to 70% cement replacement using Indian ingredients.
5. Partial replacement of sand by fly ash of coarser grain size is also a good research subject.

REFERENCES

[1]Basu Prabir C., "NPP Containment Structures: Indian Experience in Silica Fume Based HPC," *The Indian Concrete Journal*, **75** [10] 656-664 (2001)

[2]Shreeti Mavinkurve, Basu Prabir C., Kulkarni Vijay R.,"High Performance Concrete Using High Reactivity Metakaolin," *The Indian Concrete Journal*, **77** [5] 1077-1085 (2003)

[3]Kulkarni Vijay R., "Some Recent Trend in the Sphere in Concrete," In *Proceedings of Asian Concrete Forum*, Indian Concrete Institutes, pp 115-129, Pune, 2003

[4]Malhotra, V. M., and Mehta, P. K., *Fly-Ash in Concrete*, 2nd edition, CANMET MSL 94-45 (IR), Canada, 1994

[5]Mehta, P.K., and Gjrov, O.E., "Properties of Cement Concrete Containing Fly Ash and Condensed Silica Fume," *Cement and Concrete Research*, **12** 587-596 (1982)

[6]Malhotra, V.M., "Super Plasticised Fly Ash Concrete for Structural Applications," *Concrete International*, **8** [12] 28-91 (28-31)

[7]Shivasundaram, V., Carette, G.G., and Malhotra, V.M., "Mechanical Properties, Creep, and Resistance to Diffusion of Chloride Ions of Concretes Incorporating High Volumes of ASTM Class F Fly Ash from Seven Different Sources," *ACI Materials Journal* **88** 407-416 (1991)

[8]Wesche, K., *Fly ash in Concrete Properties and Performance*, International Union of Testing and Research Laboratories, Paris, France, Rilem Report 7, Chapman and Hall, London, UK, 1990

[9]Dhir, R.K., Darfour, E.S., and Munday, J.G.L., "Strength Charracteristics of Concrete Containing PFA Additive," *Silicates Industriels* **44**[1] 23-29 (1979).

[10]-------, "Specification for Fly Ash for Use as Pozzolana and Admixture," IS3812, Bureau of Indian Standards, New Delhi, 1981

[11]Subhajit Saraswati, Basu Prabir C., "High Volume Fly Ash Concrete with Indian Cement," In *Proceedings (CD-ROM), Third Quinquennial International Symposium on Innovative World of Concrete*, Indian Concrete Institute, Pune, India, 2002

[12]Basu Prabir C., Subhajit Saraswati, "Durability of High Performance Concrete: An Overview and Related Issues." In *Proceedings (CD-ROM), International Symposium on Advances in Concrete through Science an Engineering*, North Western University, Evanston, Illinois, USA, 2004

[13]Basu Prabir C., Subhajit Saraswati, "Effects of Mixing Methods on Durability of HPC Composites," In *Proceedings (CD-ROM), International Symposium on Advances in Concrete through Science an Engineering*, North Western University, Evanston, Illinois, USA, 2004

[14]Gopalan, M.K., and Haque, M.N., "Effect of Curing on The properties of fly ash Concrete," *ACI Materials Journal* **84** 14-19 (1991)

RHEOLOGICAL MEASUREMENTS AND VERY EARLY AGE VISCOELASTIC PROPERTY MEASUREMENTS OF CONCRETE

M. Neelamegam
Deputy Director
Structural Engineering Research Centre
CSIR Campus, Taramani, Chennai 600 113

N.P. Rajmane
Deputy Director
Structural Engineering Research Centre
CSIR Campus, Taramani, Chennai 600 113

J.K. Dattatreya
Assistant Director
Structural Engineering Research Centre
CSIR Campus, Taramani, Chennai 600 113

ABSTRACT

Although the plastic state of concrete is only transient in nature, rheological properties are of vital importance for ensuring consistency and uniformity of the mixture, its transportability, placement and consolidation. Measurement and modification of these properties are therefore essential for mix design and quality control. This paper presents a critical assessment of rheological measurement techniques and the modelling of properties. Some of the studies carried out at the Structural Engineering Research Centre (SERC), Chennai are also presented. A few hours after placement and consolidation, concrete starts showing viscoelastic behaviour. Such a rapidly ageing visco-elastic material cannot be easily modelled. However, the properties in this stage are preliminary measures of performance at later ages. The importance of this phase of concrete is evident from the growing record of research activities in the world regarding early age properties. Besides strength, other properties of great interest include shrinkage, creep and internal thermal behaviour. This paper critically examines some of the models for ageing viscoelastic materials and discusses some of the studies on early and very early age viscoelastic behaviour and problems associated with them.

INTRODUCTION

Portland cement when mixed with sufficient water forms a free flowing paste whose rheological characteristics are modified by the presence of fine aggregates, coarse aggregates and mineral and chemical admixtures. Although the plastic state of concrete is only transient in nature, it is of vital importance for ensuring consistency and uniformity of the mixture, its transportability, placement and consolidation. Measurement and modification of the rheological properties are, therefore essential mix design and quality control tools. The improvements in production technology of concrete, popularity of ready-mix concrete and advent of high performance and self-compacting concrete have further emphasised the importance of rheological measurements on fresh concrete. When concrete is to be hauled over long distances, the variation in rheological properties and means to retain and or restore them are also necessary. Thus, measurements and modification before placement of the mixture are common features. Rheological measurements can give much information that is of use in such industrially important areas as oil well cementing, grouting, vibratory compaction and real-mix quality control.

Conventionally, simple tools, which are generally empirical in nature, such as slump test and flow test, have been used as measures of rheological properties of fresh concrete. Now-a-days, the inadequacy of these tests especially for highly workable concrete has been realised and improved measures such as slump flow, L-flow, V-flow, O-ring, U-tube, etc., are being used in both the laboratory and field for the development of mixes and as part of quality control schemes. More objective measurements using rheometers and viscometers are used at the laboratory level mainly for mix development and modification.

This paper presents a critical assessment of rheological measurement techniques and modelling of properties for mix design and development. The Herschel-Barkley model, Tatersal's two-parameter Bingham model and their further improvements are discussed. Some of the studies carried out at SERC, Chennai are also presented. Models for ageing viscoelastic materials are also presented. Some of the studies on early age viscoelastic properties carried by various investigators are also presented.

RHEOLOGICAL MODELS FOR CONCRETE

The rheology of cement and concrete governs the ease with which these materials flow and is influenced by composition and a variety of external factors. Concrete could be considered as a granular mixture (encompassing the whole population of grains) suspended in water and neglecting air content. If a shear stress is applied to the system, the material deforms provided it is high enough to overcome the friction between the particles. The shear threshold or yield stress is governed by the number and nature of contacts between the grains and the liquid phase controls the mean distance between the grains. In addition, the strain gradient controls the fluid velocity in the pores of a granular mixture as in a Newtonian material. Therefore the resistance to deformation by applied shear stress (τ) could be expressed as

$$\tau = \tau_0 + \mu\dot{\gamma} \tag{1}$$

Where, τ = shear stress, μ = viscosity and $\dot{\gamma}$ = strain rate

This simplest way of modelling the rheological behaviour of such a two-phase material by a Bingham model was proposed by Tattersall[1] is considered to be suitable for most commonly used concretes, including high performance concrete. A rheometer is often necessary to determine these parameters. More recently, on the basis of a large experimental data, Herschel and Berkley (Larrard et al[2]) proposed a three-parameter model, which uses a power law relationship between shear stress and strain gradients in the form:

$$\tau = \tau_0 + \mu\dot{\gamma}^x \tag{2}$$

x = strain rate parameter

Limitations of these models are high degree of uncertainty of parameters and difficulties in obtaining the flow properties by change of mix parameters. For simplicity, Larrard, et al.[2] suggested a combined parameter for strain rate, viscosity and power x, which could be more useful.

RHEOLOGICAL MEASUREMENTS WITH CEMENTITIOUS PASTES IN VISCOMETERS

Chapels, et al.[3] developed an appropriate procedure to study the rheology of cement pastes as a function of time during the dormant period. They investigated model dense suspensions of alumina particles, aluminous cement pastes, and cement pastes. They recommended that use of low shear stress conditions is appropriate for study of cementitious pastes in viscometers as concrete is subjected to very small shearing.

Subverts and Reick[4] studied the effect of fly ash on the rheological properties of cement paste by using a Couette rheometer. They concluded that the flow behaviour of pure fly ash paste with sufficient flocculation does not differ from that of cement paste. They developed an expression relating solids concentration to yield stress. The effects of fly ash and cement with regard to viscosity are additively superimposed.

Ivanov and Roshavelov[5] studied the complex influence of condensed silica fume, superplasticizer and the mineral composition of cement on rheological behaviour of fresh cement pastes using design of experiments. They found that the presence of condensed silica fume significantly increases yield stress and viscosity.

ASSESSMENT OF WORKABILITY CHARACTERISTICS OF SELF COMPACTING CONCRETES

Generally for mix design of concrete and on site evaluation, simple and modified workability tests are found to be more useful for adjusting the mixture proportions of a trial mixtures and quality control. For self-compacting concretes (SCC), Okamura and Ozawa[6,7] proposed two simple measures of workability of SCCs viz., the slump flow test and V-funnel test (Figs.1 and 2). Further many additional instruments were proposed by Yonezawa, et al.[8] (L-flow tester Fig. 3) and Okamura, et al.[7] and Shinto, et al.[9] (U-tube Fig. 4). Many more modifications, such as U-funnel with unequal limbs [Fig. 5], and fillability apparatus (Fig. 6) have been developed to assess the flow, passing and fillability of SCC mixtures. Mukai, et al. (as reported by Domone, et al.[10]) employed a plexiglass model of the bottom bulb section of AASHTO type I girder (Fig.7) with a 305 mm long cross section of width varying from 152 mm to 405 mm and overall height 365 mm. Four legs of 17 mm diameter rebars were placed at 51 mm c/c spacing.

Shinto, et al.[9] employed a conical viscometer (Fig.8) consisting of outer rotating cylinder type 90 mm diameter x 90 mm height with a fixed inner cylinder of 30 mm diameter x 50 mm height for assessment of rheological properties of mortar fraction of self compacting concretes. Sedran, et al.[11] and Larrard[2] used the 'BTRHEOM' RHEOMETER (Fig.9) developed at LCPC to characterize the rheology of fresh concrete.

Investigations were carried out on rheological properties of cement pastes, cement mortars and self-compacting concrete by Dattatreya and Neelamegam[12] at SERC, Chennai. Six different SCC mixtures were developed and studied their slump flow; slump flow time and V-funnel flow times were studied. The rheological properties of the equivalent mortar extracted from the SCC mixtures were also studied. Table1 shows the rheological properties of SCC and SCC equivalent mortar mixtures.

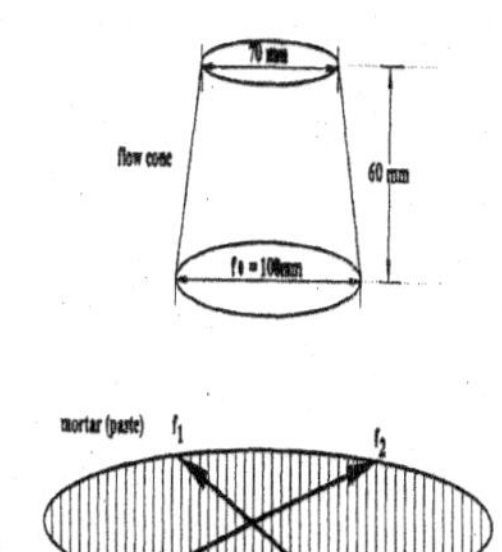

Fig.1 Slump Flow Test Apparatus for Paste/Mortar

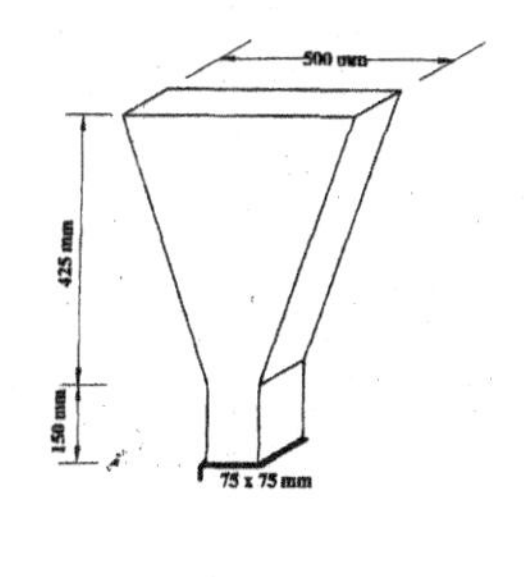

Fig.2 Schematic Diagram of the V-Funnel Apparatus for SCC

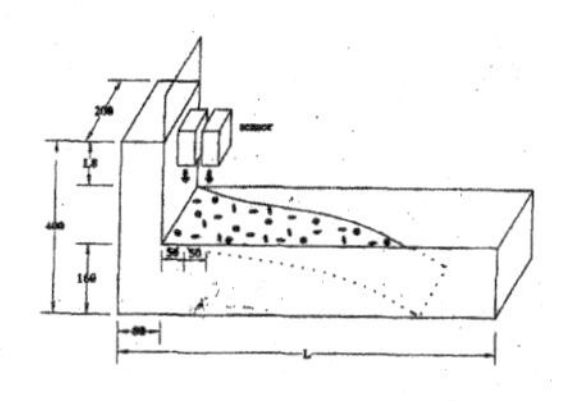

Fig.3 L-Flow Meter for SCC

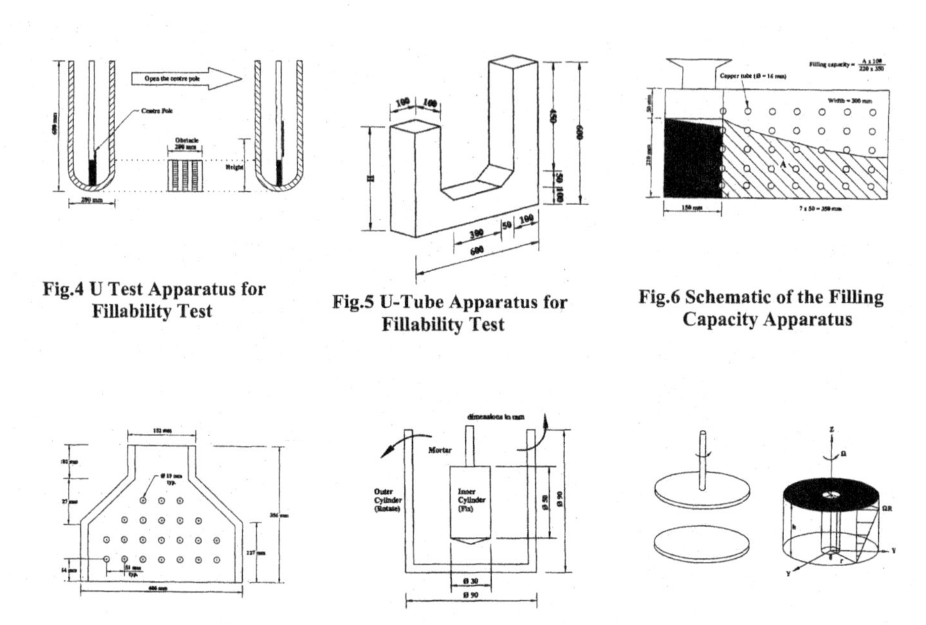

Fig.4 U Test Apparatus for Fillability Test

Fig.5 U-Tube Apparatus for Fillability Test

Fig.6 Schematic of the Filling Capacity Apparatus

Fig.7 Model AASHTO Type I Girder

Fig.8 Details of Viscometer for Mortar

Fig.9 Concrete Flow in a Parallel - Plate Rheometer

Table I Rheological properties of SCC and equivalent mortar of SCC mixtures

Mix Id.	Mineral admixture	Flow properties of SCC mixtures			Flow properties of equivalent mortar mixtures		
		Slump flow (mm)	Slump flow Time (Sec)	V-funnel flow time (Sec)	Mix Id.	Slump flow (mm)	V-funnel flow time (Sec)
A1	SF	450	6	6	---		
A2	SF	480	6	3	---		
A3	SF	650	4	2	---		
A4	SF	500	4	3	A4M	210	2.26
A5	SF	500	2	1	A5M	150	1.83
A6	SF+FA	650	3	1	A6M	235	1.50

VISCOELASTICITY

Generally, the viscoelastic behaviour of concrete is studied from creep test, where the stress is kept constant and the increase in strain over the time is recorded or the relaxation test, where the strain is kept constant and the decrease in stress over time is recorded. The behaviour of viscoelastic materials could be approximately estimated by the creation of rheological models based on two fundamental elements: the linear spring and the linear viscous dashpot in various combinations. Some of the simple combinations include the Maxwell element (spring and dashpot in series), the Kelvin element (spring and dashpot in parallel) and a standard solid (spring in series with the Kelvin element). More generalized rheological models consisting of several combinations of the Maxwell elements or the Kelvin elements could be employed for improved estimation of viscoelastic behaviour.

TIME-VARIABLE RHEOLOGICAL MODELS

Concrete changes its mechanical properties with time due to the hydration reaction. However, in the models presented so far, the elastic modulus E and the viscosity coefficient η are constant over time. Consequently they have limited success in modelling the complex response of concrete. One model, however, that accounts for this behaviour is the ageing Maxwell element. Dischinger used this model to derive a specific creep function φ (t, τ) (i.e., strain per unit stress at time t for a constant stress applied at age τ, in the following form:

$$\phi(t,\tau) = \frac{1}{E(\tau)} + \int_{\tau}^{t} \frac{d\tau'}{\eta(\tau')} \tag{3}$$

This formulation, however, implies that the creep curves are parallel , which does not match the experimental behaviour. Generally, this method underestimates creep strain. A more generalized ageing model leads to differential equation of the form:

$$\left[\frac{d^n}{d^n} + p_1(t)\frac{d^{n-1}}{dt^{n-1}} + ... + p_n(t)\right]\sigma(t) = \left[\frac{d^n}{d^n} + q_1(t)\frac{d^{n-1}}{dt^{n-1}} + ... + q_n(t)\right]\varepsilon(t) \tag{4}$$

Where the parameters p_1 .. p_n, q_1 .. q_n and also the stress σ and strain ε are functions of time *t*. Such a model involving hypoelstic springs and dashpots and considering ageing could yield the most realistic representation of creep, relaxation and other viscoelastic behaviour.

BEHAVIOUR OF VERY EARLY AGE CONCRETE SUBJECT TO CONSTRUCTION LOAD

There is no generally agreed definition of very early age properties. It is believed to be the properties of concrete within the first one or two days [Civil Engineering Association Committee, 42[13]]. In this stage, concrete behaves essentially like a viscoelastic material. If the loads are considerable, non-linearity could also come into play in addition to time dependence of properties. The measurement and modelling of very early age viscoelastic properties is gaining considerable importance with increasing adoption of fast track construction, which leads to construction loads and at times part of the dead load coming into play at very early ages. It is also possible that structures are exposed to loads at this very early age. Early age viscoelstic properties may be of relevance due to the behaviour of the material, due to mechanical influences in the construction process, or due to mix design employing use of mineral and chemical admixtures, which many times have adverse effects on very early age properties. These properties may influence the performance of the structure not only in the early age but also later on especially with regard to long term durability. Examples of such situations are early form stripping, slip form work, deformation in supporting formwork due to increasing loads during the concreting operations, water pressure on caissons at an early age, large lifts used in self compacting concrete etc. Structural analysis of very early age concrete, has not been fully established yet because mechanical properties show very complicated behaviour during the setting process of cement.

For the purpose of rationalization of concreting practice, Okamoto and Endoh[14] carried out an investigation to obtain knowledge on deformation of concrete for construction load resulting from the stripping of formwork, and to evaluate physical properties of setting process with rheological models. The objective of the investigations was to evaluate the physical properties of the setting process with rheological models and to present a simple analysis based on theory of linear viscoelasticity. They compared measured results on the problem of time dependent deformations caused by dead load of column. The conclusions obtained by the authors are as follows:

- For very early age concrete under sustained loading, a constitutive equation based on four-element viscoelastic model (consisting of one Kelvin and one Maxwell element) agrees with experimental results of creep tests.
- As the age and temperature of concrete increased, rheological constants for a four-element model at very early age increased exponentially. Properties of setting process of very early age concrete can be evaluated by viscoelasticity approach by using a creep test. However, a more generalized model accounting for age dependence could lead to better results.
- In creep under multi-stage loading, the strain due to first loading increased greatly, but the strain due to additional loading was extremely small in quantity.
- As stress ratio became more than 0.80, creep rupture occurred as is well known.
- Although calculated results by finite element analysis and simple analysis were in agreement in respect of the deformations caused by dead load of column, the measured values were higher than calculated results, because the reliability of instantaneous modulus of elasticity by creep test was unsatisfactory owing to observational error and measurement difficulties, but both of the time–history curves showed almost the same shape. Using reasonable rheological constants, deformation of very early age concrete subjected to construction load can be estimated by the analytical procedure presented with this study

EARLY AGE PROPERTIES

According to Neville[15], as a first approximation age does not influence viscoelstic behaviour, viz., creep, other than through stress –strength ratio. However, the same concrete loaded at different ages undergoes different growths in strength. Therefore, for a constant applied stress, the stress-strength ratio while under load depends on the age of loading. As a consequence creep varies with age of loading and could be more or less depending upon the stress-strength ratio. This is seen in some typical results presented in Fig.10 [Zdenek Smerda and Vladimir Kristek[16], 1988]. The creep may not be proportional to applied stress , especially at higher stresses. Some investigators have suggested that to allow for the effect of age at loading, creep coefficients could be expressed as a function of ratio of strength at any early age t and that at 90 days.

CONCLUDING REMARKS

- Although the plastic state of concrete is only transient in nature, rheological properties are of vital importance for ensuring consistency and uniformity of the mixture, its transportability, placement and consolidation. Measurement and modification of these properties are therefore essential for mix design and quality control.

- The measurement and modelling of very early age viscoelastic properties is gaining considerable importance with increasing adoption of fast track construction, which leads to construction and some times part of the dead load coming into play at very early ages.
- Limitations of these models are high degree of uncertainty of parameters and its practical utility of obtaining the flow properties by change of mix parameters.
- creep varies with age of loading and could be more or less depending upon the stress-strength ratio.
- The creep may not be proportional to applied stress , especially at higher stresses.
- Although calculated results of very early age deformations by finite element analysis and simple viscoelastic analysis agreed in respect of deformation by dead load of column, measured values were higher than calculated results, because the reliability of the instantaneous modulus of elasticity by creep test was unsatisfactory owing to by observational error, but both of the time-history curves showed almost the same shape. Inputting reasonable rheological constants, deformation of very early age concrete subjected to construction load can be estimated by analysis presented with this study

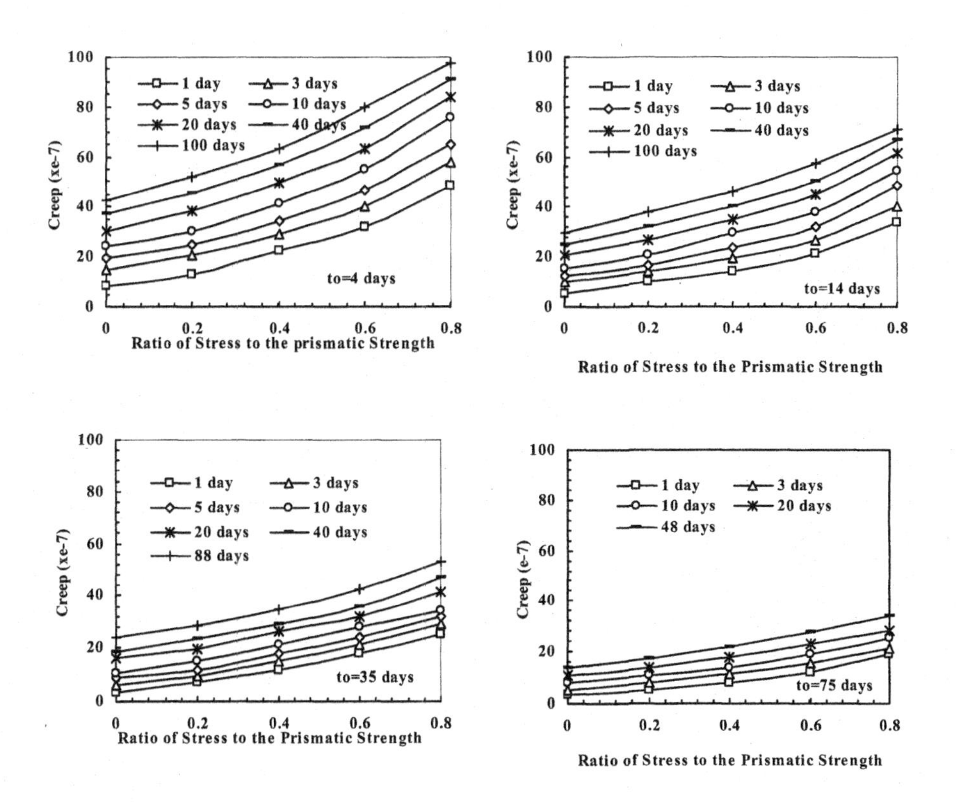

Fig.10 Effect of the Ratio of Stress σ to the Prismatic Strength R_c of Concrete on Creep in Compression in the interval [t to t_o]

ACKNOWLEDGEMENT

This paper is published with the kind permission of the Director, SERC, Chennai. The authors sincerely thank their colleagues and technical staff in the Concrete Composites Laboratory of SERC for their help and encouragement.

REFERENCES

[1]C. H. Tattersall, "Workability and Quality Control of Concrete," E & FN Spon, London, 1999

[2]F. de Larrard, "Concrete Mixture Proportioning," E & FN Spon, London, 1999

[3]J. Chappuis, " Rheological Measurements with Cement Pastes in Viscometers: A Comprehensive Approach", Rheology of Fresh Cement and Concrete, *The British Society of Rheology*, E.&F.N. SPON, 1990

[4]F. Sybertz and P. Reick, "Effect of Fly Ash on the Rheological Properties of Cement Paste," Rheology of Fresh Cement and Concrete, *The British Society of Rheology*, E.&F.N. SPON, 1990

[5]Y.P Inanov, and T.T. Roshavelov, "The Effect of Condensed Silica Fume on the Rheological Behaviour of Cement Pastes," Rheology of Fresh Cement and Concrete, *The British Society of Rheology*, E.&F.N. SPON, 1990

[6]H. Okamura, and K. Ozawa, "Design for SCC Concrete, *Library of the JSCE*," No. 25, 107-120 (1994)

[7]H.Okamura and K. Ozawa, " Self Compactable HPC in Japan", *SP169, ACI*, Detroit, 31-44 (1994)

[8]V. Yonezawa, I. Isumi, , K. Mitsui and T. Okuno, "A Study on the Flowability of High-Strength Concrete Using L-Flow test," *Proceeding of the Japan Concrete Institute*, Vol. 11, 171-176 (1989)

[9]T. Shindo, K. Yokota and K.Yokai, "Effect of Mix Constituents on Rheological Properties of Super Workable Concrete," *Proceeding of the RILEM conference on Production Methods and Workability of Concrete*, Paisley, Scotland, 263-270 (1996)

[10]P.L.Domone and H.W. Chai, "Design and Testing of SCC," *RILEM International Conference on Production Method and Workability of Concrete*, 223-236 (1996)

[11]T. Sedran, F.D. Larrard, F. Hurst and C. Contaminus, "Mix Design of SCC," *Proceedings of the International RILEM Conference*, Paisly, Scotland, June, 439-450 (1996).

[12]J.K. Dattatreya, M. Neelamegam, N.P. Rajamane and S. Gopalakrishnan, "Self Compacting Concrete with High Volumes of Cement Replacement Materials," *Innovative World of Concrete-2003, International Conference and Exhibition*, ICI Pune Centre

[13]CEA Committee, "Properties of concrete at early ages," *Materials and Structures*, Vol. 13, No. 75, pp 265-274 (1980)

[14]H. Okamoto, and T. Endoh, "Modelling in Analysis of Deformation of Very Early Age Concrete Subject to Construction Load," Rheology of Fresh Cement and Concrete, *The British Society of Rheology*, E.&F.N. SPON, 1990

[15]A.M. Neville, "Creep of Concrete: Plain, Reinforced, and Prestressed," North-Holland Publishing Company, Amsterdam, 1970

[16]Z.Smerda and V. Kristek "Creep and Shrinkage of Concrete Elements and Structures," Developments in Civil Engineering, 21, Elsevier, New York, 1988

NON-DESTRUCTIVE AND PARTIALLY-DESTRUCTIVE TEST METHODS FOR CONDITION ASSESSMENT OF CORROSION AFFECTED STRUCTURES

H.G. Sreenath
Deputy Director & Head – PSC Group
Structural Engineering Research Centre
Chennai – 600 113, India

ABSTRACT

The damage caused by reinforcement corrosion in concrete structures has been considered as one of the major durability problems affecting the service life of concrete structures. Corrosion prone as well as corrosion-affected structures requires a systematic inspection and investigation to assess the condition of concrete and embedded steel. Several Non Destructive (ND) and Partially Destructive (PD) Testing systems are available. In spite of limitations, these test methods and the measured parameters by them do help to make a reasonably reliable assessment of the corrosion-affected structure. This paper presents a systematic methodology for assessing the condition of corrosion affected reinforced concrete structures using in-situ testing methods.

INTRODUCTION

Concrete structures exposed to marine or chemical environment undergo distress during its service life due to corrosion of embedded steel reinforcement. Engineers and researchers have tried to understand the root cause of this phenomenon in concrete structures, and have suggested mandatory clauses in codes of practice under durability. It is important to note that care taken during execution of structures at site determines its durability. The research and advancements in the area of materials, instrumentation and introduction of concrete chemicals for repair and materials for enhancing the service life have opened options for the engineers. Non-Destructive Testing (ND) methods can be used as quality assurance program and take suitable measures, if required. Considering this aspect, infrastructures of national importance in India, such as, Nuclear/ Thermal Power Plants, Port complexes, etc. have made NDT mandatory. But the structures that are in service require safety audit once in 10-15 years to assess the condition of the structure and take suitable measures, if required, to avoid distress, and enhance its service life. The condition assessment of structures by systematic inspection and testing using Non Destructive Testing (ND) and Partially Destructive Testing (PD)[1] will help in planning suitable strategy either for durable repairs or for undertaking suitable methods to enhance the service life of the structure based on test results.

The paper discusses the methodology for assessing the condition of corrosion-affected structures by ND & PD and its interpretation to plan strategy for repairs, or methods to enhance service life of structures.

CONDITION ASSESSMENT OF CORROSION AFFECTED STRUCTURES

In the recent past, the damage caused by rebar corrosion in concrete structures has been considered as one of the major durability problems affecting the service life of concrete structures. Corrosion-prone as well as corrosion-affected structures require a systematic inspection and testing in order to assess the condition of concrete, presence of corrosion activity, and extent and severity of corrosion. The parameters that influence the corrosion process are: (1) the cover thickness, (2) the quality of concrete in the cover region, especially in terms of permeability, diffusivity, (3) environmental conditions, (4) pH value and chloride level in concrete, and (5) presence of cracks, etc. Further, the electrochemical parameters such as half cell potential, resistivity of concrete, and corrosion rate reflect the extent and severity of corrosion. Hence, the investigation should consider suitable tests to assess the condition of concrete, the nature and extent of corrosive environment within the concrete, and condition of steel bar in respect of corrosion. With the help of extensive research carried out world wide, several insitu testing systems have been developed, and each one has its own merits and limitations. In spite of limitations, the test methods and the measured parameters do help to make a reasonably reliable assessment if the values are interpreted correctly. The process and mechanism of corrosion of steel in general and rebar embedded in concrete in particular have been reasonably well understood and considerable data are available [2, 3].

The general procedure for condition assessment of corrosion affected reinforced concrete structures is given in Fig. 1. The results of all the ND and PD tests including electrochemical parameters and their interpretation methodology are described briefly.

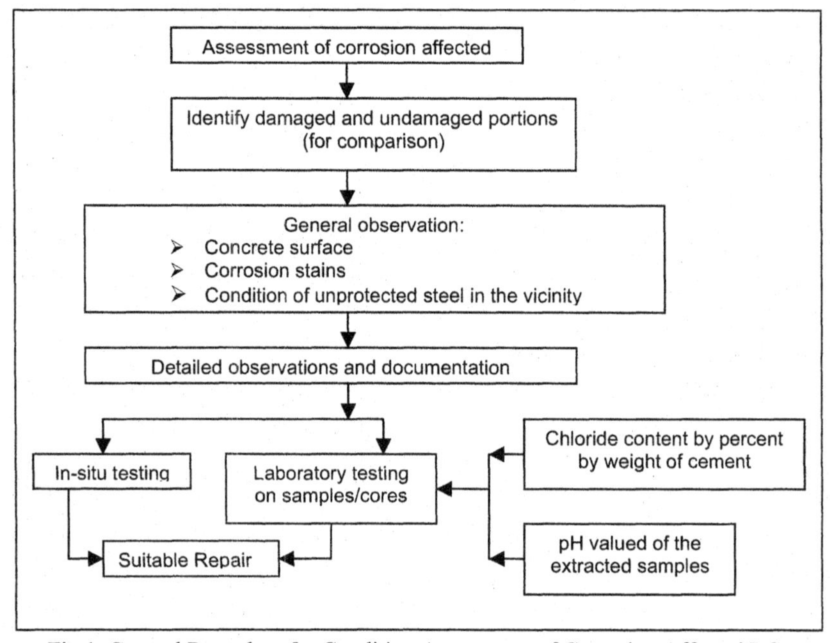

Fig.1. General Procedure for Condition Assessment of Corrosion Affected RC Structures

Visual Observation and Photographic Documentation

Visual examination is the starting point of inspection. Cracks, rust staining, and spalling are the most obvious defects that can be identified. If visual inspection of a structure suggests that a problem may be present, an in-depth examination should be carried out. This should comprise tests, which will identify the cause and extent of the problem and allow for a prediction to be made about the future projected life of the structure.

Cracking and spalling of cover concrete due to corrosion of reinforcement are shown in Fig 2 & 3. Cracking, crazing and honeycombing may be considered to be cracking. Cracking along the reinforcements indicate corrosion of reinforcement. When embedded steel has corroded sufficiently, the expansive volume of rust can cause the surface layer of concrete to become separated or delaminated from the body of the concrete. A survey to determine the extent of such areas is useful to assess the degree of deterioration of a structure. Delamination can be found on a small scale by simply striking the concrete surface with a hammer. This technique is very rapid for small areas, and needs no special equipment. Access to the surface is necessary.

Fig. 2 Cracking in a column due to Corrosion of reinforcement

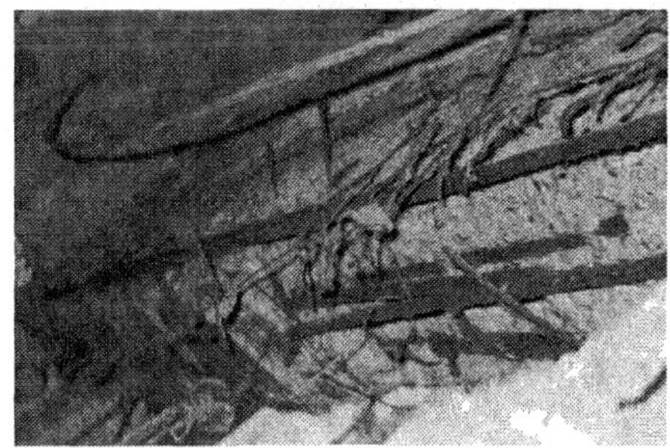

Fig. 3 Heavy spalling of cover concrete due to corrosion of reinforcement

Evaluation of physical parameters of concrete

Rebound Hammer Test: This test is normally conducted to assess the condition of cover concrete and to identify the presence of delamination. This test is carried out on the identified member in a systematic way over well-defined grid points on either sides of the member. The rebound members measured by the hammer are noted, and the test is carried out at each point as defined in ASTM- /BIS-1881 (Part-4). Guidelines for assessing quality of cover concrete are given in the Table I.

Table I. Quality of Concrete Cover from Rebound Numbers

Average rebound number	Quality of concrete
> 40	Very Good hard layer
30 to 40	Good layer
20 to 30	Fair
< 20	Poor concrete
0	Delaminated

Cover Meter Survey: The necessity to provide adequate cover thickness to control and delay corrosion needs no emphasis. Cracking, cover thickness, and concrete quality are three major interactive and inter-related parameters influencing steel corrosion. Cover is an important factor that preserves the electrochemical stability of steel in chloride contaminated concrete. A cover thickness survey is useful to determine as to what cover exists at a specific location, where damage has been identified, and elsewhere for comparison on the same structure.

The cover to reinforcement, or any other embedded steel is required to be known so that the results of other tests, such as carbonation and chloride concentration, can be correctly interpreted with respect to future performance of the concrete structure. The interpretation of cover thickness survey is given Table II.

Table II. Interpretation of Cover Thickness Survey

Test results	Interpretations
Required cover thickness and good quality Concrete.	Relatively not corrosion prone
Required cover thickness and bad quality cover concrete	Corrosion prone
Very less cover thickness, yet good quality cover concrete	Corrosion prone

Ultrasonic Pulse Velocity (UPV) Test: Ultrasonic scanning is a recognised non-destructive test method to assess the homogeneity and integrity of concrete. The test essentially consists of transmitting ultrasonic pulses of 50-54 kHz frequency through a concrete medium and receiving at the other end. The time of travel of ultrasonic pulses is measured and the pulse velocity is calculated by dividing the thickness of concrete member (path length), by measured time. Higher the integrity of the concrete, higher is the velocity. In this type of investigation, pulse velocity values can be used primarily to identify weak locations/regions of concrete from the corrosion point of view as well as to assess its present condition. The velocity values can also be combined with rebound number to make a more realistic assessment on the condition of surface concrete. The UPV scanning is also to be carried out in a systematic way on well defined grid points as done for rebound hammer test. Guidelines for classification of concrete quality based on UPV are given in Table III. As already discussed, RHV & UPV values in combination can provide information on corrosion prone areas. Guidelines are given in Table IV.

Table III. General Guidelines for Concrete Quality based on UPV

Velocity	Concrete quality
> 4.0 km/sec	Very good to excellent
3.5 - 4.0 km/sec	Good to very good, slight porosity may exist
3.0 - 3.5 km/sec	Satisfactory but loss of integrity is suspected
< 3.0 km/sec	Poor and loss of integrity exists

Core Sampling and Testing: Realistic assessment of concrete quality and strength can be assessed by core sampling and testing. The core samples can be used for:

- Strength determination
- Chemical analysis
- Excess voidage estimation
- Petrography analysis

The locations for core sampling can be selected based on the guidelines given in Table - IV. When a core is cut and taken out, it is in a wet condition. Observation of the core may give an idea of the aggregate type, size, distribution, etc. But, other features such as location and size of the reinforcement, honeycombing, cracks, and defects are observed in a dry core. Table V. shows how the condition of concrete can be interpreted after the visual examination of the core is completed.

Table IV. Identification of Corrosion prone Locations based on UPV and Hammer Readings

Test results	Interpretations
High UPV values, high Impact Hammer Nos.	Not corrosion prone
Medium range UPV values, low impact Hammer	Surface delamination, low quality of surface concrete, corrosion prone
Low UPV, high impact hammer numbers	Not corrosion prone, however, to be confirmed by chemical tests, carbonation, pH
Low UPV values, low impact hammer numbers	Corrosion prone - requires chemical and electrochemical tests

Table V. Interpretation from Core Sampling and Testing

Test results	**Interpretations**
Integral core with well distributed aggregates and mortar matrix with normal density	Not corrosion prone
Presence of surface voids and low density concrete in cover regions	Corrosion prone
Weight loss of reinforcement, if any.	Corrosion

In order to get strength directly, the core is extracted without cutting the reinforcement. This can be achieved by using rebar locator. Normally, cylinders should have length to diameter of equal to 2, in case it is different, correction needs to be made. The correction factors are given in ASTM, BSI and BIS. In case cutting of reinforcement is inevitable, correction may be made to obtain equivalent cube strength. ASTM and BSI codes provide empirical formulae.

Chemical Parameters

The results of insitu tests for concrete described in previous sections primarily help to assess the physical condition of concrete qualitatively with reference to corrosion affected, corrosion prone, and not corrosion prone locations in a member or a structure. In order to identify the presence of corrosive environment within the concrete and extent and severity of corrosion, chemical and electrochemical tests are required. The test methods and general guidelines for interpretation are described briefly.

Carbonation Test and pH Value: Carbonation test can be carried out using simple phenolphthalein test. Phenolphthalein spray is an indicator of pH value and as carbonation has its effect on pH of concrete, the change in colour indicates the extent of carbonation. The depth of carbonation is estimated based on the change in colour profile. The pH value can be determined by collecting broken samples or from core samples by analysing in the laboratory from water extracts and pH meter or RCT Kit (Fig.4)

Chloride Content: Chloride content can be determined by collecting broken samples or from the tested core samples. Primarily, the chloride up to cover thickness is of prime importance. The test consists of powdering the sample, obtaining the water extracts, and conducting standard titration experiment for determining the water soluble chloride content which is expressed by weight of concrete or by weight of cement, if the mix ratio is known.

The method gives the average chloride content in the cover region; where as the level of chloride near the steel concrete interface is of more importance.

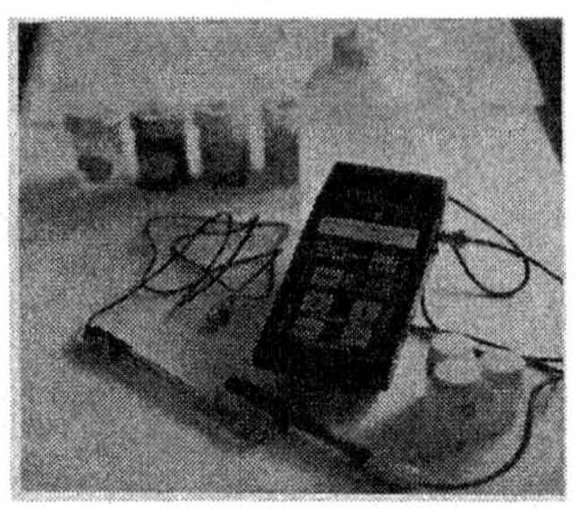

Fig.4. RCT KIT

Further, a chloride profile across the cover thickness will be a more useful measurement as this can help to make a rough estimate on chloride diffusion rate. "Rapid Chloride Test Kit" [4] can be used at the site. The test consists of obtaining powdered sample by drilling, collecting the sample from different depths (every 5 mm), mixing the sample (of about 1.5 gm weight) with a special chloride extraction liquid, and measuring the electrical potential of the liquid by chloride-ion sensitive electrode. With the help of a calibration graph relating electrical potential and chloride content, the chloride content of the sample can be directly determined. As the quantity of sample required is very little, a chloride profile for every 5 mm depth across the cover concrete up to steel concrete interface can be established. As mentioned already, the corrosive environment within concrete gets established once the pH value is lowered to 10 or less, or the chloride level reaches the threshold value of about 0.4 to 0.6% by weight of cement[5]. Table VI. gives qualitative guidelines for identification of corrosion prone locations based on pH values and chloride content.

Table VI. Guidelines for Identification of Corrosion Prone Locations based on Chemical Analysis

Test results	Interpretations
High pH values greater than 11.5 and very low chloride content	No corrosion
High pH values and high chloride content greater than threshold values (0.4% - 0.6% by weight of cement)	Corrosion prone
Low pH values and high chloride content (greater than 0.4% - 0.6% by weight of cement)	Corrosion prone
Low pH values and high chloride content	Corrosion prone

Half Cell Potential Survey: Corrosion being an electrochemical phenomenon, the electrode potential of steel rebar with reference to a standard electrode undergoes changes depending on corrosion activity. A systematic survey on well-defined grid points gives useful information on the presence or probability of corrosion activity. The gird points used for other measurements, namely, Rebound Hammer and UPV can be used for making the data more meaningful. The common standard electrodes used are (i) Copper-Copper Sulphate

Electrode (CSE) (ii) Silver-Silver Chloride Electrode (SSE) (iii) Standard Calomel Electrode (SCE). A simple field arrangement for measuring half-cell potential is shown in Fig.5.

The general guidelines for identifying the probability of corrosion based on half-cell potential values as suggested in ASTM C 876[6] are given in Table VII.

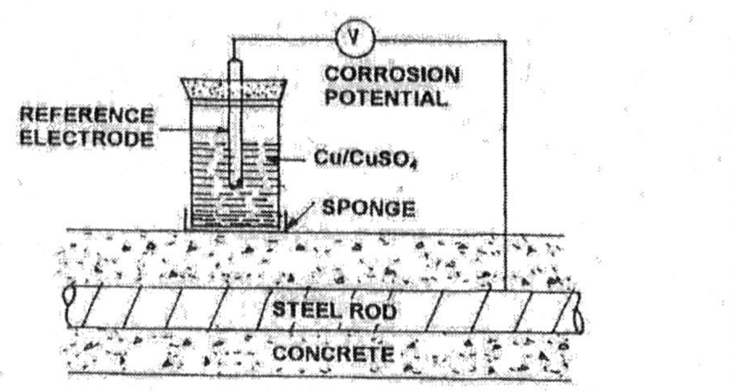

Fig. 5 Typical setup for Potential Measurement in reinforcement

It is important to realize that the potential of any metal in cement concrete environment is a function of a large number of variables such as concrete composition, pore liquid, concrete resistivity, cover thickness, degree of polarisation, etc. Hence, no quantitative conclusion can be drawn from it. The corrosion potential is determined by a balance between the anodic and cathodic reactions and the corrosion rate deals with the actual rate of the reaction.

Table VII. Corrosion Risk by Half Cell Potential

Corrosion	Potential
> 95%	More negative than - 350 mV
0%	- 200 to - 350 mV
< 5%	More positive than - 200 mV

The corrosion potential is determined by a balance between the anodic and cathodic reactions, and the corrosion rate deals with the actual rate of the reaction. Potentials do not give information on the amount of corrosion. Parameters affecting potential is shown in Fig.6.

However, a systematic "potential mapping survey" is considered to be more useful for on-site identification of the corrosion state of the reinforcements. A prerequisite of this survey is a study of the reinforcement drawings (if existing) plus resistance testing to ensure continuity of the steel cage over the location being surveyed. Potential readings are obtained over several points on a structure/structural member and over well-defined grid points. Fig. 7 shows the half-cell potential measurement made on an actual structure. Half cell contour will also help to install corrosion control and monitoring systems recently developed.

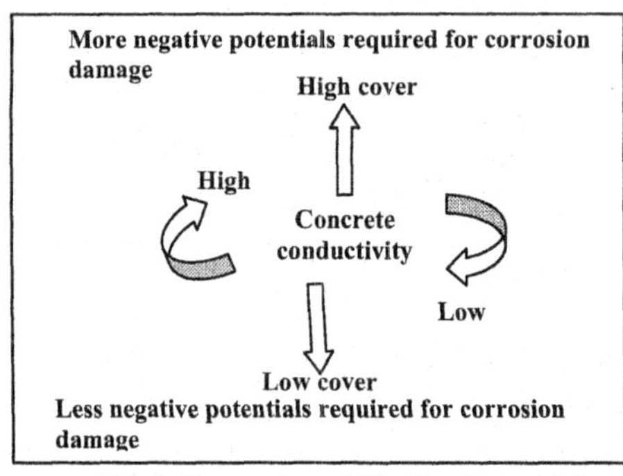

Fig.6. Parameters Affecting Potential

Fig.7. Half-cell Potential Measurement on a structure

Resistivity Mapping: The electrical resistance of concrete plays an important role in determining the magnitude of corrosion at any specific location. This parameter is expressed in terms of "Resistivity" in ohm- centimeter or kilo ohms- centimeter. The factors that govern the resistivity values are; constituents of concrete, chemical contents of concrete such as moisture, chloride level, and other ions. In view of the many influencing factors it is possible that resistivity values will vary quite significantly over a structure. Therefore, it is necessary that a systematic survey of resistivity values over well-defined grid points is suggested and a plotted map can be used to assess the corroding areas. Table VIII. Indicates the general guidelines of resistivity values based on which areas having probable corrosion risk can be identified in concrete structures. The method essentially consists of using a 4 probe technique in which a known current is applied between two outer probes and the voltage drop between the inner two elements is read off allowing for a direct evaluation of resistance R, using a mathematical conversion factor. Resistivity is calculated as $\rho = 2\ \pi x.R.x$ a, where 'a' is the spacing of probes. The principle of four-probe resistivity testing meter and Contour is shown in Fig.8 and Fig.9.

Table VIII. Corrosion Risk from Resistivity

Resistivity (ohm - cm)	Corrosion probability
Greater than 20,000	Negligible
10,000 - 20,000	Low
5,000 - 10,000	High
Less than 5,000	Very high

Fig.8 Resistivity testing meter

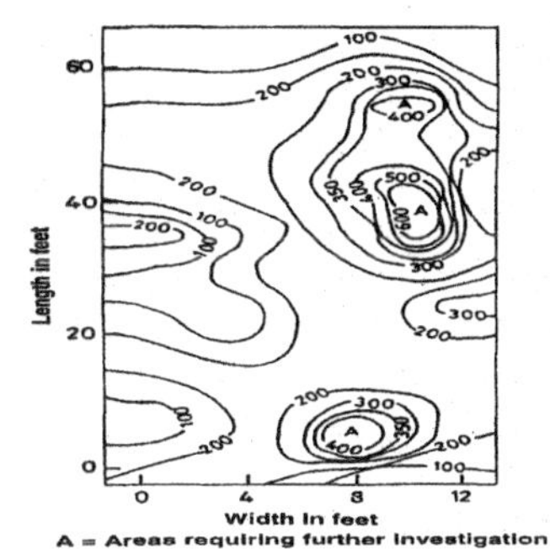

Fig.9 Resistivity contour

Table IX. gives some guidelines for a qualitative identification of corrosion prone areas based on combined results of half-cell potential and resistivity.

Table IX. Corrosion Probability based on Resistivity and Potential Mapping

Test results	Interpretations
High resistivity greater than 10,000 ohm cm and low potentials - more positive than -200 mV (CSE)	No active corrosion - relatively cathodic
Low resistivity below 10,000 ohm cm and potentials between -200 mV to -250 mV (CSE)	Initiation of corrosion activity - relatively anodic
Low resistivity about 5,000 ohm cm and potentials -200 mV to-350 mV (CSE)	Presence of corrosion activity - anodic
Low resistivity below 5,000 ohm cm and potential more negative than -350 mV (CSE)	High intensity of corrosion - fully anodic
Higher potential gradient and high conductivity	High rate of corrosion

Measurement of Corrosion Rate: In reinforced concrete structures, determination of actual rate at which the reinforcement is corroding assumes major importance. One method is by "Linear Polarisation Resistance" (LPR) technique for on-site study of corrosion rates of steel in concrete[7] commercial instrument GECOR-6 that works on linear polarization principle can be used. The application of GECOR-6 for measuring the corrosion current in the reinforcement in an actual structure is shown in Fig.10

Fig.10. GECOR-6 Instrument being used in the Structure

The techniques discussed above for assessing the condition of reinforcement embedded in concrete using half-cell corrosion potential mapping, Gecor-6 come under 'Electrochemical methods (ECM)'. There are other ECM's that have been used successfully in the field. They are Galvanostatic pulse and Electrochemical noise methods. Developments in Non-Electrochemical Methods (NECM) have given rise to acoustic emission and ultrasonics. Some of these methods can be used in isolation or in combination in the field investigation to assess the condition of reinforcement in the structural members depending on location and types of structures. Some of the other ECM's and NECM's that are popular have been described briefly.

Galvanostatic Pulse method: Galvanostatic Pulse method[8-11] has been commercialized. 'Galva Pulse' [12] (Fig.11) instrument has been used successfully in the field.

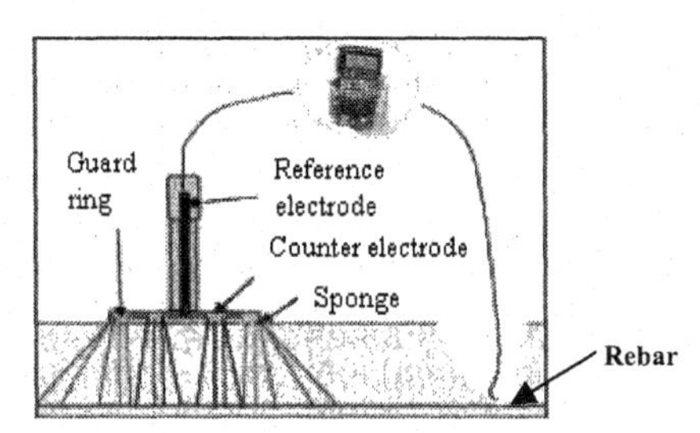

Fig.11. Galva Pulse Instrument

In this method an anodic current pulse "I" (typically in the range of 5 to 400 μA) for 5-10 seconds duration is applied to the system from a counter electrode placed over the concrete surface. A guard ring confines the current to an area "A" of the reinforcement below the central counter electrode. The reinforcement is polarized in the anodic direction compared to its free corrosion potential with SSE (Ag/AgCl) reference electrode. The resulting electrochemical potential of the reinforcement is recorded as a function of time.

When constant current is applied to the system, Ohmic drop = $I*R_p\,(\Omega)$. The polarization resistance of the reinforcement "R_p" is calculated by Stern Geary equation,

$$I_{corr} = 26/R_p \qquad (1)$$

$$\text{Corrosion rate} = (11.6 \times I_{corr} / A)$$

Where,

Corrosion rate = μm/year, I_{corr} = μA and A= Confined area of reinforcement in cm^2. The factor 11.6 is for black steel.

The instrument gives corrosion rate, polarisation potential and electrical resistance of the cover concrete. The instrument offers two modes of measurement. Half cell potential and resistance measurements that take 1-2 seconds per test (anodic and cathodic areas), while for corrosion rate, half-cell potential and electrical resistance measurements it takes 5-10 seconds (used in anodic areas).

Electrochemical Noise Method (ECN): Electrochemical noise method[13-16] has been used in structures to determine active corrosion and to study different systems of corrosion process in steel. During corrosion process, the fluctuations in corrosion potential and current generate noise in the corroding area of the rebar. Hence, this method is being used to monitor localized corrosion in environment of low conductivity, such as, those created when the corrosive electrolyte is simply covered by thin films on the surface. Page and Lambert[13] from their experiments on large numbers of steel electrodes embedded in concrete exposed to an external source of chloride ions, observed clearly linear logarithmic relationship between corrosion rate as determined by linear polarization method and the standard deviation of ECN in the form:

$$\mathrm{Log}_{10}I = 0.171 + 0.823\ \log_{10}SD$$

Hardon et. al[14] also arrived at similar conclusions. Typical block diagram for ECN measurement instrumentation[16] is shown below. (Fig.12)

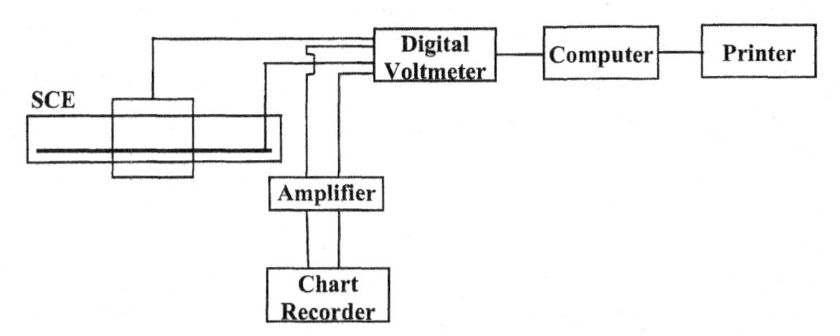

Fig.12. Typical block diagram for ECN measurement instrumentation

Acoustic Emission Technique (AE): This is one of the Non- Destructive Evaluation (NDE) methods for detecting corrosion activity, initiation of corrosion and corrosion rate in the embedded reinforcements. The primary advantage of AE over other NDE techniques is that it can directly detect the process of flaw growth. The basic principle is that during corrosion process, the corrosion products formed over the rebar swells and exert pressure to the surrounding concrete inducing microcracks. Stress waves generated during the expansion process when high enough breaks the interface layer. The growth of microcracks is directly related to the amount of corrosion products of a corroding rebar. Hence, by detecting the AE event rate and amplitude, the degree of the corrosion can be interpreted [17].Typical laboratory experimental test setup[18] is shown below. (Fig.13)

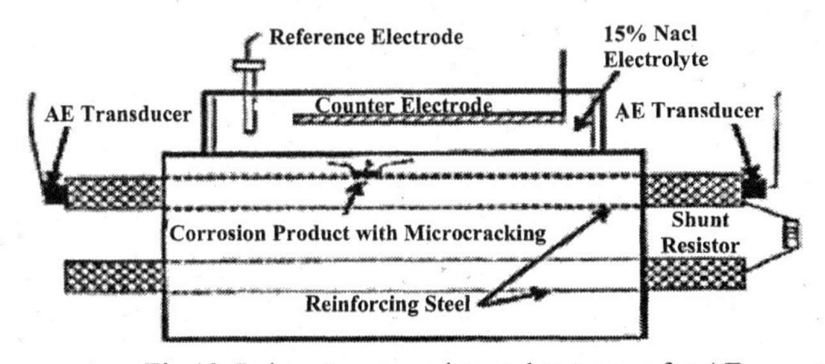

Fig.13. Laboratory experimental test setup for AE

Extensive laboratory experiments have been carried out by many researchers[19-22] using AE technique to study types of corrosion, location of corrosion/corrosion products and to quantify corrosion.

More advanced methods such as radiography and ultrasonics are being used to detect rebar corrosion and they have met with limited success due to the skill needed to analyse the data as well as the cost of the equipment. In addition, radiography uses X-ray or gamma rays that require extensive safety precautions. Considering this aspect, the methods discussed are very useful.

CONCLUDING REMARKS

A periodical measurement of rate of corrosion taking into account temperature and humidity and a statistical analysis of the data measured over a period of time may yield better and reliable assessment in predicting service life. It also becomes necessary to define the level of damage more clearly to declare a structure unserviceable. The methods described above provide useful tool in ascertaining the condition of the structure and plan suitable strategy of repairs, if required. The methods can also be used for condition monitoring of structures once in every ten years to determine any change in the properties of concrete and steel, and take suitable measures at right time, if required, to enhance its service life.

Considering the corrosion as one of the main durability problems in concrete structures, research in many parts of the world is on, for introducing non-ferrous reinforcements in structures, to completely eliminate the corrosion problems. The codal provisions for these reinforcements are yet to be established for use in practice. In structures already in service, different types of corrosion controlling and monitoring probes have been introduced, and codified that help in enhancing the service life of corrosion affected structures. In order to use these probes at right places in the structural elements, non-destructive and partially destructive tests form an important tool.

REFERENCES

[1]Malhotra V.M. - Chapter 2, p 21 “Testing hardened concrete: non-destructive methods,” *The Iowa State University Press*, AMES, IOWA and ACI, Detroit, Michigan (1976)

[2]Allan P. Crane (Editor), - “Corrosion of reinforcement in concrete construction,” Society of Chemical Industry, London, Publishers - Ellis Horwood Ltd., First Edition, 1979

[3]Schiessel P. (Editor), - “Corrosion of steel in concrete,” Report of the Technical Committee, 60- CSC RILEM, published by Chapman and Hall, 1988

[4]Operation Manual, - Rapid chloride test kit, Germann Instruments; Emdrupvej 102 DK-2400, Copenhagen NV; Denmark

[5]Page C.L, Treadway, KWJ and Bamforth PB (Editors) - Corrosion of reinforcement in concrete; Society of Chemical Industry; Elsevier Applied Science, May 1990

[6]Standard test method for half-cell potentials of reinforcing steel in concrete, ASTM C 876-80

[7]Feliv, J.A. Gonsalez, and Andrade, C – “Polarization resistance measurements in large concrete specimens; mathematical solution for a unidirectional current distribution,” Materials and Structures, RILEM V 22, 1989, pp 199-205

[8]Elsener,B., Wojtas, H., Bohni, H. “Galvanostatic Pulse Measurements – Rapid on site Corrosion Monitoring,” Proceeding of International Conference held at the University of Sheffield, 24-28 July 1994

[9]Newton, C.J., Sykes, J.M. “A Galvanic Pulse Technique for investigation of steel corrosion in concrete,” *Corrosion Science*, **28** 1051-1073 (1988)

[10]Stern, M., Geary, A.L., “Electrochemical Polarisation, I.A Theoretical analysis of Shape of Polarisation Curves,” *Journal of the Electrochemical Society*, **104** 56-63(1957)

[11]Feliu, S., Gonzales, J.A., Andrade, C., Feliu, V., “Onsite Determination of the Polarisation Resistance in a Reinforced concrete Beam,” *Corrosion Engineering*, **44** 761-765 (1988)

[12] www.germann.org

[13]Page, C.L., and Lambert, P., "Analytical and electrochemical investigations of reinforcement corrosion," Contractor Report.30.Transport and Road Research Laboratory, Crowthorne, 1986
[14]Hardon, R.G., Lambert, P., and Page, C.L., "Relationship between Electrochemical Noise and Corrosion Rate of Steel in Salt Contaminated Concrete," *Br. Corros, J.*,**23**[4] 225-228 (1988)
[15]Katwan, M.J., "Corrosion Fatigue of Reinforced Concrete," Ph.D. Thesis, University of Glasgow, (October 1988), 525-552
[16]Katwan,M.J., Hodgkiess, T., and Arthur.P.D., " Electrochemical noise technique for the prediction of corrosion rate of steel in concrete," *Materials and Structures/ Materiaux et Constructions*, **29** 286-294 (1996)
[17]Zongjin Li, Faming Li, Allan Zdunek, Eric Landis and Surendra P. Shah, "Application of Acoustic Emission Technique to Detection of Reinforcing Steel Corrosion in Concrete," *ACI Materials Journal*, January-February 1998, pp.68-76
[18]Alan D. Zdunek and David Prine, and Zongjin Li., Eric Landis.Surendra Shah, "Early Detection of Steel Rebar Corrosion by Acoustic Emission Monitoring," Paper No.547, presented at CORROSION95, the NACE international Annual Conference and Corrosion Show
[19]Seah, K.H.W., Lim, K.B., Chew, C.H. and Toeh, S.H., "The correlation of Acoustic emission with rate of corrosion," *Corrosion Science*, **34**[10] 1707-1713 (1993)
[20]Li, Zongjin and Shah, S.P., "Localization of Microcracking in concrete under uniaxial tension," *ACI Materials Journal*, **91**[4] 372-381 (1994)
[21]Maji, A.K. and Shah, S.P., "Process zone and acoustic emission measurement in concrete," *Experimental Mechanics*, **28** 27-33 (1998)
[22]Landis, Eric N., and Shah Surendra, P., "Recovery of microcrack parameters in mortar using quantitative acoustic emission," *Journal of Nondestructive Evaluation*, **12**[4], 219-232 (1993)

FIBER OPTIC INSTRUMENTS FOR MONITORING SERVICE LIFE PERFORMANCE OF IN-PLACE CONCRETE

K. Ravisankar
Deputy Director
SERC, CSIR Campus, Taramani
Chennai-600 113, India
e-mail:- kravi@sercm.org

ABSTRACT

Developments in high performance materials and innovative construction techniques have paved the way for construction of large concrete structures involving huge investments. These structures are built with certain reliability and a notion that the structure will perform as per design during their service life. However, there have been instances of damage and deterioration to the constructed systems during the active service life. Knowledge of the structure's condition could improve the service performance of the structure. The concept of monitoring service life performance of concrete structures is continuously evolving as a result of developments of new sensor and instrumentation technologies. Conventional sensors like electrical resistance strain gages have major limitations in evaluating long-term behaviour of structural members, especially in concrete structural systems. Recent developments in fiber optic sensor systems have the potential to offer advantages that can essentially eliminate conventional sensor deficiencies. Experimental studies were carried out in the laboratory to evaluate the performance of fiber optic sensors and assess their suitability for monitoring service life performance of concrete structures. In this paper, the importance and potential of fiber optic sensors for monitoring of concrete structures have been highlighted. Brief details of the laboratory studies carried out on fiber optic sensors are also covered.

INTRODUCTION

Structural health monitoring can be defined as the science of inferring the health and safety of an engineered system by monitoring its performance. Monitoring technology can play an important role in securing system integrity, minimizing maintenance cost and maintaining longevity of structures. It can also provide verification of current design analysis and suggest improvements for retrofit and future designs. Apart from this, implementing new materials and innovative construction techniques are the key factors that necessitate performance monitoring of structures. Presently, deterioration and damage are identified by visually observing the signs they exhibit as they progress. The effectiveness of visual inspection in reaching all the critical locations and in finding all the possible defects is somewhat questionable. Hence there is growing interest in developing appropriate devices and instruments for automated structural monitoring and damage assessment in the field of civil engineering. Effective monitoring, reliable data analysis, rational data interpretation and correct decision making are challenging problems.

Conventional sensors for monitoring the performance of structures are mainly limited to the application of electric and magnetic principles using electrical resistance strain gages, linear differential transformers, etc. While conventional sensors can serve well for short-term measurements, they have major limitations in evaluating long-term performance of structural system. The main problem associated with these conventional sensor techniques, stem from their response to ambient electrical noise and potential for degradation with age. Another major problem with conventional sensor system is the cabling. An embedded sensing system for a major civil structure may require large number of sensors, which means that many cables will be required to connect them to the measuring instrument. If they are embedded in concrete structures, such bulky cables may affect the integrity of the structure. These cables will also suffer from electro-magnetic interference, resulting in high noise-to-signal ratio[1].

Fiber optic sensor systems have the potential to offer many advantages over conventional sensors and it can essentially eliminate conventional sensor deficiencies and permit reliable long-term measurements and monitoring. Fiber optic sensors are extremely well suited to undertake point, quasi-distributed, and truly distributed measurements of strain and other physical parameters[2]. The most attractive feature of fiber optic sensors is their inherent ability to serve as both the sensing element and the signal transmission medium. Fiber optic sensing technology shows promise of providing superior performance to conventional sensors for measuring strain, temperature, vibration, displacement, pressure etc., under field conditions. Fiber optic sensors are now being developed to measure a number of physical and chemical parameters in not only the laboratory research environment, but also in practical field application. Due to the small size and geometric adaptability, fiber optic sensors can be easily integrated into a concrete structural system. Concrete structural systems integrated with fiber optic sensors are capable of serving a dual purpose by providing in-place, quality control of concrete at early ages and condition monitoring of structural anomalies thereafter. A significant amount of interest in fiber optic sensing techniques has been generated in the civil engineering community with regard to instrumentation of bridges and similar structures.

Fiber optic sensing is an emerging technology, which is now receiving attention in the civil engineering field. The importance and potential of fiber optic sensors for monitoring of concrete structures have been highlighted. Brief details of the laboratory studies carried out on fiber optic sensors are also covered.

INSTRUMENTATION FOR PERFORMANCE MONITORING OF CONCRETE STRUCTURES

Automatic measurement of instrumented concrete structures is becoming more common for both diagnostic system identification purposes and in field behaviour monitoring during construction. The considerations necessary to deploy a successful monitoring system include proper instrumentation, reliable signal processing and knowledgeable information processing. Since instrumentation systems include sensory devices, sensor technology is of critical importance in the development of monitoring systems.

Continuous monitoring of deflection requires very stable, reliable instrumentation such as the LVDT. Another inexpensive, alternative method is to use a digital theodolite with targets fixed to the structure which undergoes deformation and some reference objects which do not undergo any deformation. Continuous monitoring may not be possible in this case, but periodic measurements will give adequate data for analysis. Strains can be measured using vibrating wire gages or fiber optic strain gages. They can be either embedded (in the case of concrete) or can be surface mounted. The temperature sensors (either thermisters, thermocouples or fiber optic based temperature sensors) can be installed prior to concreting or after concrete is set. These can measure the temperature of concrete either inside or at the surface. Rotations are measured by tilt meters/inclinometers. Humidity influences the shrinkage and creep of concrete. Hence, the data on relative humidity may be required for analysis. Load cells (vibrating wire gage based) can be used for measuring prestressing force. The potential for corrosion of reinforcement is measured using rebar probe assemblies with reference electrodes. Accelerometers can be used to measure the dynamic behaviour of the structure.

FIBER OPTIC SENSORS

Fiber optic sensing technology is destined to become one of the core technologies of the 21st century for the construction industry. Fiber optic sensors are extremely small in diameter, very light, sensitive to strain and temperature changes, resistant to corrosion and fatigue, and capable of wide bandwidth operation[3]. They are compatible with concrete/composite materials, safe to handle, unlikely to initiate fires or explosions, and completely immune to electromagnetic interference, and have found use as strain sensors for monitoring concrete structures. Fiber optic sensors are fabricated using high strength silica and they can be classified under different categories. Localised, distributed and multiplexed sensors are based on sensing methods. Intensity, interferometric, polarimetric and spectrometric based sensors are classified according to the transduction mechanism[4].

Localized fiber optic sensors determine the measurand over a specific segment of the optical fiber, and are similar to conventional strain or temperature gages. One of the most critical parameters to be measured for structural integrity and health monitoring of a structure is the local strain. Sensing based on intensity modulation pertains to light intensity losses that are associated with straining of an optical fibers along any given portion of the length. Sensors taking advantage of this phenomenon are termed intensity or amplitude type sensors. Intensity based fiber sensors offer the advantages of ease of fabrication, robustness, and simplicity of signal processing.

Spectrometric sensors are widely used in the sensing of chemical reactions, and remote monitoring of contaminants in ground water. For structural monitoring applications, the transduction mechanism in these types of sensors is based on relating the changes in wavelength of light to the measurand of interest, i.e. strain. An example of such sensors for measuring strains is the Bragg grating type fiber sensors. The construction is based on the creation of a permanent periodic modulation of the refractive index in the core of a photosensitive optical fiber by transverse illumination with an interference pattern created by a pair of laser beams. This creates a Bragg grating in the core of the optical fiber. By virtue of its strain sensitivity, the Bragg grating is an effective sensor for many applications. The fiber optic grating acts as a wavelength-selective mirror for incoming light. The reflected portion of the light consists of a narrow spectral band while the remainder is simply transmitted through the grating. The transmitted light is simply lost or as in the serial multiplexing schemes, it is used to interrogate gratings further along in the fiber. The optical instrumentation for Bragg type sensors is highly intricate, as they require sensitive spectrometers for detecting the minute changes in the wavelength of light.

Phase modulated or Interferometric sensors cover a broad range of optical phenomena for sensing purpose. Interferometric sensors are highly sensitive for measuring strains. These sensors usually consist of a reference fiber which is isolated from the perturbation being measured, and a sensing fiber which is made extremely susceptible to the incidence of a perturbation. The output from the two fibers interferes at the photo detector, and the magnitude of the interference is dependent on the incident perturbation. An exception to a two arm Interferometric sensor is a single fiber Fabry-Perot type sensor. In this type of sensor, a cavity comprising two mirrors (reflectors) which are parallel to each other and perpendicular to the axis of the optical fiber form the localised sensing region. Here the reference and sensing optical fiber are one and the same up to the first mirror, which constitutes the start of the sensing region. A change in the optical path length between the mirrors leads to a shift in the cavity distance. Fabry-Perot cavity is formed between the air-glass interfaces of two fiber end faces aligned in a hollow core fiber (Fig.1). Changes in the separation between the two fiber end faces, known as the air gap length, cause Interferometric fringe variations.

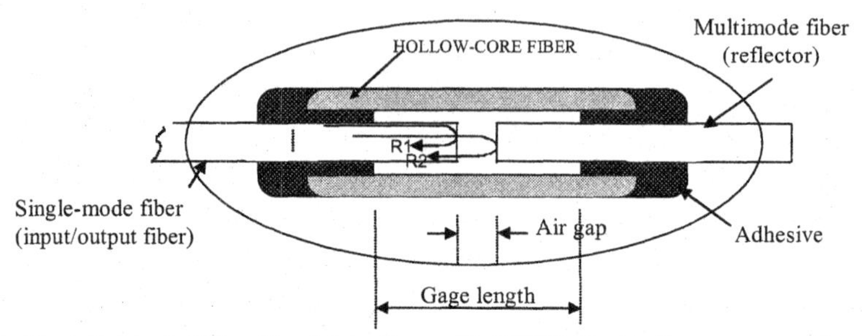

Fig.1 Extrinsic Fabry Perot Interferometric(EFPI) fiber optic sensor configuration

Polarimetric sensors form a special class of phase sensors, and take advantage of the polarization characteristics of light for transduction. Fringe shifts due to external perturbations in polarization maintaining single mode fibers are caused by the interference of two mutually perpendicular polarized waves. The sensitivity of polarimetric sensors is dependent on the polarization characteristics of the fiber, such as birefringence and the beat length. Polarimetric sensors are more sensitive than the intensity type sensors.

EXPERIMENTAL STUDIES ON PERFORMANCE ASSESSMENT OF EFPI FIBER OPTIC SENSORS UNDER DIFFERENT LOADINGS

Experimental studies were carried out in the laboratory to evaluate the performance of EFPI (Extrinsic Fabry-Perot Interferometric) fiber optic strain sensors under different loadings[5&6]. Commercially available EFPI sensors and the associated measuring equipments were used for the studies. Brief details of the laboratory studies carried out are covered here.

Development of Technique for Embedding Fiber Optic Sensors in Concrete

Embedding bare fiber optic strain sensors in concrete structures is not advisable because of their fragility. The process of placing and compacting of concrete, exerts severe stress which can damage the sensors. Hence they must be properly protected. Another important aspect of sensor embedding is the ingress/egress of the sensor lead to/from the host structure. The optical lead wires which are very fragile also need to be protected from damage at ingress/egress locations. The main problem is to protect the sensor and carriers against damage during concreting and vibration during compaction. One method of safeguarding the sensor is to provide a protective layer called encapsulation between the sensor and the host structure. The properties of this encapsulation can have a major influence on the life and functionality of the sensor. The material used for encapsulation should be compatible to the surrounding concrete to ensure complete strain transfer.

Experiments were carried out for evolving suitable procedures for embedding fiber optic sensors in concrete. Three different methods of encapsulation were attempted; (i) encapsulation of sensors by liquid epoxy monomer and subsequent polymerization, (ii) encapsulation of sensor by a pair of flat acrylic sheets, and (iii) encapsulation by a pair of epoxy sheets. In each method, two encapsulated samples were prepared and they were identically embedded in concrete cylinders and tested in an universal testing machine (UTM). Experiments carried out using liquid epoxy encapsulation were not successful due to complete lack of strain transfer in both the specimens. The details of acrylic/epoxy sheets as encapsulation, embedding in concrete cylinders and testing in a UTM are described below:

One 50mm long EFPI fiber optic strain sensor was encapsulated using two sheets of 100×10×2mm thick as shown in Fig.2. A groove was cut in one of the sheets and a 50mm long fiber optic strain sensor was bonded using epoxy cement. Then another sheet was placed over and sealed using liquid epoxy. The optical fiber cable also was protected with PVC sleeve to prevent it from mechanical damage during handling and also to protect it at the ingress/egress locations. Concrete cylinders of size 150mm dia. 300mm height were cast and one epoxy encapsulated fiber optic sensor was embedded inside at the centre of each concrete cylinder during casting. After completion of curing of concrete cylinders, four electrical resistance strain gages were bonded to the outer surface of the each cylinder. The instrumented concrete cylinder was subjected to compressive load using an UTM. The load was applied in steps and for each step, the output from embedded fiber optic strain sensor and the surface mounted electrical resistance strain gages were recorded. The strain response obtained from embedded fiber optic sensor was compared with the average of the four conventional electrical resistance strain gage responses. To check the reliability, the experiment was repeated on a second specimen in each case. From the above studies it was observed that the flat acrylic sheets as encapsulation showed lower response compared to electrical resistance strain gages while much better (closer to the average of the four electrical resistance strain gages) response was observed with epoxy sheets as encapsulation as shown in Fig.3.

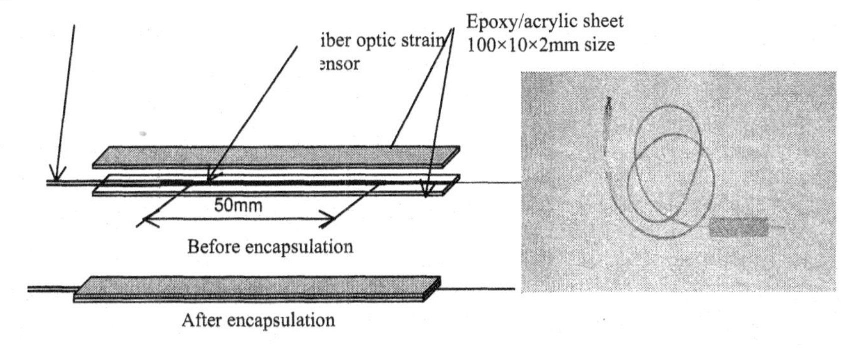

Fig. 2 Encapsulation of fiber optic sensor for embedment in concrete using acrylic/epoxy sheets

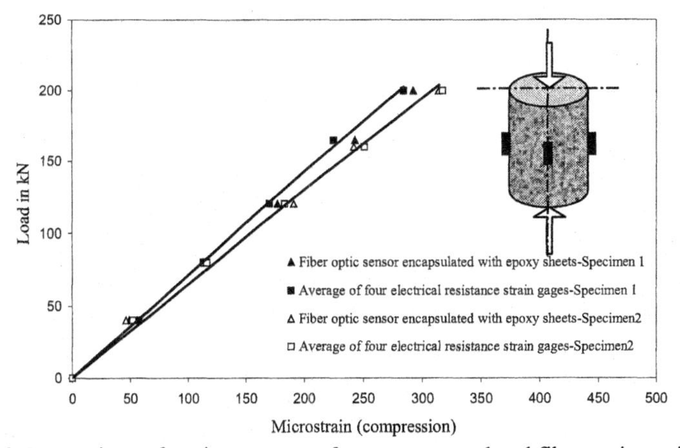

Fig. 3 Comparison of strain response of epoxy encapsulated fiber optic strain sensor with electrical resistance strain gage-specimens 1 and 2

The ability of an optical fiber sensor to monitor strain distribution in a structural material depends on the bonding characteristics between the material and sensor. Procedures were formulated for mounting EFPI sensors on re-bars of concrete structures as well as bonding it on concrete and steel structures. Details of experimental studies carried out to assess the performance of EFPI sensor bonded to a reinforcing bar of a concrete beam are described here. For this purpose, one reinforced concrete beam of size 100 x 200 x 1500 mm was cast. A fiber optic strain sensor (10 mm size) was bonded to the bottom reinforcing bar of the concrete beam and an electrical resistance strain gage was also bonded very close to it, for comparing fiber optic strain sensor response. To prevent damages to the sensors during concreting, they were suitably covered. An experimental test set-up was arranged to create a constant bending zone at the instrumented locations of the beam. The beam was subjected to four-point bending load and the load was applied in steps. Strain responses from fiber optic sensors and conventional strain gages were recorded for all loading steps. Load vs. strain plots were made and close comparison was observed between the fiber optic sensor values and conventional strain gage values. The developed techniques for mounting the sensor in or on the structure could be implemented to actual field problems also.

Performance Studies of Surface Mounting Type Fiber Optic Strain Sensor Under Static Load

Experimental studies were conducted to evaluate the performance of EFPI fiber optic strain sensors under static loading condition. In order to assess the behaviour of EFPI fiber optic sensor under axial load, experiments were conducted using standard uniaxial tensile specimens. For this study, one mild steel tensile specimen (prepared as per ASTM Standard) was instrumented with one 20 mm long surface mounted fiber optic strain sensor (kept at one surface) and two conventional electrical resistance strain gages, kept one on each surface of the specimen. The specimen was subjected to axial tension by using an UTM. The load was applied in steps of 2 kN, up to a maximum of 18 kN and the strain responses from fiber optic sensor and conventional strain gages were recorded. Three cycles of loading and unloading were carried out. Load vs. strain plots were prepared to evaluate the comparative performance of EFPI fiber optic sensor. The agreement of values between the fiber optic sensor and conventional strain gage is fairly good (Fig.4).

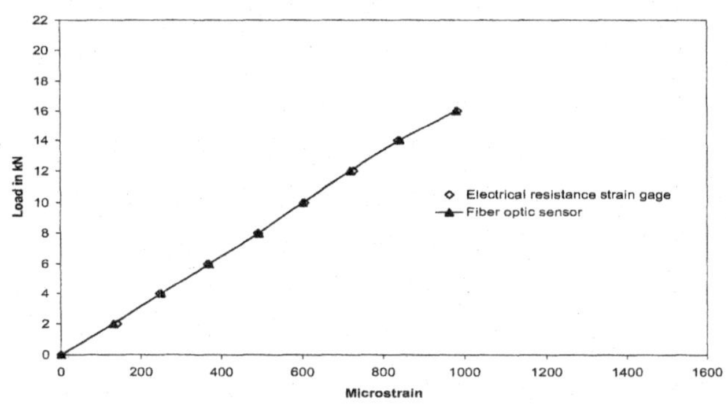

Fig.4 Comparison of fiber optic sensor with electrical resistance strain gage-steel specimen under uniaxial tension

In order to assess the behaviour of fiber optic sensors under flexural loading, a four-point bending load test was conducted. One mild steel I-beam was prepared for the bending test. The steel I-beam specimen was instrumented with two surface mounted EFPI fiber optic strain sensors (one each kept at the top and bottom flanges) and two conventional electrical resistance strain gages, bonded adjacent to the fiber optic sensor. The test set-up was designed to create a constant bending zone at the instrumented locations of the I-beam. The beam was subjected to a four-point bending load and the load was applied in steps. Strain responses from the fiber optic sensor and conventional strain gages were recorded for all loading steps. Close comparison between the fiber optic sensor values and conventional strain gage values was observed and the sensitivity was found to be nearly identical in both tension and compression.

Experiments were also conducted to assess the behaviour of fiber optic sensors, when fixed on concrete surfaces. A fiber optic strain sensor was fixed on the outer surface of a standard concrete cylinder (150 mm dia. and 300 mm long cylinder). One conventional electrical resistance strain gage was also mounted very close to the fiber optic strain sensor . The cylinder was subjected to a compressive load and the load was applied in steps. For each step loading, output from fiber optic sensor and conventional strain gage were recorded. Several loading/unloading trials were conducted. Load vs. strain variation was plotted to assess the sensitivity of the EFPI fiber optic strain sensor. A slight difference in strain response was observed between the two sensors.

Development of Temperature Compensation Procedures for Fiber Optic Strain Sensors

It is necessary to correctly interpret the influence of various parameters affecting the strain sensor output, while monitoring structures. Variation in temperature is an important factor that influences the fiber optic strain sensor output. It is important to clearly differentiate between the gage output due to thermal expansion of the structural material from that of strain due to loading.

During health monitoring of civil engineering structures such as bridges, the output from the fiber optic strain sensor may have a component of apparent strain due to temperature effects which is to be properly accounted for. This apparent strain is caused due to the temperature difference between the reference temperature and the temperature at any other time of measurement. The variation in temperature of structures between extreme seasons can be as much as 30 to 40° C, and this variation may cause an apparent strain of large magnitude. To correct the measured strain, the apparent strain must be established separately. In order to correct the temperature effects, temperature calibration was carried out from laboratory experiments on steel and concrete, using EFPI fiber optic strain sensors. A steel specimen of size 300 x 20 x 3 was prepared and one fiber optic strain sensor was bonded to the test specimen. A temperature sensor (electrical resistance type) was also bonded adjacent to a fiber optic strain sensor, to measure the surface temperature of the specimen. The instrumented test specimen was placed inside a temperature controlled oven and the temperature was raised in steps from ambient temperature to a maximum of 80°C. The temperature of the test specimen was allowed to stabilize at each stage, before measurements were carried out. Strain from the fiber optic strain sensor and temperature from temperature sensors were recorded for each temperature setting. In a similar fashion, temperature calibration was conducted for obtaining temperature correction data for concrete. While conducting temperature calibration for concrete, a temperature controlled water bath was used instead of a temperature controlled oven to eliminate the drying shrinkage effect. Also the concrete specimen was soaked in water for sufficient period to obtain moisture saturated condition. Fig. 5 shows the temperature vs. strain plot from which temperature correction coefficients for concrete can be obtained.

Performance assessment of fiber optic sensor under fatigue load

Reliable performance of the fiber optic sensor under high cycle fatigue load is an essential requirement for monitoring structures, especially bridges. Experiments were carried out to assess the performance of fiber optic sensors under fatigue load. Performance of a fiber optic sensor on a steel beam was carried out. A sinusoidal loading corresponding to a minimum of 20 MPa to a maximum of 160 MPa stress at a frequency of 5 Hz was applied to the instrumented steel specimen. The fiber optic strain measurement was consistent with the load amplitudes during a two million cycle fatigue test. In order to ascertain the reliability and repeatability of performance of fiber optic sensors under fatigue load, experiments were carried out on the second test specimen and here again the performance was found to be consistently good.

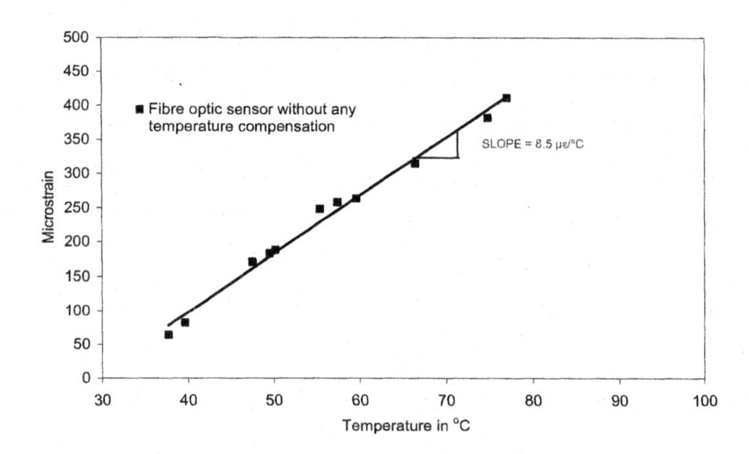

Long-Term Stability Assessment of Fiber Optic Sensors

Reliable measurement of strains over long periods is an essential requirement for performance monitoring of structures. Experiments were conducted to evaluate the long-term performance of fiber optic sensors. One concrete beam of 100 x 200 x 3200 mm was post-tensioned with three prestressing wires (7 mm dia). One of the high tensile prestressing wire was instrumented with two fiber optic strain sensors (kept diametrically opposite) at one location of the wire. The beam was additionally instrumented with two surface mounting type fiber optic sensors. Adjacent to the surface mounted fiber optic sensors, vibrating wire strain gages were also fixed for comparison. Tensioning of wires was carried out sequentially one after the other (starting from the middle wire first), by means of a hydraulic jack. At the end of prestressing the beam, the strains in the tendons as well as concrete surface strains were measured.

Continuous monitoring of the prestressed concrete beams was carried out for a period of 300 days. Strains from fiber optic sensors, and vibrating wire gages were recorded at regular intervals. The plot of strain vs. time of the fiber optic sensors bonded to the prestressing wire (Fig.6) shows stable behaviour of the gages. Relative comparison of the fiber optic sensor with the vibrating wire sensor bonded to the concrete surface was also evaluated. The response of the fiber optic sensor compares well with that of vibrating wire gages. The daily variation in the strain response is attributed to the daily temperature variations. From the analysis of measured data, it is seen that the output of fiber optic sensors is stable under long-term loading condition and compare well with that of vibrating wire sensors.

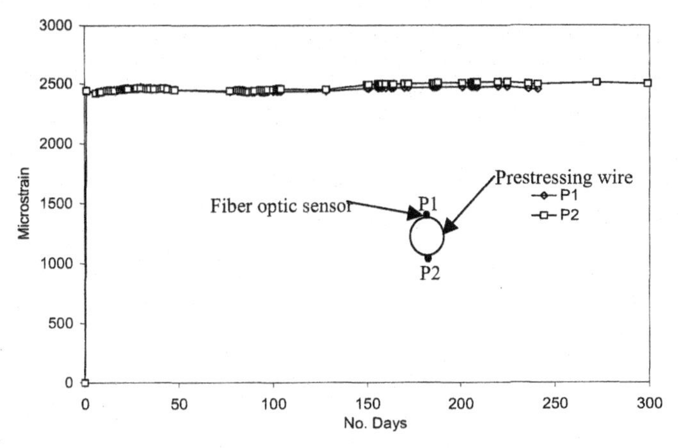

Fig.6 Strain vs time plot - fiber optic sensors bonded to a prestressing wire

SUMMARY

Civil infrastructure systems are generally the most expensive investments/assets in any country and these systems are deteriorating at an alarming rate. New tools and devices that provide feed back on the 'state of health' of constructed systems are needed[7]. In this paper, the importance and potential of fiber optic sensors for performance monitoring of concrete structures have been highlighted. Brief details of the laboratory studies carried out on EFPI fiber optic sensors are also covered. EFPI fiber optic strain sensors possess good qualities for application to civil engineering structures. Fiber optic sensing is an emerging technology, which is now receiving attention in the civil engineering field. The advantages of geometric adaptability, high sensitivity, immunity from corrosion and electro-magnetic interference and excellent long-term stability of the sensor suggest that fiber optic sensor system deserves wider application for monitoring service life performance of concrete structures.

REFERENCES

[1]D.R. Huston and P.L Fuhr, "Fiber Optic Smart Civil Structures," pp.647-65 in *Fiber Optic Smart Structures*, 1st ed. Edited by Eric Udd, John Wiley & Sons, Inc., New York, 1995

[2]R.M. Measures, "Need for Integrated Structural Monitoring (Chapter-2)" in *Structural Monitoring with Fiber Optic Technology*, 1st ed., Academic Press, California, 2001.

[3] M.B. Kodindouma and R.L.Idriss , "An Integrated Sensing System for Highway Bridge Monitoring," pp. 132-40 in *Smart Systems for Bridges, Structures and Highways*, Edited by L.K. Matthews, Proceedings of SPIE, San Diego, California, 2719 (1996)

[4]F. Ansari, "State of the Art in the Applications of Fiber Optic Sensors in Cementitious Composites," *Cement and Concrete Composites*, **19** 3-19 (1997)

[5]K. Kesavan, K. Ravisankar, T. Narayanan, S. Parivallal and P. Sreeshylam, "Temperature Calibration of EFPI Fiber optic Strain Sensors," *Experimental Techniques*, **28** [1] 31-33 (2004).

[6]S. Parivallal K. Ravisankar, T. Narayanan, K. Kesavan and P. Sreeshylam, "Fiber optic Sensors for Health Monitoring of Civil Engineering Structures," *Journal of Structural Engineering*, **31** [1] 9-14 (2004).
[7]K.P.Chong, N.J.Carino and G.Washer, "Health monitoring of Civil Infrastructures," pp.1-16 in *Health Monitoring and Management of Civil Infrastructure Systems*, Edited by S.B. Chase and A.E. Aktan, Proceedings of SPIE, Newport Beach, USA, 4337 (2001).

CONSIDERING MOISTURE GRADIENTS AND TIME-DEPENDENT CRACK GROWTH IN RESTRAINED CONCRETE ELEMENTS SUBJECTED TO DRYING

N. Neithalath, B. Pease, J.-H. Moon, F. Rajabipour, J. Weiss
Purdue University
550 Stadium Mall Drive,
West Lafayette, IN 47907

E. Attiogbe
Degussa Admixtures, Inc.,
23700 Chagrin Boulevard
Cleveland, OH 44122

ABSTRACT

Recent research has shown that the risk of early-age shrinkage cracking in concrete is influenced by many factors including the rate of shrinkage, the rate of strength development, the degree of restraint, and the extent of stress relaxation (creep). Previous research introduced an analytical model to predict residual stress development in restrained concrete elements, however this model assumed that the concrete was undamaged (i.e., uncracked) at each age before determining if the increase in shrinkage was significant enough to cause failure. This paper presents an approach to account for the role of moisture gradients on residual stress development that considers stable crack growth in restrained specimens over time. The model can be used to illustrate how alterations in binder composition (i.e., changes in ultimate shrinkage and autogenous/drying effects) change the residual stresses that develop and the resulting potential for cracking.

INTRODUCTION

Portland cement based materials experience volume changes in response to moisture, temperature, or chemical changes. The extent of these volume changes depend on numerous factors, including the material properties, temperature, relative humidity, geometry of the test specimens, and the age when the specimen is exposed to drying. In cases where these volume changes are restrained, tensile stresses can develop in the material, which can ultimately lead to cracking.

Previous research on early-age shrinkage cracking has focused on assessing the residual stresses that develop due to restraint as well as determining how likely a material may be to crack. A procedure was developed to determine the residual stresses that occur as a result of restraint in the restrained ring test [1,2]. In the analysis of the performance of various restrained shrinkage specimens, Attiogbe et al. [2] observed that the time of visible cracking (t_{cr}) is inversely related to the 'average section stress rate' ($\partial\sigma/\partial t$) at the time of cracking as described in Equation 1:

$$t_{CR} = 2.2\left(\frac{\partial \sigma}{\partial t}\right)^{-1.07} \tag{1}$$

where t_{CR} is given in days and $\partial\sigma/\partial t$ is given in MPa/day. Attiogbe et al. [2] suggested that this relationship forms a type of 'failure envelope' and as a result it may be possible to describe the behavior of a wide range of materials. Preliminary concepts were presented [2] to illustrate that for the form of Equation 1 may arise due to the fact that the penetration of the moisture gradient is related to the square root of time. This paper further explores how moisture gradients influence stable crack growth in a specimen subjected to external restraint. A simple model is developed that uses the 'crack band' concept to consider both moisture gradients and stable crack growth. The following section highlights the basic approach for treating the relationship between moisture loss, shrinkage and residual stress while the third section introduces the basic components of the modeling approach for time dependent crack growth. The final section illustrates some of the implications of this model by showing how the prediction of cracking would be influenced by different amounts of overall shrinkage and varying proportions of drying and autogenous shrinkage.

RELATING MOISTURE LOSS, FREE SHRINKAGE, AND STRESS DEVELOPMENT

Shrinkage can occur due to external drying or autogenous effects (self-desiccation/chemical shrinkage). Autogenous shrinkage describes shrinkage that occurs in sealed concrete when water is consumed by the hydration reactions causing both a reduction in the volume of the reacted products and a reduction in the internal pore pressure (i.e., relative humidity). Autogenous shrinkage increases as the water to cement ratio is reduced and is generally related to the extent of hydration that has taken place. External drying describes the process by which the moisture content inside the concrete specimen tries to equilibrate with the surrounding environment. This process is extremely slow and can take months or years, even for relatively small specimens. It can be shown that during this time the evaporation and diffusion process results in a moisture gradient near the surface of the concrete [3]. The internal stress distribution in concrete can be related to a shrinkage gradient which is related to an internal relative humidity gradient [4,5,6,7].

In order to compute the residual stresses that develop in a material, the internal moisture distribution needs to be determined. The diffusion of moisture (moisture can be expressed as water content, however relative pore pressure or humidity are frequently preferred since this eliminates the need to consider the effects of hydration directly) can be expressed using the governing differential equation

$$\frac{\partial H}{\partial(t-t_O)} = D_M \frac{\partial^2 H}{\partial x^2} + \frac{\partial H_{Auto}}{\partial t} \tag{2}$$

where H is the humidity, t is the age of the specimen, $(t-t_0)$ is time of drying, x is the distance from the drying surface, and D_M is the moisture diffusion coefficient. The second term on the right hand side of the equation refers to the change in relative humidity due to autogenous effects. The second term is frequently omitted when high water to cement ratio mixtures are investigated. While Bazant and Najjar demonstrated that the predictions using Equation 2 can be improved by considering the dependence of the moisture diffusion coefficient on local moisture conditions [8] (i.e., non-linear moisture diffusion), recent experimental studies have suggested [6,7] that for short drying times a linear approach may provide a reasonable estimate for early-age concrete and can greatly simplify the calculations since the equation can be directly solved as shown in equation 3:

$$H(x,\ t) = H_I - (H_I - H_S) \cdot \left[\text{erfc}\left[\frac{x}{2\sqrt{D_M t}} \right] \right] \quad (3)$$

where, H(x,t) is the relative humidity at a depth (x) from the drying surface at a time from the start of drying (t), erfc is the complementary error function, H_S is the relative humidity at the surface of the specimen, H_I is the relative humidity in the interior of a sealed concrete, and D_M is the aging moisture diffusion coefficient of concrete (m^2/sec) [7]. The combined effects of the time dependent diffusion coefficient and drying time have been accounted for in this paper to describe the moisture profile using a single term ($\gamma = 2\sqrt{D_M t}$).

As an example, the aging diffusion coefficient for a mortar with a water-to-cement ratio (w/c) of 0.50 is shown in Figure 1 (Type I cement). This diffusion coefficient was determined using a series of electrical measurements in slab specimens having pairs of embedded electrodes at various distances from the drying surface [9]. A series of calibration specimens were used to correlate the electrical conductivity measured at each electrode pair to the internal humidity of concrete at that point. The obtained humidity profiles (Figure 1a) can be fit to equation 3 to determine the age dependent moisture diffusion coefficient (D_M and γ) for the first 50 days of drying as shown in Figure 1b. It can be seen that γ can be approximated to begin at a value of 0.006 and to increase in a linear fashion.

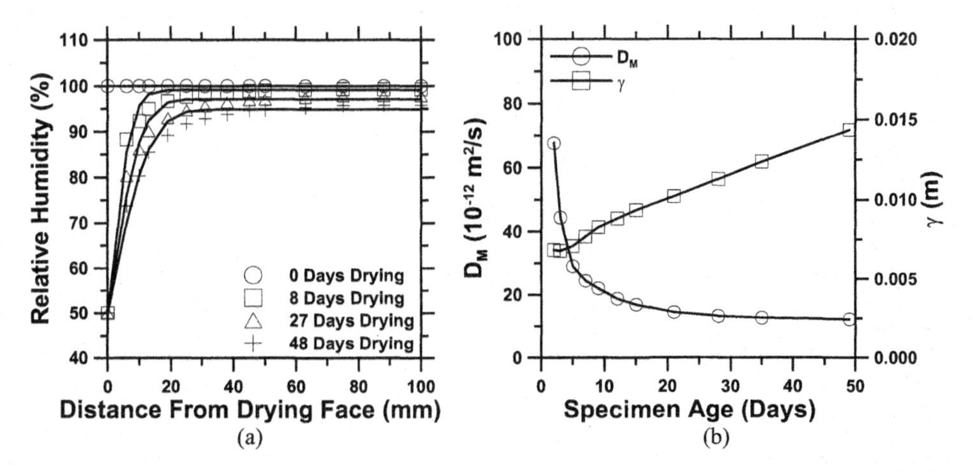

Figure 1: a) Illustration of the Fit between Equation 3 and Experimental Observations and b) an Illustration of the Time Dependent Variations in the Moisture Diffusion Coefficient and Gamma

As the internal pore pressure (i.e., relative humidity) in the specimen (the paste is truly the only shrinking portion but this relationship can be written for paste, mortar or concrete) is decreased, the specimen shrinks. While several approaches have been developed to relate the change in relative humidity to shrinkage [4], the simplest approach suggests that the free shrinkage can be approximately related to the change in relative humidity (i.e., for relatively small changes in relative humidity, H > 50%) with a linear relationship. As a result the drying shrinkage strain

(ε_{SH}(t)) can be related to the change in relative humidity (i.e., (100%-H(x,t)) through a constant that we will call the free shrinkage coefficient ($\varepsilon_{SH-\infty}$) as shown in equation 4:

$$\varepsilon_{SH}(t) = \varepsilon_{SH-\infty}(100\% - H(x,t)) \tag{4}$$

This approach can be used to describe the free shrinkage of a specimen with a moisture gradient (if changes in elastic modulus are neglected) by computing the overall shrinkage of a specimen as the average of the shrinkage at each depth across the cross section of the specimen (Equation 5).

$$\varepsilon_{Bulk}(t) = \varepsilon_{SH-avg}(t) = \int_0^D [100\% - H(x,t)] \frac{\varepsilon_{SH-50-\infty}}{D} dx \tag{5}$$

where $\varepsilon_{SH-50-\infty}$ is twice the long-term (i.e., ultimate) shrinkage of the specimen at 50% relative humidity and D is the specimen thickness.

A modeling approach was previously proposed where the cracking was assessed by comparing the values of residual stresses with age-dependent fracture resistance or the tensile strength of the material at a particular age [10]. This model considered stress development in an uncracked specimen and applied the failure criteria assuming a very small crack (i.e., on the order of 20% of the maximum aggregate size [5]) does not grow until ultimate failure occurs. This model was extended to include moisture gradients to illustrate the influence of specimen size on the age of cracking [11]. In the following section, a portion of this model is further developed to illustrate the role of moisture gradients on the stress field that develops. In addition the model is used to illustrate the potential for stable crack growth in concrete under restraint.

A MODEL FOR STABLE CRACK GROWTH DUE TO MOISTURE GRADIENTS

As previously discussed, moisture gradients can result in residual stresses at the drying surface which can result in the development of surface cracking [5-7, 11, 12]. These stresses can eventually lead to through cracking [11, 12]. To account for time dependent crack growth, a preliminary analytical modeling approach is introduced that uses the crack band concept. The stresses that develop across the depth of the specimen are estimated using the approach described in the previous section. Based on the stresses, the process of cracking in the specimen could be monitored, facilitating a convenient approach to compare the time of cracking of specimens with different mixture proportions, geometries, and drying conditions.

Figure 2 provides a schematic illustration of the model which is based on the crack band model proposed by Bazant and co-workers [14] and Hillerborg and co-workers [13]. The specimen is considered to have two distinct regions: an undamaged or bulk region and the damaged or cracked region. The ultimate shrinkage strain in the bulk and cracked region is the same as the free shrinkage of the specimen*. However the shrinkage is applied to the bulk section uniformly across the cross-section and the residual stress is computed while the shrinkage in the cracked section is divided into a series of elements along its depth and the stress in each of these layers is computed. For each layer, the shrinkage strain is calculated using

---

* It should be noted that even in the bulk section microcracking would be expected to occur (15), however this cracking is not specifically dealt with in this paper and only the crack that leads to through cracking is treated.

Equation 4 at each layer. These stresses are modeled in both the cracked and bulk sections using the step-wise model described in the following section.

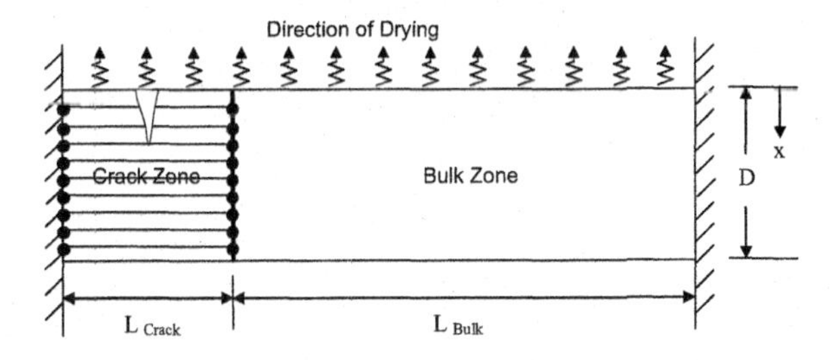

Fig.2. Schematic Illustration of the Model

To satisfy the condition of force equilibrium, the net force in the cracked section must be equal to the force in the bulk section. This necessitates that the sum of the forces in each of the cracked layers (i.e., the stresses in all cracked layers) should equal the force in the bulk zone:

$$\sum_{i=1}^{n} F_{crack-i} = F_{Bulk} \tag{6}$$

In addition, displacement compatability must be applied. The displacement in the overall system is equal to the sum of the displacements in the cracked and bulk region. If the specimen is completley restrained from shrinking (i.e., 100% restraint), the displacement continuity condition can be written by setting, in each layer, the sum of shrinkage displacement and the displacement due to the restraining stresses equal to zero as shown in equation 7

$$(\varepsilon_{Bulk} + \frac{\sigma_{Bulk}}{E_{eff-Bulk}})L_{Bulk} + (\varepsilon_{Crack-i} + \frac{\sigma_{Crack-i}}{E_{eff-Crack}})L_{Crack} = 0 \tag{7}$$

where ε_{Bulk} is the free shrinkage strain in the bulk zone (i.e., the shrinkage averaged over the entire depth of the specimen (Equation 5)), $\varepsilon_{Crack-i}$ is the free shrinkage strain in the ith layer in the cracked zone, calculated from Equation 4, σ_{Bulk} is the residual stress that develops in the bulk region, $\sigma_{Crack-i}$ is the residual stress that develops in the ith layer in the cracked zone, and E_{eff} is the effective elastic modulus which considers both the elastic and creep/relaxation effects.

If it is assumed that the stress distribution in the specimen is linearly proportional to the shrinkage strain, the stresses will also have the form of an error function. As a result the stress in an element at level *i* (corresponding to a depth x_i) can be approximately written as shown in equation 8:

$$\sigma_{Crack-i} = \frac{\varepsilon_{Crack-i} . L_{Crack} + \varepsilon_{Bulk} . L_{Bulk}}{\left(\frac{L_{Bulk}}{E_{eff-Bulk}} \frac{\int_0^D erfc(\frac{x}{\gamma})dx}{erfc(\frac{x_i}{\gamma})} \right) + \left(\frac{L_{Crack}}{E_{eff-Crack}} \right)} \tag{8}$$

This expression provides a simple and direct method for determining the stresses at any layer in the cracked region, as a function of the depth of the layer, the thickness of the specimen, the geometrical parameters of the specimen, and the effective elastic modulus of concrete.

As drying progresses, it can be assumed that the elements will start cracking as the stresses in a particular layer (determined from Equation 8) exceeds the tensile strength of the material ($f'_t(t)$). The stress in the cracked layer can be thought to be redistributed equally among all the uncracked layers. It should be noted however that an equal distribution is not an ideal method of stress redistribution and further work is investigating more realistic methods to account for post-peak stress and stress redistribution. At a certain age, the stresses in all the layers become equal to or greater than the tensile strength of the material; at this time, the specimen cracks completely (t_{cr}).

ILLUSTRATING PREDICTIONS FOR TIME TO CRACKING

Using the analytical model described in the previous section, simulations will be described in the following to illustrate some of the key aspects associated with the model predictions. For the sake of this analysis, a 75 mm thick concrete section with a length of 1 meter will be considered (this is similar to the geometry of many testing frames and the length associated with the ring specimen, i.e., $C = 2\pi R_{IC}$). The relative humidity at the surface of the specimen (H_S) was taken as 50%, whereas in the specimen interior, the humidity was taken to be 100% unless noted otherwise (100% implies no autogenous effects, however in reality autogenous effects will nearly always be present). The values for γ were similar to those shown in Figure 1 and unless otherwise stated $\varepsilon_{SH-50-\infty}$ is assumed to be 2500 $\mu\varepsilon$ which is a typical value for previously tested specimens [5,11]. The tensile strength is assumed to be 5 MPa, and the effective elastic modulus is assumed to be 15 GPa. The cracked section is assumed to be 10 mm wide and was divided into layers throughout this thickness with each layer having a thickness of 1 mm. The humidity profile, shrinkage strains, and residual stresses in each of these layers were calculated using the model.

Figure 3 shows the relative humidity profile and corresponding residual stress profile as a function of the moisture gradient parameter γ. As γ increases (Figure 1 illustrates that for one concrete this may be linear with increased drying time) the depth of the drying front begins to penetrate further into the core of the concrete. It should be noted from Figure 3b that a large portion of the residual stress that develops in the specimen is attributed to the residual stresses that are redistributed during cracking. In this case, it is assumed that in any position where the tensile strength is exceeded, no stress is transferred across the crack. However it is known that cement-based materials demonstrate some ability to transfer stresses across cracks (especially if these cracks are relatively narrow [5]). As such additional work is needed to better understand how these cracks form and how they transfer and redistribute stress.

One technique that has recently been used to assess how cracking develops is acoustic emission. Recent studies have shown that both free and restrained specimens exhibit acoustic activity (i.e., microcracking) due to differential shrinkage [15] and cracking around the aggregates [20]. It is however observed that, in addition to the acoustic activity that is related to surface microcracking, restrained specimens develop additional cracking (an increase in acoustic emission energy) which has been attributed to the development of one (or more) localized cracks [15]. While a promising relationship appears to exist between fracture energy and acoustic energy [18], more work is needed to establish this relationship for restrained specimens and to correlate this energy with the type of localized damage that is occuring in the damaged region.

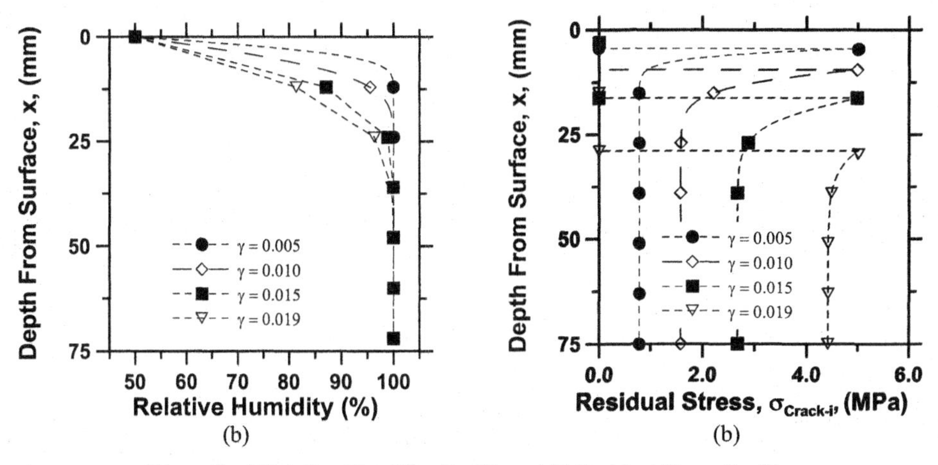

Figure 3: a) Relative Humidity Profile and b) Residual Stress Profile
($\varepsilon_{SH-50-\infty}$ is -2500 $\mu\varepsilon$, f'_t is 5 MPa, and the E' is15 GPa)

THE INFLUENCE OF A CHANGE IN THE ULTIMATE FREE SHRINKAGE

Two methods have been advocated to reduce the potential for cracking: reducing the rate of stress development and reducing the magnitude of shrinkage [5]. Recent research has shown that both of these concepts may be obtained by changing the binder with the addition of shrinkage reducing admixtures (SRA) [17,19]. The moisture profiles measured in concrete with and without SRA were measured to be similar, however the shrinkage coefficients change dramatically with the addition of SRA [11]. The change in the ultimate shrinkage coefficient ($\varepsilon_{SH-\infty}$) with the addition of SRA can be attributed to the reduction in the surface tension of the pore water [5].

To illustrate the influence of the ultimate shrinkage coefficient on the behavior of restrained elements a series of simulations were performed where only the ultimate shrinkage coefficient was varied. It can be seen that a higher shrinkage coefficient results in an earlier age of cracking and a more rapid development of a longer crack (Figure 4a). Figure 4b shows that as the ultimate shrinkage coefficient is reduced, the gamma function (recall that this was approximately linear proportional to drying time) that is required to cause failure increases following the form of a power function (similar to the form of Equation 1). It is interesting to note that when the shrinkage coefficient is reduced to a low enough value (i.e., 665 $\mu\varepsilon$ in this case) cracking will not occur even if the specimen dries out completely. This can be attributed to the fact that the gradient is not severe enough to cause a crack to form.

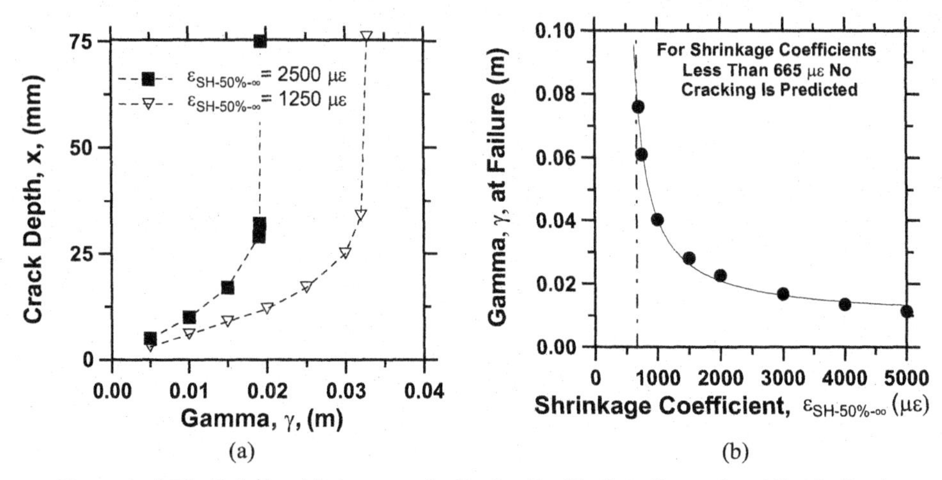

Figure 4: a) The Relationship between the Drying Profile (i.e., Gamma) and Stable Crack Growth and b) The Relationship between the Shrinkage Coefficient and Gamma at Failure

INFLUENCE OF THE RATIO OF DRYING TO AUTOGENOUS SHRINKAGE

As previously mentioned, autogenous shrinkage occurs in concrete mixtures, especially in mixtures with a low water-to-cement ratio (w/c). Autogenous shrinkage generally occurs uniformly across the cross-section, and as such, it can significantly alter the distribution of shrinkage strain and the resulting residual stresses. By altering the distribution of stress, the stable crack would be expected to grow differently. To illustrate this effect, the aforementioned model was used with the following constants ($\varepsilon_{SH-50-\infty}$ is 2500 με, f'_t is 5 MPa, and E' is 15 GPa). The internal relative humidity (H_I) and moisture distribution factor (γ) were altered to obtain the overall section shrinkage and ratio between the autogenous (i.e., shrinkage in a section of constant H_I) and drying shrinkage (total minus autogenous shrinkage) shown in Figure 5.

Figure 5a illustrates how the shrinkage strain profile differs for specimens with the same overall section shrinkage (i.e., 1250 με). It can be seen that for cases with a higher proportion of shrinkage occurring due to autogenous effects, the moisture condition is much more uniform across the cross section. While each of the shrinkage distributions shown in Figure 5a results in the same overall section shrinkage and net residual stress in the bulk region, the stress distribution and the resulting stable crack length are quite different as shown in Figure 5b. It can be seen that for specimens with a higher proportion of shrinkage due to drying, a larger stable crack develops for the same overall section shrinkage. This implies that the specimens with stronger moisture gradients have a deeper crack at the time of failure. This can also be seen in Figure 5c which shows the overall net section stress (i.e., the stress in the bulk region) to specimen strength at the time of failure. It can be seen that this ratio is lower than 1 for all cases except for the case of pure autogenous shrinkage, and this lower ratio is consistent with experimental observations described by Attiogbe et al., [2].

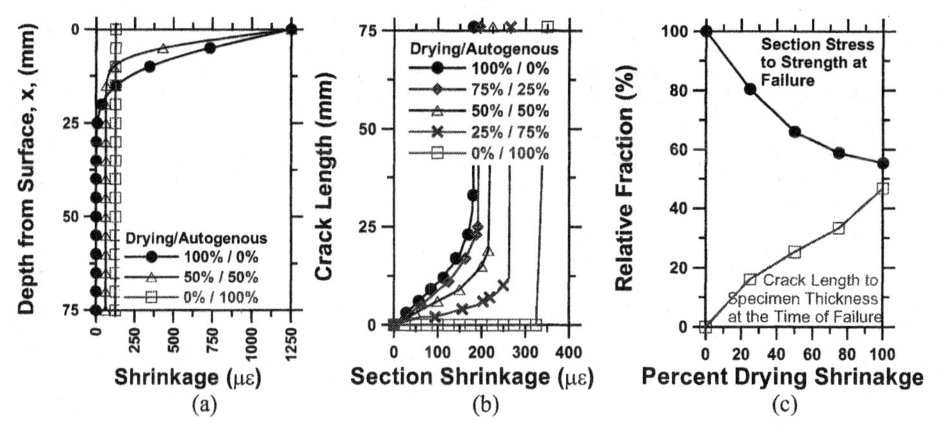

Figure 5: The Influence of the Drying to Autogenous Shrinkage Ratio: a) Shrinkage Distribution Across the Section for an Overall Section Shrinkage of 1250 με, b) Crack Length Versus Section Shrinkage and c) Section Stress at Failure ($\varepsilon_{SH-50-\infty}$ is 2500 με, f'_t is 5 MPa, and the E' is 15 GPa)

SUMMARY

This paper has described the initial steps towards the development of an analytical model that can describe the early-age shrinkage cracking behavior of concrete. Moisture diffusion has been considered using a linear solution to the Fick's second law of diffusion. The combined effects of a time dependent diffusion coefficient and the drying time have been accounted for using a single term ($\gamma = 2\sqrt{D_M t}$) which can be determined from electrical conductivity measurements. The analytical model described in this paper was shown to be able to predict the residual stress distribution in both the bulk and cracked zones. Progressive cracking in specimen was simulated and used to determine the time of through cracking. By reducing the ultimate shrinkage coefficient, the stress gradient between the surface and the core of specimen can be reduced, thereby delaying the age of through cracking. In addition, it was shown that for the same overall section shrinkage, specimens with autogenous shrinkage, demonstrate smaller stable crack growth.

ACKNOWLEDGEMENTS

The authors gratefully acknowledge support received from Degussa Admixtures, Inc. In addition, the authors gratefully acknowledge support received from the National Science Foundation under Grant No. 0134272: a CAREER AWARD. Any opinions, findings and conclusions or recommendations expressed in this material are those of the authors and do not necessarily reflect the views of the National Science Foundation (NSF).

REFERENCES

[1]A.B. Hossain and W.J. Weiss, "Assessing Residual Stress Development and Stress Relaxation in Restrained Concrete Ring Specimens," *J. Cem. Con. Comp.* Vol. 26(5) July 2004, pp. 531-540

[2]E. Attiogbe, W.J. Weiss and H.T. See, "A look at the stress rate versus time of cracking relationship observed in the restrained ring test," Advances in concrete through science and engineering, RILEM, March 2004
[3]R.W. Carlson, "Drying shrinkage of large concrete members," *Journal of the American Concrete Institute*, 1937, pp. 327-336
[4]Z.P. Bazant, "Mathematical Modeling of Creep and Shrinkage Cracking in Concrete," John Wiley & Sons, Inc., June 1989
[5]W.J. Weiss, "Prediction of Early-Age Shrinkage Cracking in Concrete," Ph.D. Dissertation, Northwestern University, Evanston, IL, December 1999
[6]J.H. Moon, F. Rajabipour and W.J. Weiss, "Incorporating Moisture Diffusion In The Analysis Of The Restrained Ring Test," Proc. 4th Int. Conf. Concrete Under Severe Conditions: Environment & Loading Vol. 2 ed. K. Sakai, O.E. Gjorv, and N. Banthia, June 2004, pp. 1973-1980
[7]J.H. Moon and W.J. Weiss, "Estimating Residual Stress in the Restrained Ring Test Under Circumferential Drying," Submitted to Cement and Concrete Composites
[8]Z.P. Bazant and L.J. Najjar, "Drying of concrete as a non-linear diffusion problem," *Cem. Conc. Res.* Vol. 1, 1971, pp. 461-473
[9]A. Schießl, W.J. Weiss, J.D. Shane, N.S. Berke, N.S., T.O. Mason and S.P. Shah, "Assessing the moisture profile of drying concrete using impedance spectroscopy," *Concrete Science and Engineering* Vol. 2, 2000, pp. 106-116
[10]W.J. Weiss, W. Yang and S.P. Shah, "Shrinkage Cracking of Restrained Concrete Slabs," J. of Engrg. Mechanics Div., ASCE, 124(7), 1998, pp. 765-774
[11]W.J. Weiss and S.P. Shah, "Restrained Shrinkage Cracking: The Role of Shrinkage Reducing Admixtures and Specimen Geometry," *Materials and Structures* Vol. 35(2), March 2002 pp. 85-91
[12]Z.C. Grasley and D.A. Lange, "Modeling drying shrinkage stress gradients in concrete," J. of Testing and Evaluation, ASTM, 2002 (httpws:..netfiles.uiuc.edu/dlange/www/)
[13]A. Hillerborg, M. Modéer and P.E. Petterson, "Analysis of crack formation and crack growth in concrete by means of fracture mechanics and finite elements," *Cem. Conc. Res.* Vol. 6. 1976, pp. 773
[14]Z.P. Bazant and B.H. Oh, "Crack Band Theory for Fracture of Concrete," *Materials and Structures Research and Testing* Vol. 16, May-June 1983, pp. 155-178
[15]T. Chariton and W.J. Weiss, "Using Acoustic Emission to Monitor Damage Development in Mortars Restrained from Volumetric Changes," Concrete: Material Science to Application, A Tribute to Surendra P. Shah, eds. P. Balaguru, A. Namaan, W. Weiss, ACI SP-206, 2002, pp. 205-218
[16]A. Hossain and W.J. Weiss, "The role of specimen geometry and boundary conditions on stress development and cracking in the restrained ring test." to appear in Cement and Concrete Research
[17]B.J. Pease, H.R. Shah and W.J. Weiss, "Shrinkage behavior and residual stress development in mortar containing Shrinkage Reducing Admixtures (SRAs)," accepted ACI SP, ACI New York 2004
[18]S. Puri and W.J. Weiss, "Assessment of Localized Damage in Concrete Under Compression Using Acoustic Emission," ASCE Journal of Civil Engineering Materials (Accepted)

[19]W.J. Weiss and N.S. Berke, "Chapter 715: Admixtures for Reduction of Shrinkage and Cracking," Early Age Cracking In Cementitious Systems: Report to RILEM Technical Committee 181-EAS ed. A. Bentur, RILEM Report 25, Bagneux, France, 2003, pp. 323-338
[20]B. Pease, A. Neuwald and W.J. Weiss, "The Influence of Aggregates on Early Age Cracking in Cementitious Systems," Role of Concrete in Sustainable Development: Proc. Int. Sym. Dedicated to Surendra Shah ed. R.K. Dhir, M.D. Newlands, and K.A. Paine, Sept. 2003, pp. 329-338

MESO-SCALE STRAIN MEASUREMENTS USING SYNCHROTRON X-RAYS

J. J. Biernacki and C. J. Parnham
Department of Chemical Engineering
[1]Tennessee Technological University, Box 5013
Cookeville, TN 38505

J. Bai
Materials Processing Center
University of Tennessee
Knoxville, TN 37996

T. R. Watkins and C. R. Hubbard
Metals and Ceramics Division
Oak Ridge National Laboratory
High Temperature Materials Laboratory
PO Box 2008 MS6064
Oak Ridge, TN 37831-6064

ABSTRACT

X-ray methods developed for the determination of residual stress in crystalline materials have been applied to study residual strains and strains due to mechanically generated stresses in portland cement paste. Synchrotron X-rays were used to make measurements of meso-scale strains in the calcium hydroxide (CH) by-product of hydrated neat portland cement paste. Mechanical stresses were applied by uniformly distributed, uniaxial in situ loading of specimens and diffraction measurements were made to establish the strain states of the calcium hydroxide phase. This essentially new application may eventually provide phase resolved strains on at least the meso- and possibly the micro-scale.

INTRODUCTION

Concrete is a ubiquitous building material used in virtually every inhabited environment on earth. Despite concrete's benefits of low initial cost, formability and field fabricability, there are numerous durability related problems that limit its practical lifespan and lifecycle cost effectiveness. The four primary mechanisms of degradation are: (1) mechanical, (2) thermal, (3) chemical and (4) drying in origin. While fundamentally different, these four families of *stressors* manifest in the generation of loads (stresses) that result in deformations (strains) that lead to the formation of cracks and ultimately to failure. The goal of this research is to develop and demonstrate experimental techniques for the study of phase resolved mechanical response due to applied loads in cement, mortar, and concrete on a meso- and eventually a micro-scale.

The x-ray diffraction experiment provides a direct measurement of the lattice spacings of each crystalline phase of a material. These spacings change due to mechanical, thermal, or chemical effects. Thus, diffraction techniques provide a direct measure of strains in individual phases and can also discriminate between tensile and compressive deformation.

Portland cement concrete is comprised of aggregate, cement paste, and water-filled or partially-filled porosity. Both the aggregate and cement paste are themselves heterogeneous in nature, aggregate typically being sand (fine) and locally available rock (coarse), and the paste being comprised of hydrated and unhydrated cement. The unhydrated cement fraction includes tricalcium silicate, dicalcium silicate, tricalcium aluminate, tetracalcium aluminoferrite, and gypsum. The hydrated cement fraction includes calcium silicate hydrate, calcium hydroxide, monosulfoaluminate and, ettringite, the respective hydration products of the anhydrous phases. While a large fraction of the hydration products are amorphous, and hence do not diffract x-rays in a useful pattern due to their lack of long-range order, by-products such as CH and monosulfate as well as unreacted clinker phases are suitably crystalline to permit x-ray interpretation. In addition, aggregate phases are typically crystalline in nature because they usually contain crystalline minerals such as calcium carbonate and silica.

Research on fracture processes in concrete and other quasi-brittle materials shows that cracks initiate as micro-scale features [1]. As damage progresses, the scale and number of defects increases. Eventually the number density and size of defects grows such that the micro- and meso-scale cracks merge to form macro flaws leading to failure. Prior research has focused on the measurement of macroscopic stress and strain measurements to interpret the cumulative effect of micro-scale degradation mechanisms. For example, macroscopic shrinkage measurements due to drying of concrete have been combined with macro-scale constitutive relationships to predict drying shrinkage cracking in constrained concrete. Until now, virtually no direct measurements of the micro-scale strain states and strain distributions have been made for cement, mortar, or concrete.

Schulson et al. [2] recently demonstrated, using neutron diffraction, that stress levels within hardened cement paste subject to thermal loads could be quantified by measuring strains in the calcium hydroxide (CH) phase. Furthermore, their work demonstrated that stress distributions on a micro-scale are anisotropic, although, they were unable to measure strain states and stresses developed upon freezing due to lack of instrumental resolution [3]. Shchukin, et al., presented an x-ray diffraction-based technique for the determination of residual stresses in cement. However, description of their method is limited and their use of a laboratory x-ray source, lack of uncertainty analysis and choice of diffraction angles place doubt on the reliability of the data they present [4].

EXPERIMENTAL

X-ray Technique

A procedure called ψ tilting was used in the present study to obtain the strains between diffracting crystallographic planes. Use of this technique is necessary since unlike powder diffraction, where the state of crystallites along different planes in the material is unimportant due to the assumptions of uniformity and random orientation distribution, this procedure and its interpretation rely critically on the orientation of the crystallites and their diffracting planes within the sample. This dependence of the ψ tilting method on the orientation of crystallographic planes is important in that it allows for the determination of strains in specific directions.

In order to fully describe the ψ tilting procedure it is necessary to define some geometric terms. Figure 1(a) illustrates the two orthogonal rectilinear coordinate systems typically used in x-ray diffraction stress measurements. The $\mathbf{S_1}$, $\mathbf{S_2}$ and $\mathbf{S_3}$ axes are associated with the sample orientation and thus constitute the sample coordinate system (sample reference frame), and the $\mathbf{L_1}$, $\mathbf{L_2}$, and $\mathbf{L_3}$ axes are associated with the orientation of the laboratory measurement and constitute the laboratory coordinate system (laboratory reference frame). The sample coordinate system is defined by the geometry of the sample with the $\mathbf{S_3}$ axis being normal to the sample surface and the $\mathbf{S_1}$ and $\mathbf{S_2}$ axes lying in the plane of the sample surface. The laboratory coordinate system describes the orientation of the planes on which diffraction measurements are made and consists of the $\mathbf{L_3}$ axis that is normal to the diffracting planes, $\mathbf{L_2}$ which is in the plane defined by $\mathbf{S_1}$ and $\mathbf{S_2}$ and makes an angle ϕ with $\mathbf{S_2}$, and $\mathbf{L_1}$ which is orthogonal to $\mathbf{L_2}$ and $\mathbf{L_3}$. Under ordinary x-ray diffraction conditions, the angles ψ (tilt) and ϕ (rotation) are both zero and the laboratory and sample reference frames are the same. In all diffraction experiments, diffracted energy is gathered *only* from crystallites oriented in the laboratory reference frame (the $\mathbf{L_i}$ coordinate system), such that a set of diffracting planes in the crystallite have a normal direction vector that bisects and is coplanar with the incident and diffracted beams (this normal vector is $\mathbf{S_3}$ when $\psi = \phi = 0^o$ and $\mathbf{L_3}$ when ψ and ϕ are not zero). This makes it necessary for the diffracting planes to be perpendicular to the θ-2θ axis in order to make measurement possible. Since diffraction measures the spacing only between these specific planes, it is desirable to load the specimen in the direction of the planar normal (in the $\mathbf{L_3}$ direction). However, because an exposed surface is required for θ-2θ scanning, the load cannot be applied in this direction since the load frame would obscure the x-rays. Instead, the load must be applied in either of the two other directions orthogonal to the exposed surface normal (along the $\mathbf{S_1}$ or $\mathbf{S_2}$ direction). If the load is applied in either of these two directions, the force vector in the $\mathbf{S_3}$ direction is zero. Any diffracting crystallographic plane aligned parallel with the $\mathbf{S_1}$-$\mathbf{S_2}$ plane, orthogonal to $\mathbf{S_3}$, will not experience an applied load, but will respond only according to the Poisson effect. A non-zero force vector is experienced in the direction of the diffraction plane normal (the measurement direction) only when the diffraction plane normal is not orthogonal to the direction of the applied force vector. To measure strains other than the Poisson's strain, the sample must be tilted so that the laboratory reference frame is no longer aligned with the sample reference frame (Figure 1(a)), thereby rotating crystallites that carry a fraction of the applied force into the diffracting orientation (Figure 1(b)). For an applied force in the $\mathbf{S_1}$ direction, the sample must be tiled in the θ direction, into the $\mathbf{S_1}$-$\mathbf{S_3}$ plane (the θ-2θ plane, about the $\mathbf{S_2}$ axis). In theory this can be done, but the geometry of the equipment used usually creates some interference with the beam path over a portion of the scanning range. Instead, the force in this case is applied in the $\mathbf{S_2}$ direction and the sample is tilted in the ψ direction (into the $\mathbf{S_2}$-$\mathbf{S_3}$ plane, about the $\mathbf{S_1}$ axis when the rotational angle ϕ is set to zero). Thus, by performing θ-2θ scans at several different ψ angles, the interplanar spacing (d-spacing) in several specific directions can be measured, allowing for the determination of elements of the strain tensor.

If polycrystallinity is assumed, the resulting strain measured in the laboratory reference frame ε'_{33} can be related to the strain in the sample reference frame ε_{33} according to the following tensor transformation [5, 6]:

$$(\varepsilon'_{33}) = \frac{d_{\varphi\Psi} - d_0}{d_0} = \varepsilon_{11}\cos^2\varphi\sin^2\psi + \varepsilon_{12}\sin 2\varphi\sin^2\psi + \varepsilon_{22}\sin^2\varphi\sin^2\psi + \varepsilon_{33}\cos^2\psi + \varepsilon_{13}\cos\varphi\sin 2\psi + \varepsilon_{23}\sin\varphi\sin 2\psi \tag{1}$$

where $d_{\phi\psi}$ = the measured strained d spacing (spacing between parallel diffracting planes under load), d_0 = the measured *un-strained* d spacing (spacing between parallel diffracting planes of stress free material), ε_{ij}=the sample strains in directions $\mathbf{S_1}$, $\mathbf{S_2}$ and $\mathbf{S_3}$, ϕ = the angle of rotation of the sample about the $\mathbf{S_3}$ axis (defined as the angular difference between the $\mathbf{L_2}$ and $\mathbf{S_2}$ axes), and ψ = the laboratory tilt angle (the difference in angle between the surface normal, $\mathbf{S_3}$, and the normal to the diffracting planes, $\mathbf{L_3}$). For these experiments the angle ϕ was held constant at zero. Even with this simplification, the mathematics of the transformation allows the observed strain ε'_{33} to be a complex function of $\sin^2\psi$ [5]. Three functional behaviors can arise: (1) linear, (2) non-linear split and (3) non-linear oscillating. Cases (1) and (2) are embodied within and can be explained by Equation (1). The linear case is probably the most documented in literature and arises from a condition of plane stress at the sample surface. Case (2) is more complicated in that phases that exhibit this type of behavior do not conform strictly with the ideal plane stress scenario, and as a result the curve is non-linear and the phenomenon of "ψ-splitting" can occur where the strain vs. $\sin^2\psi$ curves for positive and negative tilt angles are not the same. Case (3) is the most complicated of these behaviors [5, 6] and is a typical result when analyzing materials that have inhomogeneous stress/strain distributions or near surface stress gradients [5].

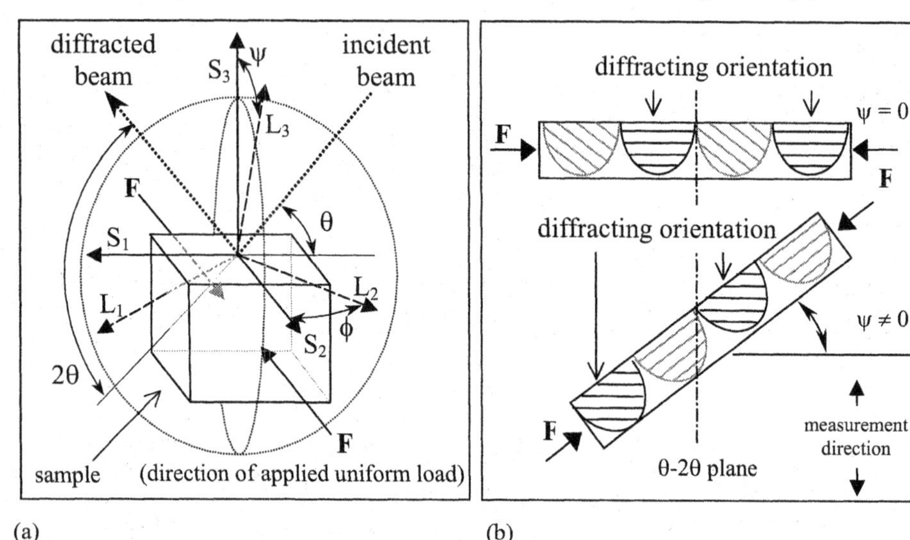

(a) (b)
Figure 1. (a) X-ray coordinate systems and (b) Cross-section illustrating the orientation of diffracting crystallites at two ψ angles.

Analysis of these functional behaviors can also be complicated by the nature of the material. Phases that are textured (texture occurs when crystal orientation is not random, but preferred on the scale of the measurement volume) or coarse-grained (materials in which the grain size is substantially large that few crystals are sampled by the x-ray beam) present a challenge in that the intensity of a given 2θ peak can change with the tilt angle ψ. So when conducting the diffraction experiment it is important to tilt to ψ angles where sufficient peak intensity is available to define the peaks in the θ-2θ scan (in order to accurately determine the 2θ location of the peak it must have an adequate height relative to the background intensity). Because texture (to varying extents) is a common condition, several computational treatments and interpretations have been formulated [7]. Interpretation for coarse-grained materials can be somewhat more illusive since the preceding discussion of strain vs. $\sin^2\psi$ behavior assumes that it is possible to measure d-spacings at all ψ tilts [5]. Coarse-grained materials are challenging since the peak intensity at a given 2θ angle can change dramatically and abruptly with ψ. Large changes in the peak intensity with ψ can occur in textured materials as well, but these changes are of a much more smooth and continuous nature. In either case it can be difficult to obtain sufficient peak intensity at arbitrary ψ tilts, so when conducting ψ tilting experiments it is necessary to have knowledge of what angles will allow for accurate measurements to be made.

Sample Preparation

Type I ordinary portland cement meeting ASTM C150 specifications was blended in the appropriate proportions with tap water to form samples of cement paste [8]. Cement paste samples were prepared using a water/cement (w/c) ratio of 0.34. The portland cement was mixed by hand with the water until a uniform paste was formed. The paste was then cast into a 0.75 inch (19.1 mm) diameter glass jar to a level of about two inches (about 50 mm). The jar was sealed and placed into a constant temperature water bath at 35 °C (95 °F). All specimens were cured in the sealed environment for at least 28 days after which time they were removed from their jars and cut into 0.4000±0.0001 inch (10.16±0.0025 mm) cubes using a diamond saw.

X-ray Experiments

X-ray strain experiments were performed using the Oak Ridge National Laboratory (ORNL) beamline X14A at the Brookhaven National Laboratory (BNL) National Synchrotron Light Source (NSLS). The ORNL facilities at NSLS are well suited for this research. The optics and highly parallel x-ray beam at X14A make it far superior to conventional laboratory x-rays and result in peak locations that are insensitive to the sample surface geometry. Beamline X14A also provides extremely high flux and high signal-to-noise ratio needed for measuring very small strains (on the order of 10^{-5}). A wavelength of 0.15472 nm (energy of 8.0136 keV, nominally equivalent to CuK_{α}) with a Ge analyzer crystal was used. The x-ray spot size was nominally 1 mm in diameter and the penetration depth expected was on the order of 100 μm, assuming a mass attenuation for the cement equal to that of concrete at 8 keV, about 40 cm^2/g [9] and a cement density of nominally 2.5 g/cm^3. Thus, while the measurement is nominally micro-scale in the depth dimension it is nominally meso-scale in the other two dimensions (surface area). The X-ray goniometer was fitted with a small load frame capable of applying up to 1000 lb_f (4448 N). θ-2θ scans were made at applied loads of 10, 200, 400 and 600 lb_f (62.5, 1250, 2500 and 3750 psi respectively) for 2θ between 129.5 and 133.5°. This range was used because it contains two well-defined and distinct peaks: one calcium hydroxide (CH) peak and one quartz (Q) peak at angles 2θ = 131.162° and 132.309°, respectively (d-spacings of 0.8496 and 0.8458 Å). These diffraction peaks are associated with (hkl) planes (2 1 4) for CH and (1 1 6) for Q*. This rather high 2θ range was selected since peak shift increases with increasing 2θ for a constant applied load (thus making the measurement more precise relative to experimental uncertainties) in accordance with Bragg's Law, Equation (2).

$$n\lambda = 2d \sin\theta \quad (2)$$

where λ is the x-ray wavelength and n is an integer. Furthermore, selection of a suitable 2θ range also depends upon identification of peaks that are nominally independent, e.g. do not overlap with peaks from other phases. X-ray scans were performed in a sequence of no-load, load, no-load, load, etc., so as to observe changes in strain states due to sequential stress cycling.

θ-2θ scans were performed at ψ angles of 0.0°, 6.8 °, 18.4 °, 27.3 °, 34.5 °, 39 °, 45.7 °, 51.6 °, 55.5 ° and 60°. These angles were chosen such that preferred orientation and coarse-grained effects of the CH phase at 130.430° 2θ were minimized. While not uniformly distributed, this series of ten angles produces a nominally spaced sequence of $\sin^2\psi$ values.

* All 2θ values given for λ = 0.15472 nm (8.0136 keV)

Diffraction peak locations (2θ values) were determined using profile fitting from which d-spacings ($d_{\phi\psi}$) were computed by applying Equation (2). Each 2θ range was fit assuming a pseudo-Voigt peak form with variable peak width, with centers of the pseudo-Voigt fits taken as the peak locations. The apparent strain was computed using the left hand side of Equation (1). All strains were referenced to powder pattern d_o-spacings generated by performing a diffraction experiment on a ground piece of representative source material. The strains, so determined, were then plotted as a function of $\sin^2\psi$ for conditions of load and no-load (Figures 2(a) through 2(e)). Confidence intervals were computed by propagating the uncertainty in 2θ determinations; plotted as error bars on Figures 2.

RESULTS

Figures 2(a) through 2(e) summarize changes in strain states as a function of tilt angle ($\sin^2\psi$) and loading conditions. This series of five plots is presented sequentially illustrating successive changes in stress states, Figure 2(a) compares the initial no-load condition compared to the first stress state of 200 lbf (1250 psi), Figure 2(b) compares the first stress state of 1250 psi to the second no-load state, Figure 2(c) compares the second no-load state to the second stress state of 400 lbf (2500 psi), etc.. In this way, changes in strain states can be compared to each preceding state individually. Note that the "no-load" stress states actually represent a stress of 62.5 psi. This negligible stress was applied to keep the sample stationary during the no-load segments (because the sample will fall out of the load frame if not confined).

The initial no-load line on Figure 2(a) represents the initial residual stress state of the CH in this sample. Assuming that the reference powder was ground finely enough that all residual stress was relieved, one could infer that the residual stress state of the CH is on the average compressive, ranging from -0.270 to 0.064 ±0.025 microstrain. The strain also appears to be a non-linear function of tilt angle.

Prior to loading, a series of replicate scans were made. After tilting the goniometer through the first ten ψ angles to complete the initial no-load experiment, four replicate scans were made by again moving the goniometer and repeating the respective θ-2θ scans. The data for the replicates, at ψ angles of 6.8°, 34.5°, 45.7° and 55.5° 2θ, are included on Figure 2(a) as shaded boxes. The replication illustrated excellent repeatability, with all four having overlapping confidence intervals and three of the four replicates being virtually indistinguishable. This repeatability experiment thereby demonstrates that changes in strain states are not an artifact of moving (tilting) the goniometer.

Upon loading to 1250 psi, the strain states appear to change, some becoming more compressive, while others become less compressive. Interpretation of this single response, however, cannot teach us anything conclusive and so the result must be interpreted in the context of subsequent stress state changes. After unloading to about 10 lb_f (62.5 psi), the strains again returned to states similar to the initial condition, compare Figures 2(a) and 2(b), thereby strengthening a conclusion that strain states, as measured by this method, are changing as a function of stress state. It bears mentioning that some of the data for this second no-load cycle was lost and could not be recreated and so is represented by the absence of data points at the four highest $\sin^2\psi$.

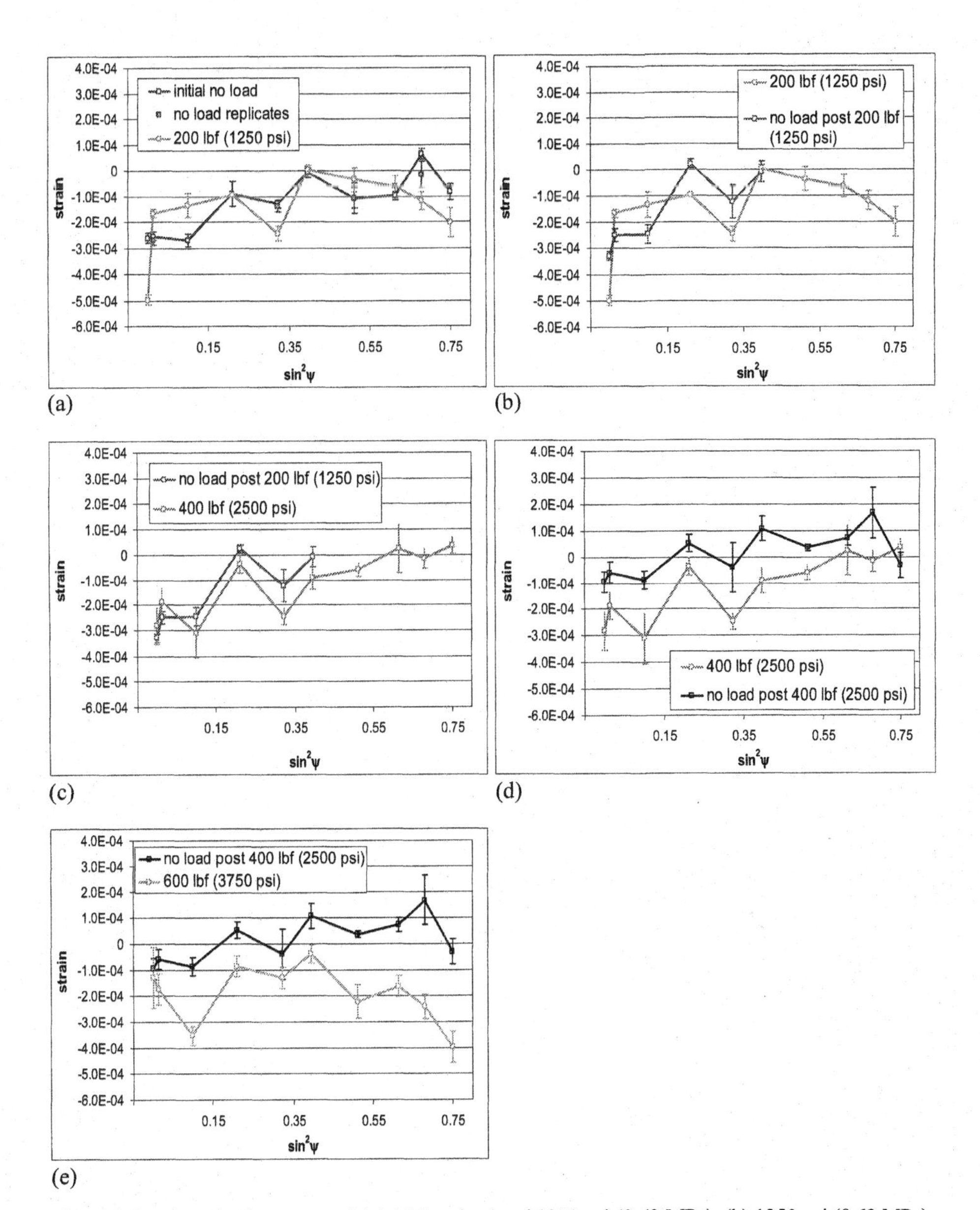

Figure 2. Load, no-load sequence: (a) initial no load and 1250 psi (8.62 MPa), (b) 1250 psi (8.62 MPa) and second no-load state, (c) second no-load state and 2500 psi (17.2 MPa), (d) 2500 psi (17.2 MPa) and third no-load state and (e) third no-load state and 3750 psi (25.6 MPa)

Figure 2(c) illustrates the next step in the sequence wherein the sample was again loaded, this time to 400 lbf (2500 psi). The net response was a generally more compressive state than either the unloaded condition or the prior 1250 psi state. Upon unloading, the strain state again shifted, on the average, to a more tensile state (Figure 2(d)). It is notable here that the net no-load stain state at this point was less compressive than the initial no-load state, indicating that residual stresses may be being relieved by load cycling, even at these relatively low loads. The final cycle was to increase the load once again, this time to 600 lb_f (3750 psi). Figure 2(e) compares the previous no-load state to this final loaded state. As one would expect, the net strain state shifts compressive and to an extent greater than either the 1250 psi or 2500 psi states.

DISCUSSION

While the above results suggest a promising application of this method, there are a number of challenges that must be addressed before the method can become generally applicable to portland cement-based composites. As discussed earlier, sensitivity to strain increases for higher 2θ values, yet peak intensity decreases and the number of suitable, non-overlapping peaks to select from is greatly reduced. This makes study of a broad range of peaks limited, even for a single crystalline phase. Furthermore, not only are portland cement-based materials textured, the constituent phases are textured to different extents, with some phases exhibiting a strong coarse-grained behavior. The hydration by-product CH, however, appears to be random enough that a judicious choice of tilt angles (ψ) can avoid gross changes in peak intensity due to either of these effects. This will not be possible in field mortar or concrete specimens that will contain aggregate crystallites that can be larger than the x-ray sample volume. It must also be pointed out that oscillations in the strain vs. $\sin^2\psi$ plot are likely real, suggesting a non-linear oscillating behavior. While outside the scope of the present study, this observation must be further verified experimentally and theoretical explanations provided, as has been done for other materials [10] Likely, the textured nature of the CH phase will be found to play a role here as well. Finally, x-ray diffraction is a surface phenomenon as far as this application is concerned. This may make interpretation and extrapolation of the x-ray-based strain information somewhat difficult, a problem that other surface techniques also face [10,11]. One alternative is to conduct similar studies using neutrons which will have a greater depth of penetration. While this may to some extent mitigate the surface constraint that x-rays have, at this time most neutron sources, unfortunately, have a much larger sampling volume (on the order of mm^3). The larger sampling volume, however, should reduce the effect of texture by averaging over a larger volume (e.g. the sampling volume increases the number of crystallites measured, which in turn increases the chances that a more random distribution of orientations will be observed).

SUMMARY AND CONCLUSIONS

X-ray methods developed for the analysis of residual stress in materials have been applied to study both residual strain states and strain states due to applied loading in portland cement paste. At this time the method has been applied to the CH hydration by-product phase in hydrated neat portland cement paste. The results, while preliminary, support the proposed hypothesis that phase resolved strain states in portland cement-based materials can be measured using synchrotron x-ray diffraction methods. And, while yet unexplained, it appears that the strain vs. $\sin^2\psi$ behavior is of the non-linear oscillating form. At this time the results are not only preliminary, but very limited. For this technique to be generally applicable to portland cement-based systems methods must be applied to overcome the complexity of working with textured and coarse-grained materials. Once developed and validated, this method may provide new opportunities for studying phase distributed residual and applied stresses in portland cement composites.

ACKNOWLEDGEMENTS

This material is based upon work supported by the National Science Foundation under Grant No. 0324616. The authors would also like to acknowledge support from the Center for Manufacturing Research at Tennessee Tech University and the High Temperature Materials Laboratory (HTML) at Oak Ridge National Laboratory (ORNL).

Any opinions, findings, and conclusions or recommendations expressed in this material are those of the authors(s) and do not necessarily reflect the views of the National Science Foundation.

REFERENCES

[1]S. P. Shah, S. E. Swartz and C. Ouyang, "Fracture Mechanics of Concrete : Applications of Fracture Mechanics to Concrete, Rock and Other Quasi-Brittle Materials," Wiley, New York (1995).

[2]E. M. Schulson, I. P. Swainson and T. M. Holden, "Internal Stress Within Hardened Cement Paste Induced Through Thermal Mismatch Calcium Hydroxide versus Calcium Silicate Hydrate," Cem. Concr. Res., 31, 1785-1791 (2001).

[3]E. M. Schulson, I. P. Swainson, T. M. Holden and C. J. Korhonen, "Hexagonal Ice in Hardened Cement, Cem. Concr. Res. 30, 191-196 (2000).

[4]E.D. Shchukin et al., "X-ray Diffraction Method for the Determination of Residual Stresses in Cement," Kolloidnyi Zhurnal, 59, 96-101 (1997).

[5]I. C. Noyan and J. B. Cohen, "Residual Stress Measurement by Diffraction and Interpretation," Springer-Verlag, New York, pp 276 (1987).

[6] V. Hauk, "Structural and Residual Stress Analysis by Nondestructive Methods," Elsevier, Amsterdam, The Netherlands, pp 640 (1997).

[7]B.D. Cullity and S.R. Stock, *Elements of X-ray Diffraction*, 3rd Ed., p.403, Prentice-Hall, Upper Saddle River, NJ (2001)

[8]*ASTM Annual Book of Standards*, Vol. 04.01 Cement; Lime; Gypsum, American Society for Testing and Materials, West Conshohocken, PA (1999).

[9]Hubble and Seltzer, "Tables of X-ray Mass Attenuation Coefficients and Mass Energy-Absorption Coefficients," NISTIR 5632, National Institute of Standards and Technology, Gaithersburg, MD (1995).

[10]Y. Xi, T.B. Bergstrom and H.M. Jennings, "Image Intensity Matching Technique: Application to the Environmental Scanning Electron Microscope," Comp. Mat. Sci. 2, 249-260 (1994)

[11].M. Neubauer, E.J. Garboczi and H.M. Jennings, "The Use of Digital Images to Determine Deformation Throughout a Microstructure: Part I Deformation Mapping Technique," J. Mater. Sci. 35, 5741-5749 (2000)

INVESTIGATION OF EARLY AGE MATERIAL PROPERTY DEVELOPMENT IN CEMENTITIOUS MATERIALS USING ONE-SIDED ULTRASONIC TECHNIQUE

K.V Subramaniam and J. Lee
City College of the City University of New York
Convent Avenue at 140th Street
New York, NY 10031

ABSTRACT

Cementitious materials exhibit a continuous change in their mechanical properties with time; there a continuous increase in the stiffness of the material which results in loss of workability before setting and a steady increase in the elastic properties of the hardening solid cementitious material after setting. An ultrasonic test setup and data analysis procedures, which provide for continuous monitoring of the hydrating cementitious materials from a very early age, are presented in this paper. The test procedures for obtaining the ultrasonic test data and the inversion subroutines for assessing the material properties of the cementitious material at different stages of hydration are discussed. The experimental test setup is used to investigate the development of shear modulus in paste, mortar and concrete samples with identical water-to-cement ratio. The observed experimental trends suggest significant differences in the kinetics of hydration and development of shear modulus in the three materials.

INTRODUCTION

The process of setting, stiffening and subsequent strength gain in concrete is produced by the hydration of cement. There is a change in the state of the material during the setting process of concrete. Immediately after mixing, cementitious materials (concrete, mortar or cement paste) resemble a viscous suspension. Conventionally, properties of concrete, such as the rheological properties, gain in stiffness through setting etc., are determined through indirect measurements performed on cement paste or mortar samples. The relative changes in the elastic and the viscous stiffness components of the cement-phase have been shown to produce a steady loss of workability in concrete [1]. The rheological measurements are however not possible after set. The increase in penetration resistance of mortar with time is currently used to quantify changes in stiffness produced by hydration, through setting. The setting of concrete is identified with fixed values of penetration resistance of mortar which is obtained by sieving concrete using a procedure which has been standardized in ASTM C 403 [2]. The penetration resistance method however does not provide for direct evaluation of setting in concrete and for assessing useful mechanical properties. After final setting, the elastic material properties of concrete are usually obtained using mechanical tests [3,4] or vibration-based measurements performed on standard specimens [5,6]. Techniques for continuously monitoring the development of material properties of concrete after casting, through setting and early stiffening are currently not available.

Direct application of properties determined from cement paste or mortar to concrete are complicated by factors which require ensuring equivalency in factors such as water-to-cement ratio, the cement content etc. Often ensuring equivalency in both the water-to-cement ratio and the cement content between cement paste, mortar and concrete is not possible. Techniques which allow for (a) direct evaluation of material properties of concrete after casting, and (b) evaluation of concrete, mortar and cement paste to determine the influence of aggregate on the development of properties in concrete from an early age, are currently under development or are not available [7-12].

A one-sided ultrasonic test procedure, which consists of continuously monitoring the waves reflected at the interface of polymethyl methacrylate (PMMA) and a cementitious material, is presented in this paper. Reflection of a stress wave at the interface of PMMA and the cementitious material produces a change in the amplitude and the phase when compared with the incident wave. The exact change in the amplitude and phase introduced by the reflection depends upon the frequency dependent material properties of the two media in contact. Since the material properties of the cementitious material change with time, monitoring the reflected waves off the PMMA-cementitious material interface provides a convenient way to assess these changes. The technique presented in this paper offers the advantage of providing continuous information through setting, stiffening and early strength gain of the cementitious material.

An experimental program which allows for systematic evaluation and comparison of the property development in cement paste, mortar and concrete with identical water-to-cement ratio, is presented in this paper. The reflected waves at the PMMA-cementitious material interface are used to monitor changes in the state of the cementitious material as reflected in its acoustic properties. The increase in the shear stiffness for the three cementitious materials, assessed using the ultrasonic technique is presented.

EXPERIMENTAL SETUP AND TEST PROGRAM

The experimental test setup for high frequency measurement of early-age response of cementitious material is shown in Figure 1 (a). An ultrasonic shear wave transducer was attached to the bottom of the mold, which holds the cementitious material, using a rapid setting epoxy. An ultrasonic pulse was introduced into PMMA, which then underwent multiple reflections between the sample-PMMA and transducer-PMMA interface. The ultrasonic signals were digitized, and transferred to the computer for further analysis. A computer program was written using Labview™ to perform the data acquisition, the data transfer to the computer, the analysis of data and the storage of results. The reflected wave signals were collected at the PMMA-water interface prior to placing the cementitious material. The reflected signals were collected and analyzed every ten minutes after placing the cementitious material in the mold.

The excitation to the shear wave ultrasonic transducer (central frequency: 1 MHz) was provided by a commercial pulser-receiver. The electrical signal from the pulser-receiver was used to excite the transducer at its resonant frequency. A sufficiently small pulse repetition rate was used to avoid interference effects between different echoes from the PMMA-cementitious material interface. An averaging procedure, which allows for eliminating random noise in the signal was adopted for the data acquisition. The test procedure comprised of averaging 100 signals in the time-domain. The electrical signals from the transducer were digitized by an oscilloscope at a sampling rate equal to 0.1 μs. The data collection by the oscilloscope was triggered by the pulser-receiver. The signal was then transferred to the computer over a GPIB interface. The time domain signal was windowed and transformed into frequency domain using the fast Fourier transform (FFT) algorithm. The amplitude and the phase angle obtained from the FFT at different frequencies were recorded and stored. The phase angle at a given frequency component was obtained by unwrapping the phase obtained from the FFT.

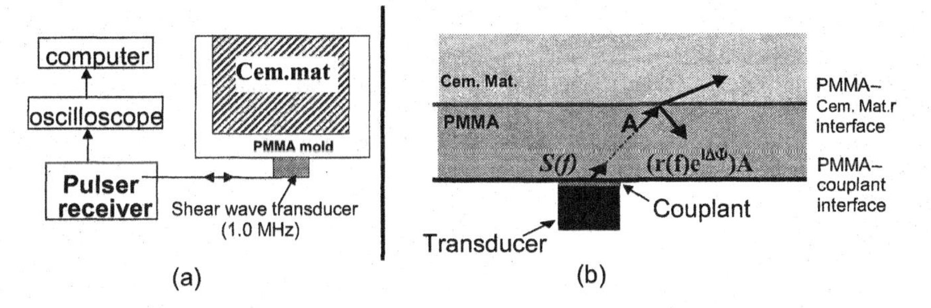

Figure 1: (a) Schematic figure of the experimental test setup for wave reflection measurements, (b) Wave reflection at the PMMA-cementitious material interface showing the change in the amplitude and phase introduced by the reflection.

Ultrasonic measurements were performed on cement paste, mortar and concrete samples with identical water-to-cement ratio. The water-to-cement ratio was kept equal to 0.4. The proportions of the constituents by weight for the mortar were: cement: water: sand = 1.0:0.4:2.0. The proportions of the constituents by weight for the concrete were: cement: water: sand: coarse aggregate = 1.0:0.4:2.0:2.0. Temperature changes in the cementitious material were also recorded using embedded thermocouples. All measurements were performed in an environmental chamber, which was maintained at 20°C.

THEORETICAL BASIS OF REFLECTION MEASUREMENTS

The choice of PMMA as the material for the mold was based on the having the ability to sensitively determine early age changes in the cementitious materials. The measurements in the test procedure introduced in this paper are based on determining the changes in the amplitude and the phase of an incident wave after one reflection at the PMMA-cementitious material interface (Figure 1(b)). Considering PMMA and the cementitious material to be visco-elastic materials, and an incident wave A(f), the reflected wave is given as

$$\text{reflected wave} = \left[r(f)e^{i\Delta\Phi}\right]A(f) = \frac{Z^*_{cem.mat.} - Z^*_{PMMA}}{Z^*_{cem.mat.} + Z^*_{PMMA}}A(f) \tag{1}$$

where r(f) is the amplitude reflection factor for an incident wave of frequency f, $Z_{cem.mat.}^*$ and Z_{PMMA}^* are the acoustic impedances of the cementitious material and PMMA, respectively, and $\Delta\Phi$ is the phase change introduced after one reflection at the PMMA-cementitious material interface. The acoustic impedance of a material is given as

$$Z = \sqrt{\rho\left(G' + iG''\right)} = \sqrt{\rho|G|} \tag{2}$$

where ρ is the density, G', G'' and $|G|$ are the storage modulus, the loss modulus and the magnitude of shear modulus of the material, respectively. In materials which exhibit frequency dependent elastic material properties, the r(f) varies with frequency. After mixing, the cementitious materials have a very low stiffness. Further, early changes in stiffness are also of a small magnitude. Therefore, sensitive detection of small changes in the acoustic impedance of the cementitious material, particularly in the early ages requires that the ratio of the acoustic impedances of the two materials be close to one. PMMA was found to be suitable for this purpose.

In the frequency domain, the stress wave signals received after a single reflection at the PMMA-water and PMMA-cementitious material interface can be represented as a simple product of terms [13]

$$F^{PMMA-water} = s(f)d_1(f)d_2(f) \text{ and } F^{PMMA-cem.mat.} = s(f)d_1(f)d_2(f)r(f) \quad (3)$$

where F is the magnitude of FFT of the captured time domain signal, $d_1(f)$ represents the geometric and material signal losses in the couplant, $d_2(f)$ is the material signal loss along the signal path in the PMMA, $S(f)$ the input source function, and $r(f)$ is the amplitude reflection factor at the PMMA-cementitious material interface. Therefore the amplitude reflection factor, r(f), can be determined by normalizing the reflected magnitude off the PMMA-cementitious material interface with the reflected amplitude off the PMMA-water interface as shown below:

$$r(f) = F^{PMMA-cem.mat.} / F^{PMMA-water} \quad (4)$$

The amplitude reflection factor, r(f), determined using Equation 4 defines the ratio of the amplitudes of incident and reflected waves at frequency f that is reflected from an interface between two materials.

The phase change at any given frequency introduced by the reflection at the PMMA-cementitious material interface can be extracted by subtracting the phase angle of the stress wave recorded after one reflection (obtained from the FFT of the time domain signal) at the PMMA-water interface from that at the PMMA-cementitious material interface. It has previously been shown that PMMA exhibits sensitivity to changes in temperature [14]. Particularly, the measured phase change is influenced by the changes in the PMMA produced by temperature. The phase change produced by varying temperature in PMMA are compensated by using the phase angle recorded after reflection at the PMMA-water interface at the same temperature as that of the mortar sample. Change of phase, ΔΦ, introduced by the reflection of stress wave at the PMMA-mortar interface can be calculated as shown below:

$$\Delta\Phi = \phi^{PMMA-cem.mat.} - \phi^{PMMA-water} \quad (5)$$

where $\phi^{PMMA-cem.mat.}$ and $\phi^{PMMA-water}$ are the phase angles obtained from the FFT of the time domain signals collected after one reflection at the PMM-cementitious material interface and the PMMA-water interface at the corresponding temperature, respectively.

A cylindrical mold with an internal diameter equal to 50mm and height equal to 50mm was used in this study. The size of the sample was chosen so as to provide a representative volume for evaluating the material and yet minimize the temperature changes produced by hydration.

EXPERIMENTAL RESULTS

Typical ultrasonic waveforms recorded in the time-domain after one reflection at the PMMA-cement paste interface are shown in figure 2(a). It can be seen that there is a progressive decrease in the amplitude of the reflected wave with time. The decrease in the wave amplitude indicates an increase in the acoustic impedance of the cement paste sample with time. After approximately 10 hours, a complete phase reversal is observed, i.e. the wave appears flipped when compared with that recorded in the first few hours. After the phase reversal, the wave amplitude continues to increase and approximately 18 hours after mixing, the reflected wave appears like a mirror image of the wave reflection at 6 hours.

The reflection factor (r) and the phase change in the stress waves ($\Delta\Phi$) introduced by reflection at PMMA-cement paste interface, captured after one reflection at the PMMA-cement paste interface at the resonant frequency of the transducer (1MHz) are plotted in Figure 2(b). The reflection factor, r(1), initially decreases in magnitude from a value close to 1.0, reaches a minimum and then starts increasing with time. The trend in the measured r(1) response corresponds with the observed trend in the reflected waveforms in Figure 2(a). The phase on the other hand exhibits a continuous increase with time. It can be seen that the minimum value of r(1) approximately corresponds in time with the a change in phase equal to 270 degrees. A 90 degree change in the measured $\Delta\Phi$ signals the initiation of the phase reversal of the reflected wave. Subsequent increase in $\Delta\Phi$ with time results in the reflected wave appearing like a mirror image of the reflected wave obtained initially.

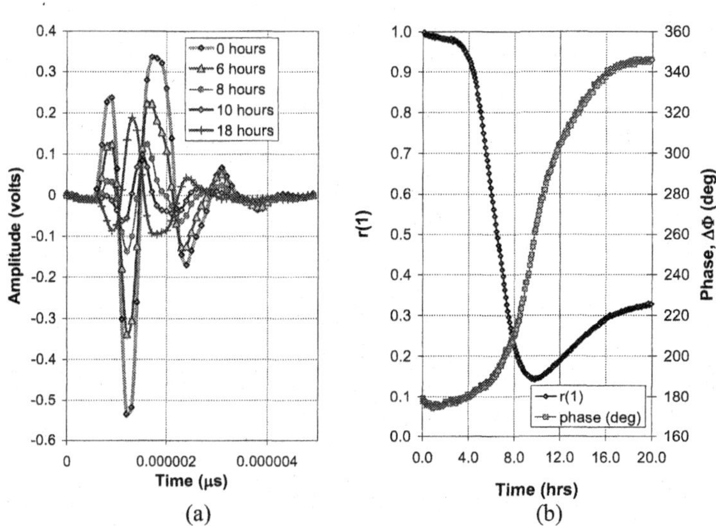

Figure 2: Ultrasonic test response collect at the PMMA-cement paste interface, (a) waveforms in time domain, (b) the amplitude reflection factor at 1 MHz (r(1)) and $\Delta\Phi$ as a function of time.

The reflection factor and the change in the phase angle at 1MHz frequency, r(1) and $\Delta\Phi$, for the cement paste, mortar and concrete specimen are shown in Figures 3(a) and (b), respectively. It can be seen that paste/concrete exhibits the fastest/slowest rate of change and that at any given time the cement paste/concrete sample has the lowest/highest value of r(1). Since the changes in the measured value of r(1) are related to changes in the acoustical properties of the cementitious material which are produced by hydration, the trends in the r(1) values indicate different rates of hydration in the three samples. Despite having the same water-to-cement ratio, the rate of change of r(1) indicates a more rapid/slow reaction in the cement paste/concrete sample. The measured phase change in mortar and concrete appear to be similar in the first 8 hours. After 12 hours the $\Delta\Phi$ of cement paste and mortar samples are similar. After approximately 10 hours the $\Delta\Phi$ of concrete is significantly lower than that of mortar and paste.

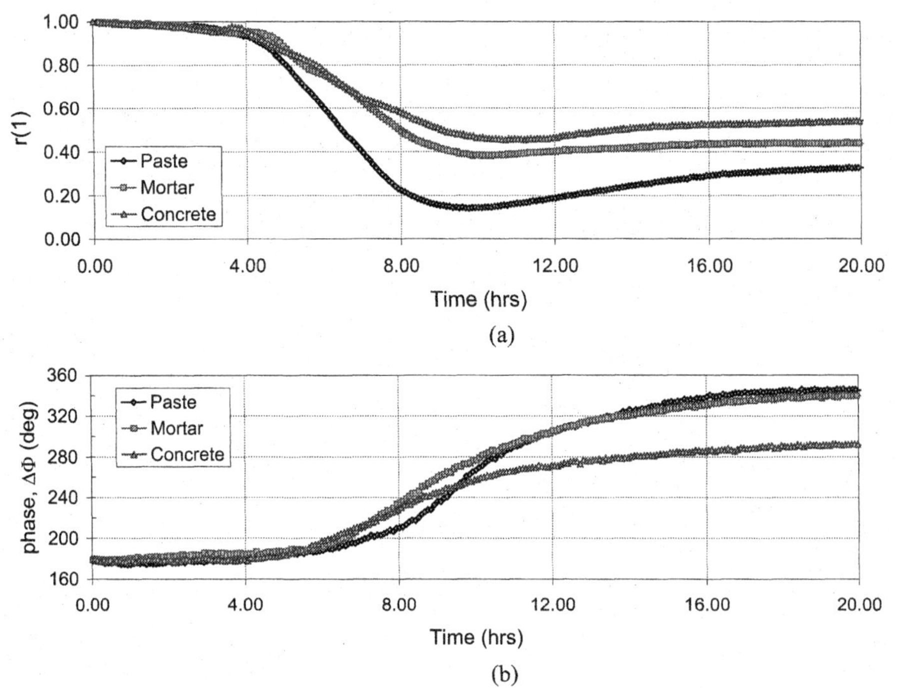

Figure 3: Ultrasonic wave response after one reflection at the PMMA-cementitious material interface, (a) r(1) as a function of time, (b) $\Delta\Phi$ as a function of time.

DISCUSSION

It has previously been shown that the for a given composition of the cementitious material, the minimum value of r(1) provides a convenient point of reference when its acoustic impedance is approximately equal to that of PMMA [15]. Since the acoustic impedance is a product of elastic stiffness and the density of the material, at the minimum point, the stiffness of mortar is a fixed percentage of the stiffness of PMMA. It was also established that the minimum point in the r(1) response of mortar corresponded in time with the occurrence of final setting time as determined using the pin penetration method [15]. The trends in r(1) for cement paste, mortar and concrete suggest that cement paste/concrete exhibits final setting earlier/later than mortar. This also suggests that the setting time determined using mortar sieved from concrete does not provide a true indication of the setting time of concrete. Further, the ultrasonic technique now provides a convenient way to determine the setting time directly in concrete.

Monitoring changes in $\Delta\Phi$ and r(1) as a function of time provides a means of assessing changes in the stiffness of the cementitious material relative to that of PMMA. The observed changes in the reflected wave amplitude and phase (r(1) and $\Delta\Phi$) can now be interpreted to provide information about the changes in the shear modulus of the cementitious material with time. Considering Equation 1, an expression for $\left|Z^*_{cem.mat.}\right| / \left|Z^*_{PMMA}\right|$ can be obtained in terms of the measured $\Delta\Phi$ and r(f) as follows

$$\left|Z^*_{cem.mat.}\right| \Big/ \left|Z^*_{PMMA}\right| = \left|\frac{1 + r(f)e^{i\Delta\Phi}}{1 - r(f)e^{i\Delta\Phi}}\right| \tag{6}$$

Considering the expression for Z* given in Equation 2, the expression for the ratio of the moduli of the acoustic impedances of the cementious material and the PMMA can be written as follows

$$\left|Z^*_{cem.mat.}\right| \Big/ \left|Z^*_{PMMA}\right| = \frac{\left|\sqrt{\rho_{cem.mat.}\left(G' + iG''\right)_{cem.mat.r}}\right|}{\left|\sqrt{\rho_{PMMA}\left(G' + iG''\right)_{PMMA}}\right|} = \frac{\sqrt{\rho_{cem.mat.}}}{\sqrt{\rho_{PMMA}}} \frac{\sqrt{\left|G_{cem.mat.}\right|}}{\sqrt{\left|G_{PMMA}\right|}} \tag{7}$$

A source function at 1MHz center frequency was used in this study. Thus considering r(1) and $\Delta\Phi$ determined at 1 MHz the shear modulus of the cementitious material can be obtained as

$$\frac{\left|G_{cem.mat.}\right|}{\left|G_{PMMA}\right|} = \left(\frac{\rho_{cem.mat.}}{\rho_{mortar}}\right)^2 \left|\frac{1 + r(1)e^{i\Delta\Phi}}{1 - r(1)e^{i\Delta\Phi}}\right|^2 \tag{8}$$

In Equation 8, r(1) and $\Delta\Phi$ are the experimentally measured changes in amplitude and phase angle after reflection at the PMMA-cementitious material interface, respectively (plotted in Figure 3). The values of r(1) and $\Delta\Phi$ change continuously with time because of the changes in the material properties of material produced by the progressing hydration of cement. Changes in $\rho_{cem.mat.}$ with time are insignificant and therefore $\rho_{PMMA}/\rho_{cem.mat.}$ can be treated as a constant in Equation 8. The densities of the cement paste, mortar and concrete were determined using hardened samples of the material.

The change in the shear moduli as a function of time for the cement paste, the mortar and the concrete samples assessed at 1MHz is shown in Figure 4. The high-frequency values of $|G_{PMMA}|$ assessed at 1MHz are currently not available. However, since G_{PMMA} is relatively invariant with time, the shear modulus of the cementitious material $|G_{cem.mat}|$., has been normalized with respect to $|G_{PMMA}|$, in Figure 4. From Figure 4, it can be seen that the observed trends in the shear modulus are similar for the cement paste and the mortar samples. There is a very gradual change in the shear modulus in the first few hours following which there is an almost exponential increase with time. The exponential increase in $G_{cem.mat.}$ is most noticeable in the cement paste and mortar samples and appears to last up to approximately 16 hours. Following this period of exponential increase the rate of increase appears decrease and approach a steady rate. The trend in change of $G_{cem.mat.}$ of the concrete sample appears to be different from those of the cement paste and mortar samples. In concrete the period of exponential increase is very brief. Following this there appears to be a hyperbolic increase with time.

The absolute values of the shear modulus at any given time are also different for the three samples. Cement paste/concrete has the highest/lowest shear modulus at any time. While in cement paste the value of shear modulus could be directly attributed to the hydration of cement paste, in mortar and concrete the shear modulus of the material depends upon the aggregate (the quantity and their relative size distribution) in addition to the progress of hydration. In mortar and in concrete, the changes in the shear modulus could however be directly attributed to hydration. It is interesting to note that while the addition of fine aggregate did not produce a significant change in the $G_{cem.mat.}$, the addition of coarse aggregate resulted in a considerable decrease of the shear modulus. Since the cementitious materials are treated as visco-elastic materials in the analytical formulation of the wave reflection at the bi-material interface, the shear modulus presented in Figure 4 contains contributions from both the elastic and the viscous components. This results presented in Figure 4 indicate that the rheological parameters obtained using cement paste may not be directly applicable for use in concrete.

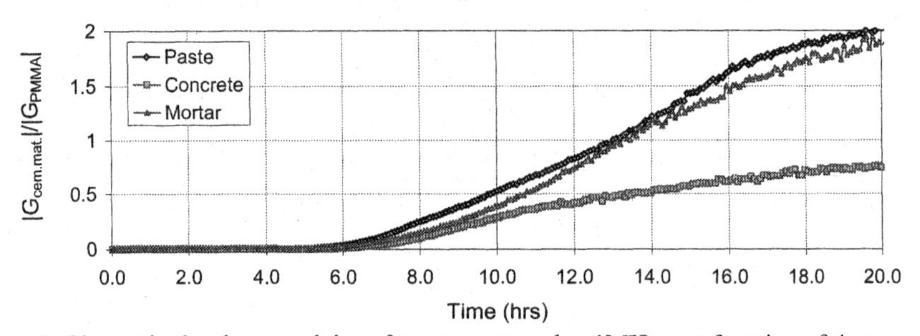

Figure 4: Change in the shear modulus of mortar assessed at 1MHz as a function of time

CONCLUSIONS

An ultrasonic technique for continuous monitoring of cementitious materials after mixing, through setting and early strength gain is reported in this paper. In the method described above, changes in stiffness of a cementitious material are monitored relative to that of PMMA. The amplitude and the phase of the waves reflected at the PMMA-cementitious material interface, were used to study the changes in the stiffness in cement paste, mortar and concrete specimens. The ultrasonic measurements are shown to be sensitive to changes in the stiffness of the cementitious material from a very early age. Based on the results presented in this paper, the following conclusions can be drawn:

1. Changes in the stiffness of the cementitious materials produce an initial decrease in the amplitude of the reflected waves at 1 MHz. The amplitude of the reflected waves reaches a minimum and then starts increasing with age. The observed increase is accompanied by a complete reversal in the phase of the wave.

2. There is a significant difference in the measured ultrasonic response from cement paste, mortar and concrete samples with identical water-to-cement ratio. The ultrasonic measurements indicate significant difference in the rate of hydration in the paste, mortar and concrete samples.
3. The shear modulus of concrete assessed at 1MHz in the early ages is significantly lower than that of paste with an identical water-to-cement ratio.

REFERENCES

[1]L.J. Struble, H. Zhang and W.G. Lei, "Oscillatory Shear Behavior of Portland Cement Paste during Early Hydration," *Con. Sci. Eng.*, 2 141-149 (2000)

[2]ASTM C 403, "Standard Test Method for Time of Setting of Concrete Mixtures by Penetration Resistance," ASTM, Philadelphia, pp. 214-218 (2000)

[3]ASTM C 39, "Test Method for Compressive Strength of Cylindrical Concrete Specimens," ASTM, Philadelphia, ASTM C 39, 18-21 (2000)

[4]ASTM C 469-94, "Standard Test Method for Static Modulus of Elasticity and Poisson's Ratio of Concrete in Compression," ASTM, Philadelphia, pp.242-245 (2000)

[5]K.V. Subramaniam, J.S. Popovics and S.P Shah, "Determining Elastic Properties of Concrete Using Vibrational Resonance Frequencies of Standard Test Cylinders," *Cem. Con. Agg.*, 22(2) 81-89 (2000)

[6]X. Jin and Z. Li, "Dynamic Property Determination for Early Age Concrete," ACI Mater. J 98(5) (2001) 365-371

[7]T.Özutürk, J. Rapoport, J.S. Popovics and S.P. Shah, "Monitoring the Setting and Hardening of Cement-Based Materials with Ultrasound," *Con. Sci. Eng.*, 11 83-91 (1999)

[8]H.W.Reinhardt, C.U.Groβe and A.T.Herb, "Ultrasonic Monitoring of Setting and Hardening of Cement Mortar – A New Device," *Mat. Stru.*, 33 580-583 (2000)

[9]J. Rapport, J. Popovics, K.V. Subramaniam and S.P. Shah, "The Use of Ultrasound to Monitor the Stiffening Process of Portland Cement Concrete with Admixtures," *ACI Mater. J.*, 97(6) 675-683 (2000)

[10]M.I. Valič and J. Stepišnik, "A Study of Hydration of Cement Pastes by Reflection of Ultrasonic Shear Waves. Part I: Apparatus, Experimental Method and Application Examples," *Kov. Zlit. Tech.*, 32(6) 551-600 (1996)

[11]A. van Beek, "Dielectric Properties of Young Concrete," PhD Thesis, Delft University, (2000)

[12]R. Zoughi, S.D. Gray and P.S. Nowak, "Microwave Nondestructive Estimation of Cement Paste Compressive Strength," *ACI Mater. J.*, 92 64-70 (1995)

[13]J.D. Achenbach, I. N. Komsky, Y. C Lee and Y. C. Angel, "Self-Calibrating Ultrasonic Technique for Crack Depth Measurement," *J. Nondes. Eva.*, 103-108 (1992)

[14]K.V. Subramaniam and J. Lee, "Ultrasonic Assessment of Early-Age Changes in the Material properties of Cementitious Materials," in Advances in Concrete through Science and Engineering, Symposium Proceedings, RILEM, March 22-24, 2004

[1] K.V. Subramaniam and J. Lee, "Monitoring the Setting Behavior of Cementitious Materials using One-Sided Ultrasonic Measurements," submitted for review to *Cem. and Con. Res*

Author Index

Achenbach, A., 137
Amde, A.M., 199
Attiogbe, E., 279

Babu, K.G., 171
Bai, J., 291
Balasubramanian, K., 21, 93
Basu, P.C., 235
Bhanumathidas, N., 183
Bharatkumar, B.H., 43
Biernacki, J.J., 149, 291
Borgnakke, C., 149
Brown, H.J., 113

Dattatreya, J.K., 247
Dinakar, P., 171

El-Shakra, Z., 123

Garci Juenger, M.C., 211
Gopalakrishnan, S., 21, 93
Gopalaratnam, V.S., 123

Hansen, W., 149
Hockenberry, T., 137
Hubbard, C.R., 291

Jain, A.K., 165
Jansen, D.C., 101
Jupe, A.C., 223

Kalidas, N., 183
Kaushik, S.K., 73
Krishnamoorthy, T.S., 83
Kurtis, K.E., 223

Lakshmanan, N., 83
Lee, J., 301
Lee, W.K., 101
Likhitruangsilp, V., 55
Livingston, R.A., 199
Lopez de Murphy, M., 137

Mihashi, H., 123
Moon, J.-H., 279
Morton, J.H., 113

Naaman, A.E., 55
Naik, N.N., 223
Neelamegam, M., 247
Neithalath, N., 279

Parameswaran, V.S., 93
Parnham, C.J., 291
Pease, B., 279
Peng, Y., 149
Peter, J.A., 21

Raghuprasad, B.K., 43
Rajabipour, F., 279
Rajmane, N.P., 247
Ravisankar, K., 269
Reddi, S.A., 33
Riding, K.A., 211

Shah, S.P., 3
Shridhar, R., 193
Sreenath, H.G., 255
Stock, S.R., 223
Stutzman, P., 223
Subramaniam, K.V., 301
Suresh, G.A.B., 183

Voigt, T., 3

Watkins, T.R., 291
Weiss, J., 279
Williams, K., 199
Wongtanakitcharoen, T., 55

Keyword Index

Activated belite-rich cement, 165
Alkali resistant glass fiber, 73

Beneficiated fly ash, 211

Cement blends, 149, 165, 171, 183, 193, 199, 211
Compact reinforced concrete, 93
Confinement-analogy, 123
Confocal microscopy, 223
Construction industry-India, 21, 33, 73, 93
Corrosion affected structure assessment, 255
Crack width modeling, 101

Defining high performance, 3, 21, 33, 43
Delayed ettringite formation, 199

Electron microscopy with x-ray imaging, 223

FaL-G, 183
Fiber dispersion, 3
Fiber optic sensors, 269
Fiber reinforced concrete, 33, 55, 73, 83, 93, 101, 123, 137, 165
Fiber reinforced plastics, 73
Fiber reinforced polymer rebar, 101
Flexural performance of reinforced concrete, 123
Fly ash in concrete, 21, 33, 43, 171, 183, 211, 235
Fracture characteristics, 43

Gergely-Lutz model, 101
Glass fibers, 73, 101

Hardwire©, 55
High silica modulus cement, 165
Hooked-end steel fibers, 137
Hybrid composites, 55
Hybrid fiber reinforced cement composites, 3
Hybrid fiber systems, 73
Hydration kinetics, 149
Hydro electric power projects, 33

Kraft pulps, 113

Laser scanning confocal microscopy, 223

Maryland State Highway projects, 199
Moisture gradients, 279
Monitoring service life performance, 269
Multi-component cementitious systems, 183
Mumbai Pune expressway, 33

Non-destructive test methods, 255
Non-reactive Frederick limestone, 199
Nuclear containment structures, 33

One-sided ultrasonic testing, 301
Optical microscopy, 223

Partially destructive test methods, 255
Performance monitoring, 269
Polycarboxylate admixtures, 193
Potassium carbonate admixture, 199
PVA fibers, 137

Ready mixed concrete applications, 113
Rheological measurement techniques, 247

Self compacting concrete, 21, 33, 73, 193, 247
Shrinkage cracking behavior, 3
Silica fume blended cements, 33, 165
Slag based concrete, 43
Slurry infiltrated fibrous concrete, 93
Slurry infiltrated mat concrete, 93
Steel fiber reinforced concrete, 73, 83, 93, 137
Steel fiber reinforced shotcrete, 93
Stereo microscopy, 223

Structural Engineering Research Centre, Chennai, India, 21, 247
Supplementary cementitious materials, 149
Sustainable development, India, 235
Synchrotron X-rays, 291

Tala hydroelectric project, 33
Three-dimensional meshes, 55
Time-variable rheological models, 247
Twisted polygonal steel fiber, 137

Underwater concrete, 33

Very early age materials, 247, 301
Viscoelasticity, 247

Water permeability, 3
Water retaining structures, 33
Wood pulp fibers, 113

X-ray microtomography, 223

9 781574 981995